Genes in the Environment

Special Symposia of The British Ecological Society

14 Integrating Ecology and Evolution in a Spatial Context
Silvertown & Antonovics, 2001

13 Ecology of Arctic Environments
Woodin & Marquiss, 1997

12 Primary Succession on Land
Miles & Walton, 1993

Annual Symposia of The British Ecological Society

42 Dispersal Ecology
Bullock, Kenward & Hails, 2002

41 Ecology: Achievement and Challenge
Press, Huntly & Levin, 2001

40 The Ecological Consequences of Environmental Heterogeneity
Hutchings, John & Stewart, 2000

39 Physiological Plant Ecology
Press, Scholes & Barker, 2000

37 Dynamics of Tropical Communities
Newbery, Prins & Brown, 1997

36 Multitrophic Interactions in Terrestrial Systems
Gange & Brown, 1997

33 Genes in Ecology
Berry, Crawford & Hewitt, 1992

32 The Ecology of Temperate Cereal Fields
Firbank, Carter, Darbyshire & Potts, 1991

31 The Scientific Management of Temperate Communities for Conservation
Spellerberg, Goldsmith & Morris, 1991

30 Toward a More Exact Ecology
Grubb & Whittaker, 1989

Genes in the Environment

The 15th Special Symposium of the British Ecological Society held at St Catherine's College, Oxford, 17–19 September 2001

EDITED BY

ROSIE S. HAILS
NERC Centre for Ecology and Hydrology
CEH Oxford
Mansfield Road
Oxford

JOHN E. BERINGER
Senate House
University of Bristol
Tyndall Avenue
Bristol

AND

H. CHARLES J. GODFRAY
NERC Centre for Population Biology
Imperial College
Silwood Park
Ascot

Blackwell Publishing

and published for them by
Blackwell Science Ltd
a Blackwell Publishing company

350 Main Street, Malden, MA 02148-5018, USA
108 Cowley Road, Oxford OX4 1JF, UK
550 Swanston Street, Carlton South, Melbourne, Victoria 3053, Australia
Kurfürstendamm 57, 10707 Berlin, Germany

First published 2003

Library of Congress Cataloging-in-Publication Data

British Ecological Society. Special Symposium (15th: 2001: St. Catherine's College, Oxford)
Genes in the environment: the 15th Special Symposium of the British Ecological Society held at St. Catherine's College, Oxford, 17–19 September, 2001/
edited by Rosie S. Hails, John Beringer, and H. Charles J. Godfray.—1st ed.
p. cm.
Includes bibliographical references and index.
ISBN 0-632-06507-9 (alk. paper)—ISBN 0-632-06508-7 (pbk.: alk. paper)
1. Ecological genetics—Congresses. 2. Population genetics—Congresses.
3. Molecular genetics—Congresses. I. Hails, Rosemary. II. Beringer, John.
III. Godfray, H.C.J., 1958– IV. Title.

QH456.B748 2001
576.5′8—dc21

2002071228

A catalogue record for this title is available from the British Library.

Set in 10 on $12\frac{1}{2}$ pt Minion
by SNP Best-set Typesetter Ltd., Hong Kong
Printed and bound in the United Kingdom
by MPG Books Ltd, Bodmin, Cornwall.

For further information on
Blackwell Publishing, visit our website:
http://www.blackwellpublishing.com

Contents

List of Contributors

M. J. Alexander
NERC Centre for Ecology and Hydrology, CEH Dorset, Winfrith Technology Centre, Dorchester DT2 8ZD, UK

D. A. Ashford
Department of Biology, University of York, York YO10 5YW, UK

R. J. Atkinson
Department of Biology, University of Stirling, Stirling FK9 4LA, UK

M. Begon
Population and Evolutionary Biology Research Group, School of Biological Sciences, University of Liverpool, Nicholson Building, Brownlow Street, Liverpool L69 3GS, UK

R. Benmayor
Molecular Ecology and Biocontrol Group, NERC Centre for Ecology and Hydrology, CEH Oxford, Mansfield Road, Oxford OX1 3SR, UK and Population and Evolutionary Biology Research Group, School of Biological Sciences, University of Liverpool, Nicholson Building, Brownlow Street, Liverpool L69 3GS, UK

J. E. Beringer
Senate House, University of Bristol, Tyndall Avenue, Bristol BS8 1TH, UK

L. M. Birkle
Department of Biology, University of York, York YO10 5YW, UK

C. M. Brasier
Forest Research Agency, Alice Holt Lodge, Farnham, Surrey GU10 4LH, UK

G. Brown
School of Biology, St Andrews University, Bute Building, St Andrews KY16 9TS, UK

K. W. Buck
Department of Biological Sciences, Sir Alexander Fleming Building, Imperial College of Science, Technology and Medicine, London SW7 2AZ, UK

S. Burke
School of Biosciences, University of Birmingham, Birmingham B15 2TT, UK

R. K. Butlin
Centre for Biodiversity & Conservation, School of Biology, University of Leeds, Leeds LS2 9JT, UK

G. R. Carvalho
Molecular Ecology & Fisheries Genetics Laboratory, Department of Biological Sciences, University of Hull, Hull HU6 7RX, UK

L. Chistoserdova
Department of Chemical Engineering Box 351750, University of Washington, Seattle, WA 98195, USA

J. I. Cooper
NERC Centre for Ecology and Hydrology, CEH Oxford, Mansfield Road, Oxford OX1 3SR, UK

J. S. Cory
Molecular Ecology and Biocontrol Group, NERC Centre for Ecology and Hydrology, CEH Oxford, Mansfield Road, Oxford OX1 3SR, UK

G. Csóka
Hungarian Forest Research Institute, Mátrafüred Research Station, 3232 Mátrafüred, POBox 2, Hungary

L. J. Crawford
Department of Biological Sciences, Sir Alexander Fleming Building, Imperial College of Science, Technology and Medicine, London SW7 2AZ, UK

A. C. Darby
Department of Biology, University of York, York YO10 5YW, UK. Current address: Centre for Tropical Veterinary Medicine, University of Edinburgh, Easter Bush Veterinary Centre, Midlothian EH5 9RG, UK

A. E. Douglas
Department of Biology, University of York, York YO10 5YW, UK

M-L. Edwards
NERC Centre for Ecology and Hydrology, CEH Oxford, Mansfield Road, Oxford OX1 3SR, UK

T. M. Embley
Department of Zoology, Natural History Museum, London SW7 5BD, UK

J. Gockel
Department of Biology, University College London, Darwin Building, Gower Street, London WC1E 6BT, UK

H. C. J. Godfray
NERC Centre for Population Biology, Imperial College of Science, Technology and Medicine, Silwood Park, Ascot, Berkshire SL5 7PY, UK

D. B. Goldstein
Department of Biology, University College London, Darwin Building, Gower Street, London WC1E 6BT, UK

A. J. Gray
NERC Centre for Ecology and Hydrology, CEH Dorset, Winfrith Technology Centre, Dorchester DT2 8ZD, UK

C. M. Griffiths
NERC Centre for Ecology and Hydrology, CEH Oxford, Mansfield Road, Oxford OX1 3SR, UK. Current address: School of Biological and Molecular Sciences, Oxford Brookes University, Gipsy Lane, Oxford OX3 0BP, UK

R. S. Hails
NERC Centre for Ecology and Hydrology, CEH Oxford, Mansfield Road, Oxford OX1 3SR, UK

G. Hall
NERC Centre for Ecology and Hydrology, Windermere Laboratory, Far Sawrey, Cumbria LA22 0LP, UK

S. E. Hartley
NERC Centre for Ecology and Hydrology, CEH Banchory, Hill of Brathens, Banchory, Kincardineshire AB31 4BW, UK. Current address: School of Biological Sciences, University of Sussex, Falmer, Brighton BN1 9QG, UK

M. J. Hatcher
Centre for Biodiversity & Conservation, School of Biology, University of Leeds, Leeds LS2 9JT, UK

L. Hauser
School of Aquatic and Fishery Sciences, University of Washington, 1122 NE Boat Street, Box 355020, Seattle, WA 98195, USA

R. B. Hitchman
NERC Centre for Ecology and Hydrology, CEH Oxford, Mansfield Road, Oxford OX1 3SR, UK

D. J. Hodgson
NERC Centre for Ecology and Hydrology, CEH Oxford, Mansfield Road, Oxford OX1 3SR, UK

P. Hunter
Horticulture Research International, Wellesbourne, Warwick CV35 9EF, UK

P. Ineson
NERC Centre for Ecology and Hydrology, CEH Merlewood Research Station, Grange-over-Sands LA11 6JU, UK

A. M. E. Jones
NERC Centre for Ecology and Hydrology, CEH Dorset, Winfrith Technology Centre, Dorchester DT2 8ZD, UK. Current address: Department of Agricultural Sciences, Imperial College at Wye, Wye, Kent TN25 5AH, UK

W. J. Kennington
Department of Biology, University College London, Darwin Building, Gower Street, London WC1E 6BT, UK

M. E. Lidstrom
Department of Chemical Engineering Box 351750 and Department of Microbiology Box 35242, University of Washington, Seattle, WA 98195, USA

A. J. McCarthy
Environmental and Molecular Microbiology Group, School of Biological Sciences, University of Liverpool, Liverpool L69 7ZB, UK

M. R. Macnair
School of Biological Sciences, University of Exeter, Hatherly Laboratories, Prince of Wales Road, Exeter EX4 4PS, UK

N. McNamara
NERC Centre for Ecology and Hydrology, CEH Merlewood Research Station, Grange-over-Sands LA11 6JU, UK

R. McVean
Population and Evolutionary Biology Research Group, School of Biological Sciences, University of Liverpool, Nicholson Building, Brownlow Street, Liverpool L69 3GS, UK

A. E. Magurran
Environmental and Evolutionary Biology, Bute Building, University of St Andrews, St Andrews, Fife KY16 9TS, UK

J. B. Mallet
The Galton Laboratory, Department of Biology, University College London, 4 Stephenson Way, London NW1 2HE, UK

R. A. Meade
Environmental and Molecular Microbiology Group, School of Biological Sciences, Life Sciences Building, University of Liverpool, Liverpool L69 7ZB, UK

T. R. Meagher
Division of Environmental and Evolutionary Biology, Sir Harold Mitchell Building, University of St Andrews, St Andrews, Fife KY16 9TH, UK

E. Mitchell
NERC Centre for Ecology and Hydrology, CEH Dorset, Winfrith Technology Centre, Dorchester DT2 8ZD, UK

S. Morris
Department of Biological Sciences, University of Warwick, Coventry CV4 7AL, UK

C. L. Moyes
NERC Centre for Ecology and Hydrology, CEH Dorset, Winfrith Technology Centre, Dorchester DT2 8ZD, UK

J. C. Murrell
Department of Biological Sciences, University of Warwick, Coventry CV4 7AL, UK

K. E. Mutch
Department of Biology, University of York, York YO10 5YW, UK

L. A. Mutch
Department of Biology, University of York, York YO10 5YW, UK

M. Naylor
NERC Centre for Ecology and Hydrology, CEH Oxford, Mansfield Road, Oxford OX1 3SR, UK

D. B. Nedwell
Department of Biological Sciences, University of Essex, Wivenhoe Park, Colchester CO4 3SQ, UK

C. van Oosterhout
Molecular & Evolutionary Ecology Research Group, Department of Biological Sciences, University of Hull, Hull HU6 7RX, UK

D. W. Pallett
NERC Centre for Ecology and Hydrology, CEH Oxford, Mansfield Road, Oxford OX1 3SR, UK

M. Paoletti
Department of Biological Sciences, Sir Alexander Fleming Building, Imperial College of Science, Technology and Medicine, London SW7 2AZ, UK

L. Partridge
Department of Biology, University College London, Darwin Building, Gower Street, London WC1E 6BT, UK

R. W. Pickup
NERC Centre for Ecology and Hydrology, CEH Windermere Laboratory, Far Sawrey, Cumbria LA22 0LP, UK

R. D. Possee
Molecular Biology of Baculoviruses Group, NERC Centre for Ecology and Hydrology, CEH Oxford, Mansfield Road, Oxford OX1 3SR, UK

J. I. Prosser
Department of Molecular and Cell Biology, University of Aberdeen, Institute of Medical Sciences, Foresterhill, Aberdeen AB25 2ZD, UK

A. S. Pullin
School of Biosciences, University of Birmingham, Birmingham B15 2TT, UK

S. Radajewski
Department of Biological Sciences, University of Warwick, Coventry CV4 7AL, UK

A. F. Raybould
NERC Centre for Ecology and Hydrology, CEH Dorset, Winfrith Technology Centre, Dorchester DT2 8ZD, UK. Current address: Syngenta, Jealott's Hill International Research Centre, Bracknell, Berkshire RG42 6EY, UK

D. S. Reay
Department of Biological Sciences, University of Essex, Wivenhoe Park, Colchester CO4 3SQ, UK

J. P. Reeves
NERC Centre for Ecology and Hydrology, CEH Oxford, Mansfield Road, Oxford OX1 3SR, UK

A. Rokas
Laboratory of Molecular Biology, University of Wisconsin, 1525 Linden Drive, Madison, WI 53706, USA

S. M. Sait
Population and Evolutionary Biology Research Group, School of Biological Sciences, University of Liverpool, Nicholson Building, Brownlow Street, Liverpool L69 3GS, UK

J. R. Saunders
Environmental and Molecular Microbiology Group, School of Biological Sciences, Life Sciences Building, University of Liverpool, L69 7ZB, UK

B. Shorrocks
Centre for Biodiversity & Conservation, School of Biology, University of Leeds, Leeds LS2 9JT, UK

A. Smithson
School of Biological Sciences, University of Exeter, Hatherly Laboratories, Prince of Wales Road, Exeter, EX4 4PS, UK

G. N. Stone
University of Edinburgh Institute of Cell, Animal and Population Biology, The Kings Buildings, West Mains Road, Edinburgh EH9 3JT, UK

D. E. Taneyhill
Centre for Biodiversity & Conservation, School of Biology, University of Leeds, Leeds LS2 9JT, UK

C. D. Thomas
Centre for Biodiversity & Conservation, School of Biology, University of Leeds, Leeds LS2 9JT, UK

D. J. Thompson
Population and Evolutionary Biology Research Group, School of Biological Sciences, University of Liverpool, Nicholson Building, Brownlow Street, Liverpool L69 3GS, UK

M. I. Thurston
Department of Zoology, University of Oxford, South Parks Road, Oxford OX1 3PS, UK

M. Tristem
NERC Centre for Population Biology, Imperial College of Science, Technology and Medicine, Silwood Park, Ascot, Berkshire SL5 7PY, UK

A. J. Vanbergen
NERC Centre for Ecology and Hydrology, CEH Banchory, Hill of Brathens, Banchory, Kincardineshire AB31 4BW, UK

C. Vassiliadis
Division of Environmental and Evolutionary Biology, Sir Harold Mitchell Building, University of St Andrews, St Andrews, Fife KY16 9TH, UK

J. A. Walsh
Horticulture Research International, Wellesbourne, Warwick CV35 9EF, UK

K. F. A. Walters
Central Science Laboratory, Sand Hutton, York YO41 1LZ, UK

A. D. Watt
NERC Centre for Ecology and Hydrology, CEH Banchory, Hill of Brathens, Banchory, Kincardineshire AB31 4BW, UK

G. Webster
Department of Molecular and Cell Biology, Institute of Medical Sciences, University of Aberdeen, Foresterhill, Aberdeen AB25 2ZD, UK. Current address: School of Biosciences, Cardiff University, Main Building, Park Place, Cardiff CF10 3TL, UK

C. B. Whitby
Environmental and Molecular Microbiology Group, School of Biological Sciences, Life Sciences Building, University of Liverpool, Liverpool L69 7ZB, UK

R. J. Wilson
Centre for Biodiversity & Conservation, School of Biology, University of Leeds, Leeds LS2 9JT, UK

I. Wynne
The Galton Laboratory, Department of Biology, University College London, 4 Stephenson Way, London NW1 2HE, UK

J. P. W. Young
Department of Biology, University of York, York YO10 5YW, UK

A. Zézé
Department of Biology, University of York, York YO10 5YW, UK

History of the British Ecological Society

The British Ecological Society is a learned society, a registered charity and a company limited by guarantee. Established in 1913 by academics to promote and foster the study of ecology in its widest sense, the Society currently has around 5000 members spread around the world. Members include research scientists, environmental consultants, teachers, local authority ecologists, conservationists and many others with an active interest in natural history and the environment. The core activities are the publication of the results of research in ecology, the development of scientific meetings and the promotion of ecological awareness through education. The Society's mission is:

To advance and support the science of ecology and publicize the outcome of research, in order to advance knowledge, education and its application.

The Society publishes four internationally renowned journals and organizes at least two major conferences each year plus a large number of smaller meetings. It also initiates a diverse range of activities to promote awareness of ecology at the public and policy maker level in addition to developing ecology in the education system, and it provides financial support for approved ecological projects. The Society is an independent organization that receives little outside funding.

British Ecological Society
26 Blades Court
Deodar Road, Putney
London SW15 2NU
United Kingdom
Tel.: +44 (0)20 8871 9797
Fax: +44 (0)20 8871 9779
E-mail: info@BritishEcologicalSociety.org
URL: www.britishecologicalsociety.org

The British Ecological Society is a limited company, registered in England No. 15228997 and a Registered Charity No. 281213.

Preface

This special symposium volume describes research on the application of molecular techniques to solve ecological questions, much of which was funded by a Natural Environment Research Council initiative EDGE (Ecological Dynamics and Genes). This is a rapidly developing area of ecology, and spans a very diverse range of topics, including community dynamics in soils and water, gene flow and spatial dynamics, and the evolution of pathogenic and symbiotic relationships.

The study of gene flow can provide insights into the adaptation of species from diverse taxa, including fungi, insects and plants, to new ecological conditions. For example, a degree of interspecific gene flow has been detected between two Dutch elm disease fungi, *Ophiostoma* spp., which have recently come into contact through range expansion. The introgression of favoured genes confers resistance to viruses that reduce the pathogenicity of this fungus. A study of the latitudinal body size cline of *Drosophila melanogaster* has investigated the roles of gene flow and selection in maintaining the cline, and the influence of the potentially asymmetrical nature of gene flow. Recent molecular advances have allowed for the first time the study of virus epidemiology in both wild and cultivated plants, as illustrated by two chapters in this volume. From a study of the silver-studded blue butterfly, *Plebejus argus*, and the brown argus, *Aricia agestis*, the population dynamics of these populations lead to spatial patterns of genetic variation that are often opposite to what may be predicted on the basis of their dispersal behaviour. These studies illustrate how both molecular methods and mathematical models can combine to explain and predict patterns in genetic structure.

Other studies have illustrated how the direct analysis of environmental DNA from microorganisms can provide insights into the dynamics of ammonia-oxidizing bacteria, or the impact of land use on the soil methane sink. The beauty of PCR techniques is that it is no longer necessary to culture the bacteria in the laboratory (a great advantage given that less than 10% were culturable). This step forward has now been more fully exploited by the development and application of the novel technique of stable isotope probing (SIP), whereby stable isotopes are fed to the microbial populations and incorporated into genetic material. This allows the distinction to be made between the active and inactive bacteria, making the study of their dynamics a real possibility.

Many of these studies lay the groundwork and pose the questions that lead us into the interdisciplinary area of genomics. The interface of molecular biology and ecology is in its infancy, but is set to be one of the fastest growing areas of ecology over the next decade. We are in a period of transition. In the past there has been a rather limited range of techniques, dictating which ecological questions can be tackled. The recent explosion of techniques will now create a research environment in which the ecological questions themselves can drive the direction of research. The challenge will be to match the appropriate techniques with the most important questions.

Rosie S. Hails
John E. Beringer
H. Charles J. Godfray

Part 1

Gene flow and spatial dynamics

Chapter 1

Estimating rates of gene flow in endemic butterfly races: the effect of metapopulation dynamics

D. E. Taneyhill, J. B. Mallet, I. Wynne, S. Burke, A. S. Pullin, R. J. Wilson, R. K. Butlin, M. J. Hatcher, B. Shorrocks and C. D. Thomas*

Introduction

Variation within species is the raw material of evolutionary change. The spatial pattern of genetic variation is fundamental to the formation of races and new species, and depends in turn on the environment: through variation in the strength of natural selection and through the population dynamics of the organisms concerned. Ecological dynamics, which consist of changes in abundance and distribution, influence genetic variation by determining the rates of gene flow between spatially separated populations. While the relative importance of allopatry and limited gene flow in producing new species has been long debated (Mayr 1942; Ehrlich & Raven 1969; Endler 1977), the contemporary view seems to be that low levels of gene flow and speciation events are necessary for new adaptations to persist, as opposed to being ephemeral in evolutionary time (Futuyma 1987). Estimating rates of gene flow is thus central to understanding how a species' spatial distribution will determine its amount of genetic diversity.

There are two special aspects of ecological dynamics that affect genetic variation and with which we are concerned in the present chapter. One is metapopulation structure: many plant and animal populations actually consist of a set of subpopulations connected by relatively low levels of interpopulation dispersal. A second, related aspect is the **non-equilibrium** character of such small local subpopulations. Since the classical models of population genetics generally assume a constant population size, standard methods of estimation may not be appropriate for local populations that experience frequent extinctions followed by recolonization and population expansion.

In our project we investigated the effect of a metapopulation structure on the spatial patterns of genetic diversity in two species of butterflies, the silver-studded blue *Plebejus argus* and the brown argus *Aricia agestis*. This chapter focuses mainly on the effect of metapopulation dynamics on the estimation of rates of gene flow from molecular genetic data. We present a short introduction to recent theoretical

* Correspondence author: det2@york.ac.uk

developments in the field, including new methods that have been developed for this research. Indirect estimates of gene flow rates from both the traditional and the more recent methods are compared. We begin with a brief introduction to metapopulation dynamics and genetics, followed by descriptions of the study species.

Metapopulations dynamics and genetics

Spatial structure is currently at the forefront of theoretical ecology (Hanski & Simberloff 1997). While spatial structure was recognized as an important aspect of population biology from the early 1950s (e.g. Andrewartha & Birch 1954), it is only in the last decade that the metapopulation concept began to dominate ecological thinking in regard to spatially fragmented habitats (Hanski & Gilpin 1997; Hanski 1999).

A metapopulation was originally defined by Richard Levins (1969) as a set of local populations, which are discrete patches within a matrix of habitat, and which are large enough to support viable, panmictic populations of the focal species. These patches may exchange migrants, but generally at a sufficiently low rate that interpatch migration has no appreciable effect on local population dynamics (Hanski & Simberloff 1997). Local patches may go extinct, and the migration between patches may effect recolonization of such empty habitat patches. These forces of extinction and recolonization produce what den Boer (1968) termed a virtual governor; i.e. a mechanism, unseen at the scale of local population dynamics, which could stabilize a collection of small subpopulations that were subject to frequent extinctions. Note that a fundamental assumption of metapopulation dynamic theory is the separation of processes in regard to timescale: local population dynamics occur on a much faster timescale than do extinction and recolonization (Taneyhill 2000). In this way metapopulation dynamics is distinguished from **landscape ecology**, where space is continuous and where local and regional dynamics are not so separated (Turner 1989; Hanski 1999).

Metapopulations are often classified into several broad categories. In Levins' original conception, the local patches were assumed to be more or less equal in size, and recolonizing propagules could arrive from any occupied patch. At the other end of a continuum is a mainland–island, or MacArthur–Wilson type metapopulation, in which there is assumed to be one large, stable, essentially immortal population that supports small satellite subpopulations and which serves as the sole source of recolonists. Schematic representations of these types of metapopulations are shown in Figure 1.1.

Empirical work during the past four decades has shown that metapopulations are ubiquitous in nature, although there is debate concerning which type is most common (Harrison & Taylor 1997). Butterflies were the first organisms shown to inhabit metapopulations; the early work of Paul Ehrlich and colleagues on the bay checkerspot *Euphydryas editha bayensis* showed that the Jasper Ridge, California population was in fact composed of several subpopulations linked by very low levels of interpatch dispersal (Ehrlich 1986). The bay checkerspot is a particularly relevant example, because the Jasper Ridge metapopulation has recently gone extinct, and

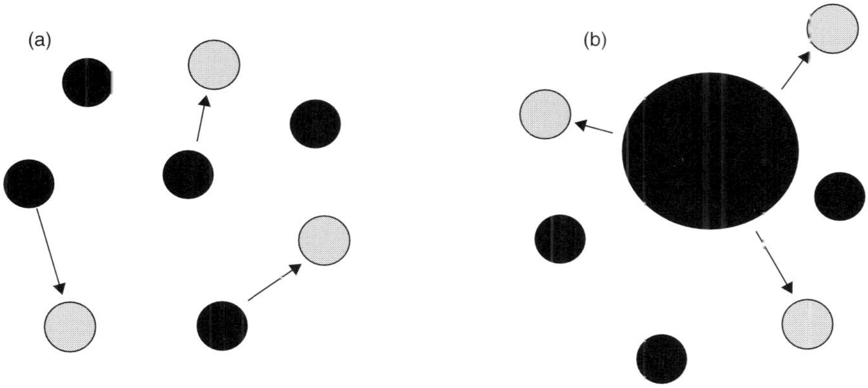

Figure 1.1 Representations of two types of metapopulations. Filled circles represent occupied patches; dotted circles represent extinct patches. (a) A Levins-type metapopulation, in which patch sizes are roughly the same size, and recolonization of extinct patches may be from any other patch. (b) A mainland–island-type metapopulation, with one large, stable population serving as the source of recolonists.

probably will not be recolonized without human intervention. Many species of British butterflies are considered to inhabit metapopulations, often due to the effects of human-based habitat fragmentation (Thomas & Hanski 1997), making it important to be aware of metapopulation theory when planning a conservation strategy.

Interestingly, the concept of a spatially structured population exchanging migrants was developed in population genetic theory, well before the metapopulation concept became established in ecology. This is mostly due to the work of Sewall Wright, who developed his famous 'shifting balance' theory of evolutionary dynamics in the 1930s (Wright 1931). Wright referred to local populations as **demes**, panmictic units that exchanged genes through interpatch migration. What ecologists would consider as a Levins-type metapopulation, as defined above, became known as the **island model** in population genetics. The crucial difference between the ecological and population genetic views was that demes were assumed to be stable, i.e. not subject to local extinctions or to fluctuations in size. In Wright's theory, the stochastic component involved not extinction but changes in gene frequencies within local demes via random genetic drift. Variants of Wright's theory in terms of spatial population structure were developed, the most important of which were the **stepping stone model**, in which gene flow occurred only between neighbouring populations (Kimura & Weiss 1964; Maruyama 1970; Latter 1973), and the **isolation by distance model**, where space was assumed to be continuous (Wright 1943, 1969). Slatkin (1977) added the metapopulation effect, so that local demes could go extinct and be recolonized. Genetic models of structured populations with extinction and recolonization were subsequently elaborated (Wade & McCauley 1988; Whitlock & McCauley 1990). A thorough review of metapopulation genetic theory may be found in Hastings and Harrison (1994).

If metapopulation genetic theory is to be useful, a critically important parameter to estimate is the rate of gene flow between local subpopulations. This is often directly proportional to (but not identical to) the rate of dispersal between subpopulations. In this chapter we shall adhere to a fairly common abuse of language and use the term **migration rate** to mean the number of migrants being exchanged between subpopulations each generation.

Methods for estimating the migration rate are classified into two groups, direct and indirect (Neigel 1997). Direct techniques usually involve a mark–release–recapture scheme. Such methods are straightforward, but do have several drawbacks: (i) observations may be biased toward shorter movements, since short distance dispersers are easier to track; (ii) the reproductive success of migrants may not be the same as that of residents; and (iii) in many cases direct observations are not feasible due to the logistical difficulty of tracking small numbers of marked individuals within large populations (Roderick 1996). The alternative, indirect approach is to use molecular genetic data, and estimate the level of gene flow by fitting statistics drawn from the data to the theoretical values from population genetic models (Slatkin 1985).

By far the most widely used indirect method for estimating migration rates is via Wright's (1950) *F*-statistics. We briefly mentioned Wright's island model above, defining it as a set of spatially discrete local populations, each of constant (effective) size *N*, exchanging a fraction *m* of migrants each generation. Wright's measure of genetic divergence for a collection of subpopulations, F_{ST}, is defined as the ratio of the genetic variance among subpopulations σ_q^2 to the total genetic variance of the whole population. For a single locus with two alleles having frequencies $\bar{q}$ and $(1 - \bar{q})$ in the collection of populations the formula is:

$$F_{ST} = \frac{\sigma_q^2}{\bar{q}(1-\bar{q})} \tag{1.1}$$

With more than two alleles, F_{ST} is generally calculated by weighted average methods described in Weir and Cockerham (1984).

Under the island model of population structure, and assuming that there is no natural selection affecting the gene locus in question, at equilibrium the relationship between F_{ST}, the migration fraction *m* and the population size *N* is:

$$F_{ST} = \frac{1}{1+4Nm} \tag{1.2}$$

so that the effective number of migrants per generation, Nm (sometimes denoted as M (Slatkin 1985)) may be estimated as:

$$Nm = \frac{1}{4}\left(\frac{1}{F_{ST}} - 1\right) \tag{1.3}$$

For genes with haploid inheritance, such as in the mitochondrial DNA, one simply substitutes the factor of 2 for each factor of 4 in the above equations.

While equation (1.3) yields a very simple estimator of the level of dispersal, and a large number of studies have now been published in which Nm is estimated by this method (Slatkin 1985; Roderick 1996), several authors have recently raised serious concerns about its suitability as an estimate of the actual number of migrants per generation. Most of the objections to Nm centre on the assumptions inherent in the underlying model. The populations are assumed to be large and stable; there is no mutation or natural selection affecting the gene frequencies; and the collection of populations is assumed to have reached a migration-drift equilibrium. Bossart and Pashley Prowell (1998) and Whitlock and McCauley (1999) suggest that these assumptions are almost never true for real populations. Mallet (2001) also points out that Nm as estimated from F_{ST} cannot truly be a number of migrants, because it is a simple transformation of a dimensionless number (the ratio of two genetic variances). The computer simulations of fluctuating populations reported by Taylor *et al.* (2000) suggested that Nm may often be a poor estimator of the true migration rate for populations not at equilibrium.

How do indirect estimates of the migration rate compare to direct estimates? In some cases there seems to be good agreement; for example, differences between the two were negligible for one metapopulation of our study species *P. argus* in the Dulas Valley of North Wales (Lewis *et al.* 1997). In other cases there may be very large differences between direct and indirect estimates. In particular, comparisons of the two often reveal what is known as **Slatkin's Paradox**: direct estimates suggest very low values of Nm, which would predict high values of F_{ST}, but measures of F_{ST} at marker loci are quite low ($<<0.1$), which would imply high rates of migration and gene flow. Perhaps the most well known example comes from the checkerspot butterfly *Euphydryas editha*, which differs very little in allozyme frequencies between populations in the Rocky Mountains and the Sierra Nevada in the western USA, despite the fact that dispersal between these areas is virtually non-existent (Slatkin 1987).

There are three explanations that have been offered for Slatkin's Paradox.

1 Balancing selection may be keeping gene frequencies from drifting to fixation (e.g. for allozyme loci). However, values of F_{ST} are often similar for multiple unlinked loci, and it is unlikely that balancing selection could be affecting many loci in the same way (Mallet 2001).

2 Low values of F_{ST} might arise from frequent recolonizations and extinctions, with long-distance founders providing the observed spatial homogeneity of gene frequencies (Slatkin 1987). Note that this second explanation has at the core of it the metapopulation dynamics with which we are concerned in this chapter, so we will return to this point in the discussion of our results. Whitlock and McCauley (1999), however, point out that extinction and recolonization could also **increase** F_{ST}, if recolonizations result in severe bottlenecks of the newly founded populations.

3 The estimates of dispersal by direct methods may be in error, if they miss the occasional long-distance dispersers.

Note that **2** and **3** are somewhat related in that it is the effect of the long-distance dispersers that is 'missed' in both cases (Mallet 2001).

We shall discuss these issues in the light of our genetic data. First, we introduce the study species.

The silver-studded blue *Plebejus argus* and the brown argus *Aricia agestis*

The silver-studded blue butterfly *Plebejus argus* Linnaeus (Lycaenidae) is widely distributed throughout most of Europe; in Britain the species has suffered range decline during the 20th century, and now occurs mainly in heathlands in southern England, and in limestone grasslands in North Wales (Thomas 1985; Thomas *et al.* 1998; Asher *et al.* 2001). The butterfly is often found in large numbers in suitable habitat, with dense clusters of male butterflies common. This species is very sedentary compared to other species of lycaenid butterflies, and thus is thought to be a poor disperser (Thomas 1985; Lewis *et al.* 1997); movements of butterflies more than approximately 400 m are virtually never observed in mark–release–recapture studies. In our study populations in North Wales, *P. argus* has a fragmented spatial distribution because it occurs primarily within calcareous grasslands which, in turn, depend on the distribution of limestone outcroppings (Figure 1.2a). The Great Orme's Head near Llandudno in North Wales supports a large, well-established population of approximately 600 000 adults (Dennis 1977), and this population has served as the source of several colonizations, both natural and from deliberate introductions (Thomas & Harrison 1992).

A natural colonization event occurred after 1983 in the area known as Happy Valley, 1 km northeast of the Great Orme population. Surveys of the Happy Valley site in several years before 1983 had detected no butterflies; by 1990 the newly founded population numbered approximately 1000 adult individuals (Thomas & Harrison 1992). Because of the sedentary nature of the butterfly—the distance between parent and satellite populations, 1000 m, is well beyond the maximum observed dispersal distance—it is thought that the Great Orme–Happy Valley system is an example of a mainland–island type metapopulation, with a very low level of migration from the parent to the satellite population.

A deliberate introduction of *P. argus* was made in the Dulas Valley in North Wales in 1942, at the sight known as Rhyd-y-foel (Marchant 1956). From this founder population the butterfly has subsequently colonized 19 surrounding areas, some of which have themselves undergone extinction and recolonization. Each of the subpopulations in the Dulas Valley is of small to moderate size, with estimated maximum peak population densities in the range 50–2000 adults (Lewis *et al.* 1997). The Dulas Valley system is thus thought to be a fair approximation to a Levins-type metapopulation.

Like *P. argus*, the brown argus *Aricia agestis* (Lycaenidae), has experienced population decline in the 20th century due to destruction and fragmentation of suitable habitat. While approximately 40% of the populations of *A. agestis* had gone extinct by the early 1980s (Thomas & Webb 1984), in southern England the butterfly has now

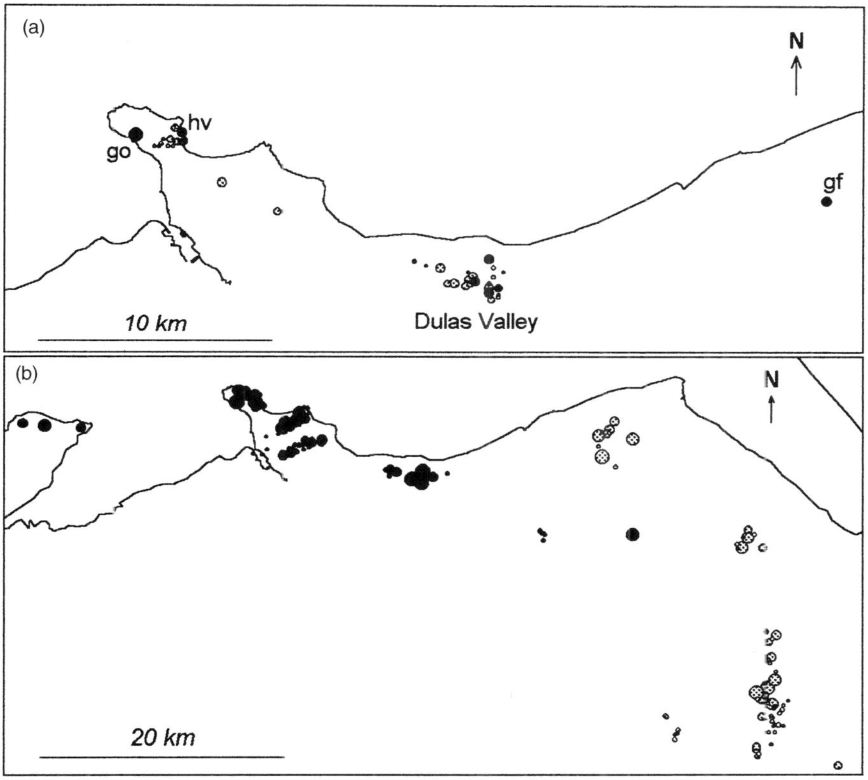

Figure 1.2 (a) The locations of sampling populations of the silver-studded blue *Plebejus argus* in North Wales. gf, Graig Fawr (a human-introduced population of *P. argus*); go, Great Orme; hv, Happy Valley. (b) The same area of North Wales at a larger spatial scale, indicating the locations of populations of the brown argus *Aricia agestis*. Univoltine populations are shown stippled, with bivoltine populations shown in black.

re-established its numbers by spreading to areas where alternative host plants (Geraniaceae) grow (Asher *et al.* 2001). In North Wales it remains restricted to small areas of limestone grassland. Populations of *A. agestis* tend to be small in comparison with those of other lycaenid butterflies, with sizes in the range of 50–100 adults in small habitat patches in Dorset (Bourn & Thomas 1993). The butterfly is a superior disperser compared to other species of lycaenids (particularly in comparison to *P. argus*), with long-range colonizing movements of several kilometres being observed (Asher *et al.* 2001). With its small populations and high dispersal capability, the brown argus is thought to be subject to frequent local extinction and recolonization.

In the study populations of *A. agestis* in North Wales, the butterfly occurs in two types across a gradient in the thermal microclimate. There is a univoltine form of the butterfly (i.e. one generation of adults per year) found in the eastern part of the study range at higher elevations, as well as a bivoltine form found at lower elevations near the coast and in the Great Orme area (Figure 1.2b). Different forms generally do not

Table 1.1 Frequencies of mtDNA haplotypes from the silver-studded blue *Plebejus argus* for the Great Orme (GO) – Happy Valley (HV) system, and the Rhyd-y-foel (RYF) metapopulation.

	Haplotype frequency							
	GO	HV	RYF	T	BW	GG	PN	MM
n	81	85	42	84	50	47	83	90
Allele								
A	—	0.082	—	—	—	—	—	—
B	0.123	0.118	0.071	0.036	0.140	0.170	0.060	0.056
C	0.420	0.365	0.118	0.786	0.620	0.574	0.663	0.433
D	0.160	0.188	0.595	0.119	0.160	0.170	0.157	0.100
E	0.173	0.176	0.095	0.060	0.020	0.043	0.048	0.378
F	—	0.024	0.048	—	0.020	0.021	0.036	—
G	0.025	0.012	—	—	—	0.021	0.012	0.033
H	0.012	—	—	—	—	—	—	—
I	0.037	—	—	—	—	—	—	—
J	0.012	0.024	—	—	—	—	—	—
M	0.012	—	—	—	—	—	0.012	—
S	—	—	—	—	—	—	0.012	—
T	—	—	0.024	—	—	—	—	—
U	0.025	—	—	—	—	—	—	—
V	—	—	—	—	—	—	—	—
A6	—	—	0.095	—	—	—	—	—
A7	—	—	0.048	—	—	—	—	—
A8	—	—	0.024	—	—	—	—	—
A9	—	—	—	—	—	—	—	—

Estimated values for Nm are: Great Orme–Happy Valley, 250; Rhyd-y-foel system, 12.

co-occur within the same metapopulation. Voltinism appears to depend on regional rather than local climatic conditions (Wilson 1999).

The molecular genetic data that we have analysed for *P. argus* consists of mitochondrial DNA restriction haplotypes, from an approximately 780 bp fragment of the mtDNA control region. Collecting methods and molecular genetic techniques are described in Brookes *et al.* (1997). Sample sizes and the observed frequencies of alleles for eight populations are listed in Table 1.1.

For *A. agestis*, 24 enzymes, representing approximately 31 loci, were screened for polymorphism; seven polymorphic loci used in this study are listed in Table 1.2. In addition, mitochondrial DNA haplotypes from the cytochrome b region were amplified for subsets of the individuals screened for allozymes.

For both data sets, values of F_{ST} (see Tables 1.1 and 1.2 for *P. argus* and *A. agestis* respectively) were computed by the method of Weir and Cockerham (1984). Calculated values of F_{ST} were used to estimate Nm using equation (1.3). For *P. argus*, the factor of 4 in equation (1.3) is replaced by 2, as the values of F_{ST} were calculated from mtDNA allele frequencies. Median and maximum observed dispersal distances for both species (Table 1.3) were estimated using mark–release–recapture (methods detailed in Thomas 1985; Wilson 1999).

Table 1.2 Standardized gene frequency variance F_{ST} among univoltine, bivoltine and all populations of the brown argus *Aricia agestis* in North Wales for seven polymorphic enzyme loci.

	F_{ST}		
Locus	Bivoltine	Univoltine	All
Pgi	0.0821	0.0684	0.0969
Pgm	0.0976	0.0849	0.0943
Me	0.0188	0.0464	0.0679
Got-f	0.2651	0.0420	0.1484
G6pd	0.1156	—	0.1450
α Gpd	0.0251	0.0121	0.0337
Mdh-f	0.0965	0.0655	0.0836
All	0.0952	0.0630	0.0983

Table 1.3 Median and maximum dispersal distances recorded in mark–release–recapture experiments for *Plebejus argus* and *Aricia agestis* in North Wales, with estimated mean number of immigrants per generation Nm from molecular genetic data.

	P. argus		
	Dulas Valley	Great Orme	*A. agestis*
No. recaptured	294	10	596
Median distance between captures (m)	15	15	45
Maximum distance (m)	160	50	960
Estimated Nm	12 (11)	250	2.25

For the Dulas Valley system, the estimate of Nm is compared to the estimated migration rate (shown in parentheses) from a mark–release–recapture study.

Results from the genetic analyses show patterns that are in some ways opposite to what one might expect given the dispersal behaviour of the two species. For the Dulas Valley metapopulation of *P. argus*, there is very close agreement between direct estimates of migration rate via mark–release–recapture and indirect estimates using F_{ST}. However, the Great Orme–Happy Valley metapopulation of *P. argus* has a rather high estimate for the migration rate (250 butterflies per generation), despite the observations that the distance between the two populations is well beyond the maximum observed dispersal distance, and that the Happy Valley site was unoccupied for several years before its founding around 1984. Since the Happy Valley population grew rapidly after it was founded, this may be a clear case of the second explanation for Slatkin's Paradox as discussed above, i.e. that extinctions and recolonizations result in populations that are not at genetic equilibrium. The estimates of Nm for the brown argus *A. agestis* are lower than those for *P. argus*, despite the fact that it is a superior disperser.

The results make it clear that more realistic population models may be necessary in order to yield accurate estimates of the migration rate when dealing with non-equilibrium populations. In the following sections, we describe alternative, more complex methods of estimation, including techniques developed for this project which are not based on the assumption of population equilibrium. We then apply these methods to our data, with the goal of explaining the patterns discussed above.

Gene genealogies and the coalescent

The estimate Nm of the migration rate from F_{ST} is a function of one dimensionless number, the ratio of two genetic variances (equation 1.1). The accuracy of this estimator will depend on how well the relevant populations fit the assumptions of the underlying model. Wright, in fact, originally defined his F-statistics as **inbreeding coefficients**, which measured the probability that alleles sampled from the various subpopulations were descended from a common ancestor. F_{ST} may be defined as the probability that two alleles, drawn at random from a set of subpopulations, are identical by descent. This suggests that looking at the family tree, or **genealogy** of a sample of genes from one or more populations, might be very useful in population genetic analysis.

In 1982, J.F.C. Kingman introduced a stochastic process called the **n-coalescent.** The idea of the coalescent is to trace the genealogy of a genetic sample by looking backward in time; i.e. to determine the pattern of ancestral relationships among the sampled genes. A fundamental idea is that in a large, stable population, all of the genes in the population can be traced back in time to a single most recent common ancestor (MRCA) (Kingman 1982a,b). Different alleles in the sample are produced by mutations that have accumulated in the descendent lineages.

To understand the coalescent process, consider a sample of two alleles drawn at random from a haploid Wright–Fisher population of size N. The probability that these two alleles have descended from a common ancestor in the previous generation (i.e. that they have the same parent) equals $1/N$ (in a Wright–Fisher population, all individuals in the previous generation are equally likely to be parents). The probability that the two sampled genes do **not** share a common ancestor in the previous generation—or in other words, that their lineages do not **coalesce** one step back in time—is thus $1 - 1/N$. For a sample of n individuals, this probability, denoted $P\{n\}$, is

$$P\{n\} = \prod_{j=1}^{n-1}(1 - j/N) \approx 1 - \frac{\binom{n}{2}}{N} \tag{1.4}$$

with the approximate relation on the right becoming more accurate as the population size N increases (Hudson 1990). The central idea of the coalescent is that this relation continues as we move back in time toward the most recent common ancestor of the genes in the sample: if we have n separate ancestors one generation back (corresponding to no coalescences), then the same relation applies for the grandparents' generation two steps back in time, and so on. Thus, if we could view the genealogy of

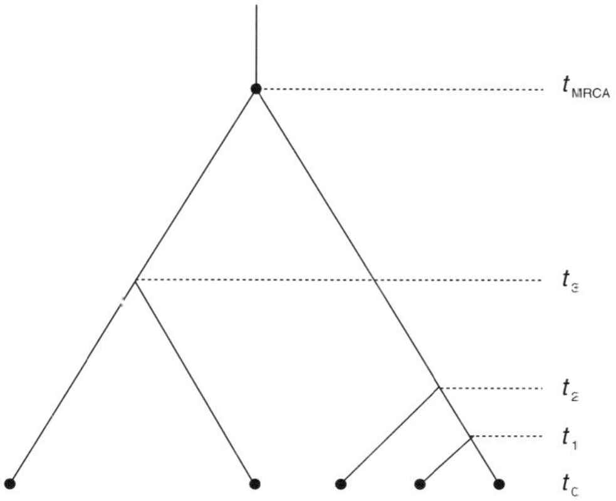

Figure 1.3 The coalescent process, tracing the genealogy of a sample of five genes from the present (t_0) to the time of the most recent common ancestor (t_{MRCA}). The times refer to coalescent events in order, moving backward in time to the most recent common ancestor.

a sample of alleles, we would see the lineages coalescing at rate $(n/2)/N$. In the formal coalescent process, time is continuous and is scaled in terms of the population size, so that the rate of coalescence is $\binom{n}{2}/N$. Note that this is a limit result as the population size N becomes infinite (meaning, for example, that there is negligible probability of two or more coalescences in the same generation). The coalescent process for a single population is illustrated in Figure 1.3.

The coalescent process yields the probability of a gene genealogy. Mutations occur along the coalescent tree; since (neutral) mutations are assumed to be rare, random events, they may be assumed to follow a Poisson distribution, so that the times between mutation events will be exponentially distributed with the mean equal to the product of the mutation rate and the number of lineages present at time t. Thus if μ is the rate parameter for mutation, and defining θ as the limit of $\theta = 4N\mu$ for large N and small μ, the lineages in the genealogy are coalescing and changing into new forms at rate $n(n + \theta - 1)/2$ (Griffiths 1980; Tavaré 1984).

In order to include migration, the coalescent may be extended to a set of populations that exchange migrants; this is the so-called **structured coalescent** (Takahata 1988; Notohara 1990; Hey 1991; Nath & Griffiths 1993, 1996; Herbots 1997; Wilkinson-Herbots 1998; Bahlo & Griffiths 2000). In a structured coalescent lineages not only coalesce but also leave the population as they migrate backward in time, so that in a collection of populations the number of lineages present at (backward) time t may increase or decrease, again until only one most recent common ancestor is left. To make the connection between F_{ST} and the coalescent, note that F_{ST} may be expressed in terms of the probability of identity by descent:

$$F_{ST} = \frac{f_0 - \bar{f}}{1 - \bar{f}} \tag{1.5}$$

where f_0 is the probability that two genes drawn at random from the same subpopulation are identical by descent, and $\bar{f}$ is the analogous probability for two genes drawn at random from the collection of subpopulations. Slatkin (1991) showed that, assuming negligible mutation, this expression could be written in terms of mean coalescence times:

$$F_{ST} = \frac{E(T) - E(T_0)}{E(T)} \tag{1.6}$$

where $E(T)$ is the average coalescence time of two genes drawn from the same subpopulation, and $E(T_0)$ is the analogous average coalescence time for two genes drawn at random from the collection of subpopulations. Of course, as we stated earlier, the collection of subpopulations is assumed to be stable and to have reached genetic equilibrium, and each subpopulation is assumed to be large relative to the sample size.

If the genealogy of a genetic sample is known, then the simple coalescent or its structured analogue can be used to compute the likelihood of the data given the parameters. Since the genealogy of the sample is unknown, however, the genealogy must be considered as missing data, and the likelihood calculated by averaging over all possible genealogies. If P represents the vector of parameters (including the population size, the mutation rate, and the interpopulation migration rate(s)), then the probability of the data given the model parameters, $P\{D|P\}$, is

$$P\{D|P\} = \sum_G P\{D|G\}P\{G|P\} \tag{1.7}$$

The probability of the data given the genealogy, $P\{D|G\}$, may be computed using standard phylogenetic methods (Swofford *et al.* 1996), and the probability of the genealogy given the parameters, $P\{G|P\}$, is given by the coalescent.

Equation (1.7) implies that one must sum over G, the set of all possible genealogies. However, there are an infinite number of possible genealogies, even in a sample of only two alleles (Beerli & Felsenstein 1999). The way around this difficulty is to average over a finite sample of genealogies that are drawn in proportion to their likelihood, in the same manner that a parametric mean can be estimated by the mean of a finite sample. This is accomplished via Markov Chain Monte Carlo (MCMC), a method by which samples may be drawn from any likelihood distribution, no matter how complicated (Gilks *et al.* 1995).

A Markov chain is a discrete-time stochastic process that moves among a set of states. With each pair of states is associated a transition probability, the probability of moving from a given state X, say, to the new state Y; the chain must also specify the probability of moving from Y to X (which is not necessarily the same as vice versa). For example, one simple Markov chain is the random walk, where the states are the integers and where a particle may move one step to the left or one step to the right. Such Markov chains may be easily implemented on the computer.

Consider now a Markov chain such as a random walk between two fixed, reflecting barriers. If a particle is at some initial state, we allow it to move among states by choosing a new state at random, where the only possible new states are one step to the left and one step to the right (except at the two reflecting barriers, where there is zero probability of a step beyond the barrier). As time increases, the proportions of moves that the particle makes to each state will converge to fixed values: we say that the Markov chain has a **limiting distribution**, or **invariant distribution**. Not all Markov chains have invariant distributions: for example, if there are no barriers in our random walk example, the particle will eventually 'walk away to infinity', and thus the proportion of time spent in each state will in fact approach zero. However, many Markov chains do have invariant distributions, and the principle of Markov Chain Monte Carlo is to construct such a chain having our target distribution as its invariant distribution.

An appropriate Markov chain can be constructed via the Hastings–Metropolis algorithm (Metropolis *et al.* 1953; Hastings 1970). Given that the Markov chain is in state X, we propose a new state Y, using a proposal distribution denoted as $Q(Y|X)$. The chain moves to Y with probability

$$P\{Y|X\} = \min\left\{1, \frac{\pi(Y)Q\{X|Y\}}{\pi(X)Q\{Y|X\}}\right\} \tag{1.8}$$

where $\pi(\cdot)$ is the likelihood of the state, and $Q(\cdot|\cdot)$ is our proposal distribution that specifies the conditional probability of moving from one state to another. If the probability of moving to the proposed state is independent of the current state, or if the conditional probabilities are equal in both directions (i.e. if $Q(Y|X) = Q(X|Y)$), then the Q terms in equation (1.8) divide out, so that the expression simplifies to

$$P\{Y|X\} = \min\left\{1, \frac{\pi(Y)}{\pi(X)}\right\} \tag{1.9}$$

which is called the Metropolis algorithm. Note that with equation (1.8) if the likelihood of the new state Y is higher than that of the current state X, then the chain **always** moves to the new state; if it is lower, then it moves to the new state with probability equal to the ratio of the likelihoods. After specifying an initial state, the Markov chain is allowed to run for a burn-in period of sufficient length such that subsequent states are chosen according to the target invariant distribution.

To estimate migration rates using the coalescent, one may compute the likelihood of the genetic sample data given the parameters of migration and mutation rates, by drawing genealogies via MCMC and using the coalescent to compute the likelihood of the data for each genealogy. The likelihood of the genealogy G, including migration events, given the model parameters P is

$$P\{G|P\} = \prod_{i=1}^{2}\left(\left[\frac{2}{\theta_i}\right]^{v_i} M_i\right)\prod_{j=1}^{T} p_j \tag{1.10}$$

where $\theta_i = 4N_e^i\mu$, the product of the effective population size N_e in population i and the mutation rate μ, M is the rate of migration into patch i divided by the mutation rate, v_i is the total number of coalescences in subpopulation i, and where the p_j are given by

$$p_j = \exp\left(-u_j \sum_{i=1}^{2}\left(k_{ij}M_i + \frac{k_{ij}(k_{ij}-1)}{\theta_i}\right)\right)$$

where u_j is the time interval between coalescent or migration events, and k_{ij} is the number of lineages in the sample in population i during time interval j (Beerli & Felsenstein 1999). An alternative MCMC approach based on the structured coalescent uses a recursion scheme originally developed by Griffiths and Tavaré (1994), and is implemented in Nath and Griffiths (1996) and Bahlo and Griffiths (2000).

Branching processes with immigration

The coalescent methods are based on the assumption of large, stable population sizes. For small, stochastically growing populations, a more realistic model may be needed in order to obtain accurate estimates of the migration rate. The mathematical model upon which our analysis is based is called a Galton–Watson branching process with immigration. We first explain what a branching process is.

In the *Educational Times* of 1 April 1873, Sir Francis Galton, the noted cousin of Charles Darwin, posed a mathematical puzzle concerning the survival and extinction of surnames within a population. The motivation of the problem was the disappearance of surnames in the British peerage, a subject of some controversy at the time. Given, say, one initial ancestor, what is the chance that the family name will survive through n successive generations? Obviously, survival of the surname in the first generation depends on the founder having one or more sons, since daughters will not pass on the surname. The survival of the surname in the next generation then depends on the likelihood that male grandchildren are produced by the founder's sons, and so on. How likely is the surname to eventually go extinct? A solution was offered by the Rev. Henry William Watson, and from their joint paper of 1874 the basic branching process takes its name of the **Galton–Watson Process**.

A schematic representation of the Galton–Watson process is shown in Figure 1.4. Beginning with one or more initial ancestors, the members of the population reproduce according to a given offspring probability model. The offspring in turn produce their own offspring, and this may be repeated *ad infinitum*. Individual lineages may die out when descendants leave no offspring, but the clan comprising all the descendants of the initial ancestor may persist. The mathematical techniques for analysing branching processes allow us to find the probability that the process will eventually go extinct, along with the distribution and expected number of lineages alive in each generation.

In the case of a spatially structured population, we can modify the Galton–Watson process to include immigration. We focus on one individual local subpopulation, with the goal of estimating the average number of immigrants per generation (emigration from the population can be included within the offspring distribution). Ini-

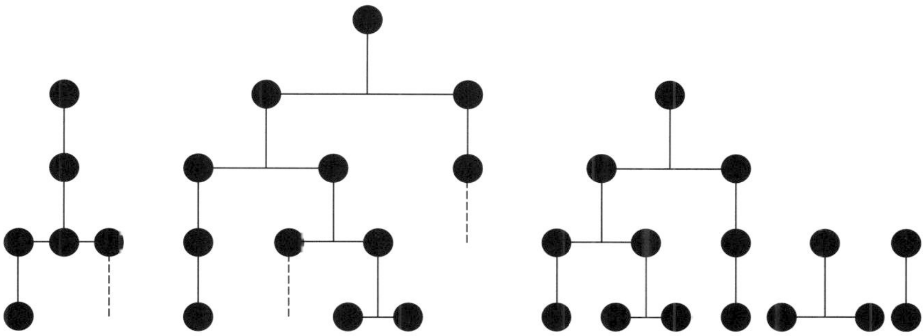

Figure 1.4 The Galton–Watson branching process with immigration. The filled circles represent individuals, which produce offspring each discrete time step. At each time step a random number of individuals immigrate into the population, thus forming their own independent branching process. Individual lineages may die out, as indicated by the dashed lines, when individuals produce no offspring.

tially the patch may be assumed to be empty. At each generation after founding a random number of immigrants arrive; the initial immigrants are the recolonizers/ founders of the subpopulation. In each successive generation the number of immigrants is a discrete random variable, as is the number of offspring produced by the members of the population. Note that the independence of reproduction assumption inherent in the branching process technically means that the individuals reproduce asexually; in terms of genetic data it means that the branching process is most applicable to genes with haploid inheritance such as those of the mitochondrial DNA, although diploid genes may also be used if mating is random and there is no bias in the transmission of gametes. A most important assumption is that mutation is negligible on the timescale of interest, so that for each allele at the gene locus, there is a corresponding branching process with immigration. While it is possible to include mutation in the Galton–Watson process, this makes the mathematics much more complicated. This is one aspect of a metapopulation that makes the analysis easier; with frequent local extinctions the age of a subpopulation may often be such that mutation has had no effect on the allele frequencies. Related to this is the assumption that the subpopulation is still in a period of unrestricted growth, i.e. that non-linear, density-dependent effects are not strongly influencing population dynamics.

The branching process with immigration is a realistic model of a recently founded, growing population within a metapopulation. The reason that it has not been more widely applied is that it is mathematically much less tractable than the standard migration models in population genetics. To apply the process to our genetic data from *P. argus* and *A. agestis* we will use several methods, one based on the diffusion approximation to the branching process with immigration, and the others employing the Markov Chain Monte Carlo techniques described in the previous section.

The diffusion approximation to the branching process with immigration takes place in continuous time, and is controlled by three parameters: λ, the instantaneous birth rate; μ, the instantaneous death rate, and γ, the instantaneous immigration rate. Briefly, the process evolves as follows. In a small interval of time Δt, an individual may immigrate into the population with probability $\gamma\Delta t$. If there are i individuals alive at time t, then one individual in the population may give birth to a new individual with probability $i\lambda\Delta t$, and similarly an individual may die (or emigrate) with probability $i\mu\Delta t$. As usual, we assume that there is no mutation, so that the various allelic types form their own independent clans. In terms of the discrete branching process with immigration described above, the continuous-time process approximates a discrete process with a **geometric** offspring distribution (Harris 1963).

Under the continuous-time process, it may be shown that the probability of a sample from a single gene locus with d alleles $\mathbf{n} = (n_1, n_2, \ldots, n_d)$, given the frequencies of the immigrant alleles from the parent population(s) $(p_1, p_2, \ldots, p_d)$ is given by:

$$P(n_1, n_2, \ldots, n_d) = \binom{N+\theta-1}{N}^{-1} \prod_{i=1}^{d} \binom{n_i+\theta p_i-1}{n_i} \tag{1.11}$$

where $\theta = \gamma/\lambda$, or the ratio of the immigration rate to the birth rate, and N is the total sample size. This distribution is known as the **compound multinomial-Dirichlet** (Rannala & Hartigan 1996). There are two very interesting properties of the distribution (equation 1.11). First, note that there is only one free parameter, θ, which is actually the ratio of the rate of immigration to the birth rate. Thus, a growing subpopulation cannot be distinguished from a declining, or **sink**, subpopulation on the basis of a genetic sample. Second, note that the distribution is invariant to time: t does not appear in the distribution, meaning that it is valid for a subpopulation of any age and size.

The sampling formula (equation 1.11) is extremely convenient, as it leads to an explicit expression for the likelihood function. However, our true interest is in the parameters γ and λ separately, since what we really wish to know is the migration rate. To this end we discuss two methods based on the discrete branching process.

If the age of the subpopulation of interest is known, then a simple first method of estimation is via implicit statistical analysis (Diggle & Gratton 1984; Moilanen 1999). We integrate the probability of the sample vector $\mathbf{n}$ given the branching process with immigration at generation k over all possible branching processes of this age, given the parameters λ and γ, weighted by the likelihood of the branching process given the two parameters. Of course, as in the genealogies discussed in the section on coalescent methods, there are too many branching processes to sum over all of them, but with the age of the population given we can sample them in proportion to their likelihood by simulating the process on the computer, and averaging over a finite sample. We then use MCMC to move among points in parameter space, i.e. sets of values of λ and γ. The likelihood of the data given the parameters is then given by an equation analogous to equation (1.7):

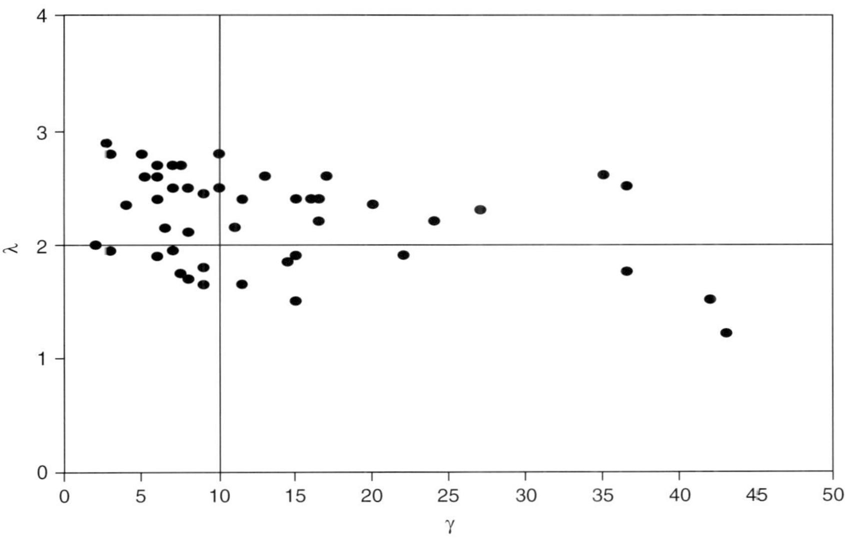

Figure 1.5 Results from a sensitivity analysis of the branching process method (first method, using implicit statistical analysis) of estimating the migration rate γ and population growth rate λ. Fifty simulated data sets were constructed, using a one-locus system with five alleles, having allele frequencies 0.35, 0.25, 0.20, 0.15 and 0.05. Simulated populations were allowed to grow for six generations after founding. For the simulations, the values of γ and λ were 10.0 and 2.0, repectively (shown as crossed lines on the graph). Most estimates are reasonably close to the true values, but as with estimates using F_{ST}, there are often outliers showing upward bias.

$$L_P(\mathbf{n}) = \frac{1}{M}\sum_{j=1}^{M} P\{\mathbf{n} \mid BP_j\} P\{BP_j \mid P\} \tag{1.12}$$

where M is the number of branching processes sampled.

Results from a sensitivity analysis of this method (D. E. Taneyhill, in preparation) show that estimates of γ, the number of immigrants per generation, are often strikingly similar to those obtained from F_{ST} via equation (1.3) (Figure 1.5). In particular, there may be upward bias when there is little difference between the proportions of the alleles in the sample and those of the alleles in the parent population, in the same manner that F_{ST} may be biased for growing populations. This is because the sampled branching processes will contain many with large amounts of migration and large final population sizes.

To improve the estimates, we may make use of knowledge of the current population size, thus avoiding any branching processes that have unrealistic numbers of migrants. However, we cannot simulate such processes on the computer, because only a vanishingly small proportion of such simulated processes would match the target population size. To solve this problem we can instead sample branching processes via Markov Chain Monte Carlo. We specify a starting process, defined in terms of the number of immigrants in each generation and the sizes of their clans of descendents in the current generation. We then construct a Markov chain designed to move among processes, using the Hastings–Metropolis algorithm. In our scheme, new processes are proposed by, for example, deleting one migrant and assigning its clan members to other clans, or by introducing one new migrant and deleting members of other clans. Assuming a Poisson number of immigrants each generation with parameter γ, and a geometric offspring distribution with mean population growth rate λ, and gene frequencies of immigrants $\mathbf{p}$, it may be shown (Taneyhill, in preparation) that the likelihood of each process given the parameters is

$$P(BP|\lambda,\gamma,\mathbf{p}) = P\left(\mathbf{M},\sum_{j=1}^{M} Z_j \,|\, \lambda,\gamma,\mathbf{p}\right)$$

$$= \prod_{k=1}^{d}\left\{\frac{e^{-p_k\gamma^n}(p_k\gamma)^M}{m_{k1}!m_{k2}!\dots m_{kn}!}\prod_{j=1}^{n}\lambda^{m_{kj(n-j)}}\left[\frac{1-s_0}{\lambda^j - s_0}\right]\left[\frac{\lambda^j - 1}{\lambda^j - s_0}\right]^{\sum(Z_{m_{jk}}-1)}\right\} \qquad (1.13)$$

where $\mathbf{M}$ is the matrix of immigrants m_{jk} whose entries are given by the number of immigrants j in generation k; M is the total number of immigrants that have arrived up to generation n; and s_0 is the probability that a clan ever goes extinct, which can be calculated explicitly (Harris 1963). The addition of the population size to the data allows for more accurate estimates of γ in simulated data sets (Figure 1.6).

Estimates of migration rates from branching process models

For the silver-studded blue *P. argus*, estimates of the migration rate in the Great Orme–Happy Valley system (Table 1.4) were much lower compared to the value estimated from F_{ST}, and more realistic given the dispersal behaviour of this species. The continuous-time estimate of about 15 butterflies per generation compares well with the discrete time estimates of $\gamma \approx$ 10–12 butterflies per generation. With an estimated population growth rate λ of about 1.9, these parameter values were fully adequate, in computer simulations, to preserve the genetic diversity observed in the Happy Valley population.

For the Dulas Valley system, the agreement was extremely good between the indirect estimates from allozymes (12 butterflies per generation) and from direct mark–release–recapture (11 butterflies per generation, assuming an effective population size about one-half of the census size; see Lewis *et al.* 1997). The branching process estimate of θ, 28.4, is in very good agreement; interpreting $\theta/2$ as Nm yields an estimated 14 effective immigrants per generation.

For the brown argus *A. agestis* there was good agreement between estimates computed from F_{ST} and those from the continuous-time branching process (Table 1.5). The estimates of the effective numbers of migrants per generation were very low for many loci, especially in the bivoltine populations, where the estimated number (two or less for the majority of the loci) implies significant genetic divergence between populations.

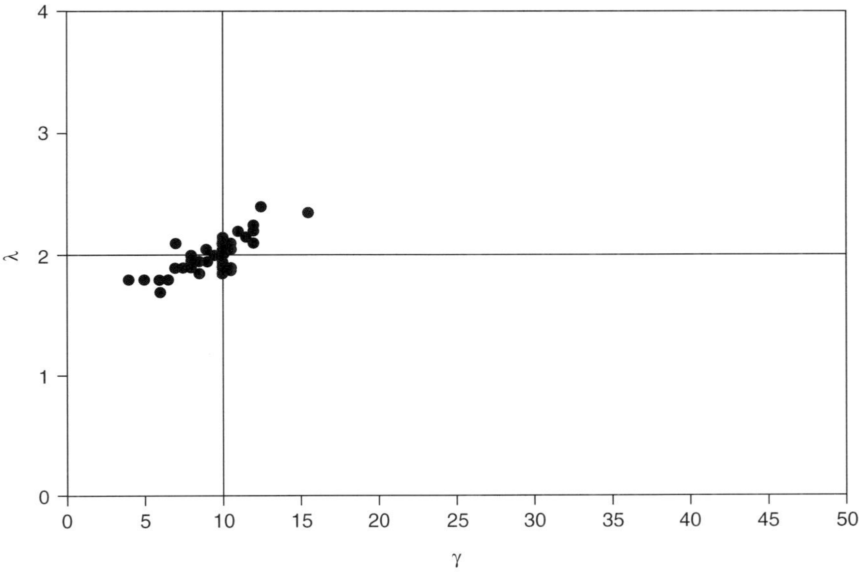

Figure 1.6 Results from a sensitivity analysis of the branching process method (second method, using Markov Chain Monte Carlo selection of branching processes). The same 50 simulated data sets were used as in Figure 1.5 (data sets were constructed using a one-locus system with five alleles, having allele frequencies 0.35, 0.25, 0.20, 0.15 and 0.05). Simulated populations were allowed to grow for six generations after founding. For the simulations, the values of γ and λ were 10.0 and 2.0, respectively (shown as crossed lines on the graph). The estimates are clustered near the true values; the greater accuracy with this method is due to the use of knowledge of the population sizes.

Table 1.4 Estimates of the migration rate (number of migrants per generation) for the silver-studded blue *Plebejus argus* in two metapopulations in North Wales. Estimates compared from three methods: F_{ST}, the continuous-time branching process and the discrete-time branching process.

	F_{ST}	BP (cont.)	BP (discrete)
Great Orme–Happy Valley	250	5	10–12
Dulas Valley	12	14	—

Conclusions

The application of metapopulation concepts to our data has allowed us to understand spatial patterns of genetic variation which, on the surface, seem at odds with the biology of the two species of butterflies. The silver-studded blue *P. argus* is a poor disperser, yet at least one set of genetic data (in the Great Orme–Happy Valley system) displays Slatkin's Paradox: F_{ST} is very low, which would imply a fairly high amount of long-distance dispersal. Our knowledge of the recent recolonization and

Table 1.5 Estimates of the number of immigrants per generation in *Aricia agestis* in North Wales, for seven allozyme loci. Estimates are from the continuous-time branching process using equation (1.11) in the text, estimated as $\theta/4$, with estimates from F_{ST} in parentheses. Cases where the likelihood function did not have a maximum are denoted (—), as are those for which F_{ST} could not be calculated.

Locus	Univoltine	Bivoltine
Pgi	2.25 (3.4)	2.0 (2.8)
Pgm	4.0 (2.69)	1.96 (2.31)
Me	5.4 (5.14)	14.69 (13.05)
Got-f	5.6 (5.7)	1.00 (0.69)
G6pd	– (–)	1.05 (1.91)
α-Gpd	– (20.4)	9.0 (9.7)
Mdh-f	2.2 (3.57)	– (2.34)

rapid growth of the Happy Valley subpopulation has given us the chance to make more realistic estimates. Analysis of the data from *P. argus* in the Dulas Valley has shown that within about 50 generations of founding, the metapopulation had moved close enough to genetic equilibrium so that it closely resembled Wright's island model, and thus both of the techniques of inference that we have employed—F_{ST} and branching processes—agree well with each other and with the direct estimates obtained from mark–release–recapture studies. The Dulas Valley system is indeed a fortunate occurrence: the subpopulation sizes are small enough, the rate of migration high enough, and the rate of population extinction apparently low enough so that all of the methods of estimation converge.

The spatial genetic patterns seen for allozyme loci in the brown argus *A. agestis* seem to be a clear example of the point made by Whitlock and McCauley (1999) concerning Slatkin's Paradox: extinction and recolonization could also have the opposite effect, i.e. cause high values of F_{ST}, if the founding populations are small enough to induce genetic bottlenecks. Thus the small populations of *A. agestis* apparently generate founder effects after recolonization, with the resulting high levels of genetic divergence and high values of F_{ST}. There are several factors that could add to this effect. There may be very low **effective** numbers of founders during recolonization, for example if most of the lineages that are founded during a recolonization event originate from only a very few females. A second possibility is that gene frequencies are affected by natural selection due to local adaptation to patch conditions; in the North Wales populations the thermal gradient in microclimate may be affecting frequencies of allozyme alleles. Our results support the idea of the voltinism forms as local races of the same species, and the estimates of the effective numbers of migrants per generation are low enough so that genetic isolation is indeed possible.

While the vast majority of studies involving indirect estimates of gene flow have been based on F_{ST}, very few of the authors of those studies have verified that their populations conform to the underlying population genetic model. Since metapopulations seem to be ubiquitous (Harrison & Taylor 1997), we suggest that metapopulation dynamics may often cause inaccurate estimation of interpopulation migration rates. Our results show the need for application of non-equilibrium methods of estimation. With the techniques that we have discussed in this chapter now available, we recommend that researchers consider the effects of non-equilibrium population dynamics whenever indirect methods are used to infer rates of migration and gene flow.

Acknowledgements

This work was funded by the Natural Environment Research Council's thematic programme EDGE (Ecological Dynamics and Genes).

References

Andrewartha, H.G. & Birch, L.C. (eds) (1954) *The Distribution and Abundance of Animals.* University of Chicago Press, Chicago.

Asher, J., Warren, M., Fox, R., Harding, P., Jeffcoate, G. & Jeffcoate, S. (2001) *The Millennium Atlas of Butterflies in Britain and Ireland.* Oxford University Press, Oxford.

Bahlo, M. & Griffiths, R.C. (2000) Inference from gene trees in a subdivided population. *Theoretical Population Biology* **57**, 79–95.

Beerli, P. & Felsenstein, J. (1999) Maximum likelihood estimation of migration rates and effective population numbers in two populations using a coalescent approach. *Genetics* **152**, 763–773.

Bossart, J.L. & Pashley Prowell, D. (1998) Genetic estimates of population structure and gene flow: limitations, lessons and new directions. *Trends in Ecology & Evolution* **13**, 202–206.

Bourn, N.A.D. & Thomas, J.A. (1993) The ecology and conservation of the brown argus butterfly *Aricia agestis* in Britain. *Biological Conservation* **63**, 67–74.

Brookes, M.I., Graneau, Y.A., King, P. *et al.* (1997) Genetic analysis of founder bottlenecks in the rare British butterfly *Plebejus argus. Conservation Biology* **11**, 648–661.

den Boer, P. (1968) Spreading of risk and stabilization of animal numbers. *Acta Biotheoretica* **18**, 165–194.

Dennis, R.L.H. (1977) *The British Butterflies: Their Origin and Establishment.* E.W. Classey, Oxon, UK.

Diggle, P.J. & Gratton, R.J. (1984) Monte Carlo methods of inference for implicit statistical models. *Journal of the Royal Statistical Society, Series B* **46**, 193–227.

Endler, J.A. (1977) *Geographic Variation, Speciation, and Clines.* Princeton University Press, Princeton, NJ.

Ehrlich, P. (1986) *The Machinery of Nature.* Simon and Schuster, New York.

Ehrlich, P. & Raven, P. (1969) Differentiation of populations. *Science* **165**, 1228–1232.

Futuyma, D.J. (1987) On the role of species in anagenesis. *American Naturalist* **130**, 465–473.

Gilks, W.R., Richardson, S. & Spiegelhalter, D.J. (1995) Introducing Markov Chain Monte Carlo. In: *Markov Chain Monte Carlo in Practice* (eds W. R. Gilks, S. Richardson, & D. J. Spiegelhalter), pp. 1–19. Chapman & Hall, London.

Griffiths, R.C. (1980) Lines of descent in the diffusion approximation of neutral Wright–Fisher models. *Theoretical Population Biology* **17**, 37–50.

Griffiths, R.C. & Tavaré, S. (1994) Simulating probability distributions in the coalescent. *Theoretical Population Biology* **46**, 131–159.

Hanski, I. (1999) Metapopulation dynamics. *Nature* **396**, 41–49.

Hanski, I. & Gilpin, M. (eds) (1997) *Metapopulation Biology: Ecology, Genetics, and Evolution.* Academic Press, San Diego, CA.

Hanski, I. & Simberloff, D. (1997) The metapopulation approach, its history, conceptual domain, and application to conservation. In: *Metapopulation Biology* (eds I. Hanski & M. Gilpin), pp. 5–26. Academic Press, San Diego, CA.

Harris, T.E. (1963) *The Theory of Branching Processes.* Prentice-Hall, New York.

Harrison, S. & Taylor, A. (1997) Empirical evidence for metapopulation dynamics. In: *Metapopulation Biology* (eds I. Hanski & M. Gilpin), pp. 27–42. Academic Press, San Diego, CA.

Hastings, A. & Harrison, S. (1994) Metapopulation dynamics and genetics. *Annual Review of Ecology and Systematics* **25**, 167–188.

Hastings, W.K. (1970) Monte Carlo sampling methods using Markov chains and their applications. *Biometrika* **57**, 97–109.

Herbots, H.M. (1997) The structured coalescent. In: *Progress in Population Genetics and Human Evolution* (eds P. Donnely & S. Tavaré). *IMA Volumes in Mathematics and its Applications*, Vol. 87, pp. 231–255, Springer-Verlag, Berlin.

Hey, J. (1991) A multi-dimensional coalescent process applied to multi-allelic selection models and migration models. *Theoretical Population Biology* **39**, 30–48.

Hudson, R.R. (1990) Gene genealogies and the coalescent process. In: *Oxford Surveys in Evolutionary Biology* (eds D. J. Futuyma & J. Antonovics), Vol. 7, pp. 1–44. Oxford University Press, Oxford.

Kimura, M. & Weiss, G. H. (1964) The stepping stone model of population structure and the decrease of genetic correlation with distance. *Genetics* **49**, 561–576.

Kingman, J.F.C. (1982a) The coalescent. *Stochastic Processes and Their Applications* **13**, 235–248.

Kingman, J.F.C. (1982b) On the genealogy of large populations. *Journal of Applied Probability* **19a**, 27–43.

Latter, B.D.H. (1973) The island model of population differentiation: a general solution. *Genetics* **73**, 147–157.

Lewis, O.T., Thomas, C.D. & Hill, J.K. (1997) Three ways of assessing metapopulation structure in the butterfly *Plebejus argus. Ecological Entomology* **22**, 283–293.

Levins, R. (1969) Some demographic and genetic consequences of environmental heterogeneity for biological control. *Bulletin of the Entomological Society of America* **15**: 237–240.

Mallet, J. (2001) Gene flow. In: *Insect Movement: Mechanisms and Consequences* (eds I. P. Woiwod, D. R. Reynolds & C. D. Thomas), pp. 337–360. CAB International, Wallingford, UK.

Marchant, A.J. (1956) *Plebejus argus* L. (Lycaenidae) *caernensis* Thompson at Llandulas and Rhyd-y-Foel. *Entomologist* **89**, 195–211.

Maruyama, T. (1970) Effective number of alleles in a subdivided population. *Theoretical Population Biology* **1**, 273–306.

Mayr, E. (1942) *Systematics and the Origin of Species.* Columbia University Press, New York.

Metropolis, N., Rosenbluth, A.W., Rosenbluth, M.N., Teller, A.H. & Teller, E. (1953) Equations of state calculations by fast computing machines. *Journal of Chemical Physics* **21**, 1087–1092.

Moilanen, A. (1999) Patch occupancy models of metapopulation dynamics: efficient parameter estimation using implicit statistical inference. *Ecology* **80**, 1031–1043.

Nath, H.B. & Griffiths, R.C. (1993) The coalescent in two colonies with symmetric migration. *Journal of Mathematical Biology* **31**, 841–851.

Nath, H.B. & Griffiths, R.C. (1996) Estimation in an island model using simulation. *Theoretical Population Biology* **50**, 227–253.

Neigel, J.E. (1997) A comparison of alternative strategies for estimating gene flow from genetic markers. *Annual Review of Ecology & Systematics* **28**, 105–128.

Notohara, M. (1990) The coalescent and the genealogical process in geographically structured populations. *Journal of Mathematical Biology* **29**, 59–75.

Rannala, B. & Hartigan, J.A. (1996) Estimating gene flow in island populations. *Genetical Research, Cambridge* **67**, 147–158.

Roderick, G.K. (1996) Genetic structure of insect populations: gene flow, phylogeography, and their uses. *Annual Review of Entomology* **41**, 325–352.

Slatkin, M. (1977) Gene flow and genetic drift in

a species subject to frequent local extinctions. *Theoretical Population Biology* **12**, 253–262.

Slatkin, M. (1985) Gene flow in natural populations. *Annual Review of Ecology & Systematics* **16**, 393–430.

Slatkin, M. (1987) Gene flow and the geographic structure of natural populations. *Science* **236**, 787–792.

Slatkin, M. (1991) Inbreeding coefficients and coalescence times. *Genetical Research, Cambridge* **58**, 167–175.

Swofford, D., Olsen, G., Waddell, P. & Hillis, D. (1996) Phylogenetic inference. In: *Molecular Systematics* (eds D. Hillis, C. Moritz & B. Mable), pp. 407–514. Sinauer Associates, Sunderland, MA.

Takahata, N. (1988) The coalescent process in two partially isolated diffusion populations. *Genetical Research, Cambridge* **52**, 213–222.

Taneyhill, D.E. (2000) Metapopulation dynamics of multple species: the geometry of competition in a fragmented habitat. *Ecological Monographs* **70**, 495–516.

Tavaré, S. (1984) Line-of-descent processes, and their application in population genetics models. *Theoretical Population Biology* **26**, 119–164.

Taylor, B.L, Chivers, S.J., Sexton, S. & Dizon, A.E. (2000) Evaluating dispersal estimates using mtDNA data: comparing analytical and simulation approaches. *Conservation Biology* **14**, 1287–1297.

Thomas, C.D. (1985) The status and conservation of the butterfly *Plebejus argus* (Lepidoptera: Lycaenidae) in northwestern Britain. *Biological Conservation* **33**, 29–51.

Thomas, C.D. & Hanski, I. (1997) Butterfly metapopulations. In: *Metapopulation Biology* (eds I. Hanski & M. Gilpin), pp. 359–401. Academic Press, San Diego, CA.

Thomas, C.D. & Harrison, S. (1992) Spatial dynamics of a patchily distributed butterfly species. *Journal of Animal Ecology* **61**, 437–446.

Thomas, C.D., Hill, J.K. & Lewis, O.T. (1998) Evolutionary consequences of habitat fragmentation in a localized butterfly. *Journal of Animal Ecology* **67**, 485–497.

Thomas, J.A. & Webb, N.R. (1984) *Butterflies of Dorset.* Dorset Natural History and Archaeological Society, Dorchester, UK.

Turner, M.G. (1989) Landscape ecology: the effect of pattern on process. *Annual Review of Ecology & Systematics* **20**, 171–197.

Wade, M.J. & McCauley, D.E. (1988) Extinction and recolonization: their effects on the genetic differentiation of local populations. *Evolution* **42**, 995–1005.

Weir, B.S. & Cockerham, C.C. (1984) Estimating F-statistics for the analysis of population structure. *Evolution* **38**, 1358–1370.

Whitlock, M.C. & McCauley, D.E. (1990) Some population genetic consequences of colony formation and extinction: genetic correlations with founding groups. *Evolution* **44**, 1717–1724.

Whitlock, M.C. & McCauley, D.E. (1999) Indirect measures of gene flow and migration: $F_{ST} \neq 1/(4Nm + 1)$. *Heredity* **82**, 117–125.

Wilkinson-Herbots, H.M. (1998) Genealogy and subpopulation differentiation under various models of population structure. *Journal of Mathematical Biology* **37**, 535–585.

Wilson, R.J. (1999) *The spatiotemporal dynamics of three Lepidopteran herbivores of* Helianthemum chamaecistus. PhD thesis, University of Leeds.

Wright, S. (1931) Evolution in Mendelian populations. *Genetics* **16**, 97–159.

Wright, S. (1943) Isolation by distance. *Genetics* **28**, 114–138.

Wright, S. (1950) The genetical structure of populations. *Annals of Eugenics* **15**, 323–354.

Wright, S. (1969) *Evolution and the Genetics of Populations. The Theory of Gene Frequencies*, Vol. 2. University of Chicago Press, Chicago, IL.

Chapter 2

Virus transmission and gene flow between two species of the Dutch elm disease fungi, *Ophiostoma ulmi* and *O. novo-ulmi*: deleterious viruses as selective agents for gene introgression*

K. W. Buck, C. M. Brasier,[†] *M. Paoletti and L. J. Crawford*

Introduction

The concept of a biological species for sexually reproducing organisms implies reproductive isolation and lack of exchange of genetic information between individuals of different species. However genetic isolation of species is not absolute and there is an increasing awareness that interspecific hybridization can play a significant role in ecological and evolutionary processes (Neuffer *et al.* 1999; Burke & Arnold 2001). Interspecific hybrids may be of lower or higher fitness compared to the parents (Arnold 1997; Arnold *et al.* 2001). Even when a primary hybrid is of relatively low fitness, backcrossing to one of the parents can give rise to introgressants in which a portion of the genome of one species has become stably incorporated into the genome of another species and there are many examples of gene introgression in animals and plants (Arnold 1992; Riesberg & Wendel 1993).

In the fungi, there are relatively few reports of interspecific hybridization occurring in nature (Brasier 1995). However interspecific hybridization is probably more common than has been realized and the availability of molecular fingerprinting and sequencing methods may lead to many more examples being discovered. Some recently reported examples of interspecific hybridization in basidiomycete and oomycete plant pathogens, in which the hybrids have acquired new properties, such as new or increased host range, or increased aggressiveness, (reviewed by Brasier 2000a, 2001), are shown in Table 2.1. In this chapter, we review recent work on gene flow between the Dutch elm disease pathogens, *Ophiostoma ulmi* and

† Correspondence author: c.brasier@forestry.gov.uk

Table 2.1 Examples of interspecific hybridization in oomycetes and basidiomycetes.

Hybrid genus	Parents	Host	Region	Reference
Phytophthora	*P. cactorum, P. nicotianae*	*Primula, Spathiphyllum*	Netherlands	Man in't Veldt *et al.* 1998
Phytophthora	*P. cambivora, P. fragariae?*	Alder	Europe	Brasier *et al.* 1999
Melampsora	*M. medusae, M. occidentalis*	Poplar	New Zealand, North America	Newcombe *et al.* 2000
Heterobasidion	*H. annosum* 'S' and 'P' species	Conifers	North America	Gonthier *et al.* 2001

O. novo-ulmi and evidence that introgression of certain 'beneficial' genes from *O. ulmi* to *O. novo-ulmi* is driven by the resistance they confer to transmission of deleterious mycoviruses. Implications for the use of viruses as biological control agents for Dutch elm disease are also considered and comparisons made with hypoviruses which cause hypovirulence in the chestnut blight fungus, *Cryphonectria parasitica.*

The two 20th century pandemics of Dutch elm disease

The 20th century saw two worldwide outbreaks of Dutch elm disease, caused by different species of ascomycete fungi in the genus *Ophiostoma* (Brasier 1986a, 1988a,b, 1990a, 1991, 2001; Brasier & Buck 2002). The first pandemic, caused by *O. ulmi*, began about 1910 in northwest Europe and quickly spread to other parts of continental Europe, into southwest Asia and overseas to Great Britain and to North America (Figure 2.1a). *O. ulmi* is relatively weakly pathogenic to the European elm species, *Ulmus carpinifolia* (smooth-leaved elm), *U. glabra* (Wych elm), *U. laevis* (white elm) and *Ulmus procera* (English elm), and in Europe this first pandemic had died down by the 1940s having caused losses of 10–40% of elms in different countries. The second pandemic was due to the appearance, around the 1940s, of a new far more aggressive species, *O. novo-ulmi*, which causes almost 100% mortality in mature elms and which can be distinguished from *O. ulmi* by many other important biological and molecular properties (Brasier 1991; Bates *et al.* 1993a,b; Pipe *et al.* 1995, 1997; Jeng *et al.* 1996). Two subspecies of *O. novo-ulmi* have been recognized, subsp. *novo-ulmi* and subsp. *americana*, which differ in their geographical ranges and in several other biological and molecular properties (Brasier & Kirk 2001). The colony morphologies of *O. ulmi* and the two subspecies of *O. novo-ulmi* are compared in Figure 2.2. Evidence indicates that subsp. *novo-ulmi* first appeared in the Moldova–Ukraine region in eastern Europe, and then spread across Europe and into southwest and central Asia. Subsp. *americana* emerged around the southern Great Lakes region of North America and spread across the North American continent and into Canada. It was imported into Britain during the 1960s and spread to westward Europe. In some parts of Europe, the geographical ranges of the two subspecies now

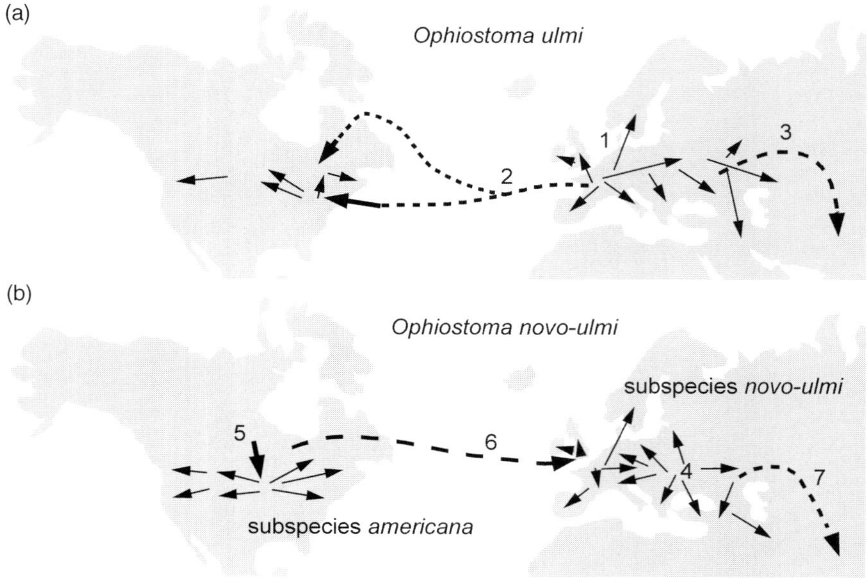

Figure 2.1 Geographical spread of *Ophiostoma ulmi* and *O. novo-ulmi* in the first and second pandemics of Dutch elm disease. Solid arrows show natural spread from probable sites of initial introduction; dashed arrows show subsequent movement via additional importation events. (a) Spread of *O. ulmi*: 1, its initial spread from a site in northwest Europe about 1910; 2, importation to North America in the 1920s; 3, introduction to Tashkent from Krasnodar in the late 1930s. (b) Spread of *O. novo-ulmi* subspecies *novo-ulmi* and *americana*: 4, original centre of appearance of subsp. *novo-ulmi* in Europe in the Romania–Moldavia region, probably in the 1940s; 5, original centre of appearance of subsp. *americana* in the Great Lakes region of North America, probably also in the 1940s; 6, importation of subsp. *americana* from the Toronto area to Britain about 1960; 7, introduction of subsp. *novo-ulmi* to the Tashkent area in the 1970s. (Adapted from Brasier 1990a, 2001.)

overlap (Figure 2.1b). As a result of the current pandemic, billions of elm trees are believed to have died from the disease worldwide, including over 25 million elms in Britain.

The annual cycle of Dutch elm disease

Dutch elm disease is characterized by an annual cycle of alternating pathogenic and saprotrophic phases (Brasier 1986a,b, 1988a). The Dutch elm disease fungi are transmitted from tree to tree by bark beetles, principally *Scolytus scolytus* and *S. multistriatus*, which colonize the bark of diseased trees for breeding purposes. The pathogenic phase is initiated in the spring when newly hatched beetles, carrying spores of *O. ulmi* or *O. novo-ulmi*, emerge from breeding galleries (Figure 2.3) in the bark of diseased trees. The beetles fly to healthy elm trees where they cut feeding grooves in the crotches between twigs, resulting in infection of the trees by the fungus, initially in a hyphal form. The fungus invades the xylem (sapstream) of the tree, where it

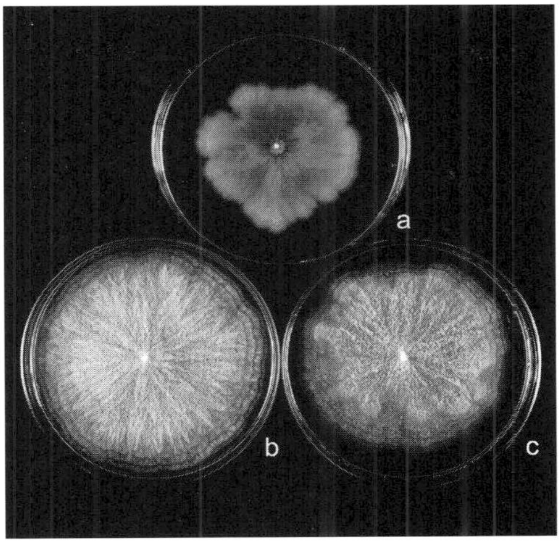

Figure 2.2 Colony morphologies of the Dutch elm disease fungi: (a) *Ophiostoma ulmi*, cause of the first pandemic; (b,c) *O. novo-ulmi* subsp. *americana* and *novo-ulmi* respectively, cause of the second, current pandemic.

Figure 2.3 A piece of dying elm bark from a tree suffering from Dutch elm disease, showing the breeding galleries produced by elm bark beetles. The pathogens, *Ophiostoma ulmi* and *O. novo-ulmi*, multiply in and around the galleries during the saprotrophic phase of the disease. Grey areas: areas occupied by spreading colonies of *O. ulmi* and *O. novo-ulmi*. White areas: remaining uncolonized bark.

multiplies in a yeast-like form and secretes a number of toxins, including a protein hydrophobin, cerato-ulmin (CU) (Temple & Horgen 2000; Wosten 2001). The water-conducting vessels of the xylem then become blocked and the tree dies. The fungus migrates from the xylem to the bark to initiate the saprotrophic phase. In

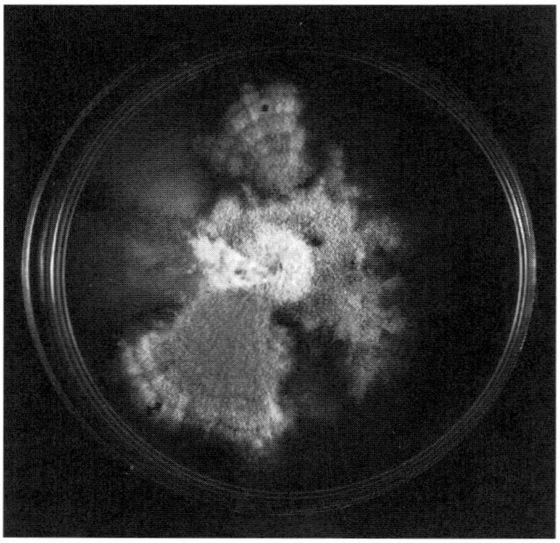

Figure 2.4 A culture of *O. novo-ulmi* infected with a deleterious mitochondrial virus, showing a typical irregular, unstable growth pattern.

the autumn the bark of the newly killed tree is invaded by bark beetles which can carry additional fungal spores acquired from other diseased elms. Hence in the bark, a variety of genotypes of *O. ulmi* or *O. novo-ulmi* is introduced. In the saprotrophic phase, the fungus grows in the beetle breeding galleries in a mycelial form (see Figure 2.3), producing a network of interwoven hyphal strands. Asexual spores (conidia) and sexual spores (ascospores) are also produced in successive waves (Webber & Brasier 1984; Webber *et al.* 1988). In the spring, newly hatched beetles emerge and fly off to initiate a new disease cycle in healthy trees.

Virus diseases of *O. ulmi* and *O. novo-ulmi*

O. ulmi and *O. novo-ulmi* themselves exhibit diseases, which are caused by transmissible agents. The diseases of *O. novo-ulmi* are characterized by slow hyphal growth, degenerate colony morphology (Figure 2.4), reduced conidial germination efficiency and reduced ability to cause disease in elms (Brasier 1983). The determinants of the diseases have been called disease-factors (d-factors) and named d^1, d^2, d^3, etc. in order of discovery (Brasier 1986b). A large body of correlative evidence has indicated that d-factors are double-stranded (ds) RNA viruses (Buck & Brasier 2002). There appear to be at least two types of viruses which cause degenerative diseases in *O. novo-umi.* The first type of virus, exemplified by the d^2-factor, replicates in the mitochondria and is associated with mitochondrial defects, such as reduction of cytochrome oxidase and production of small DNA 'plasmids' derived from the mitochondrial DNA (Rogers *et al.* 1987; Charter *et al.* 1993; Abu-Amero *et al.* 1995). These mitochondrial viruses belong to the genus *Mitovirus* of the family Narnaviridae. Mitoviruses consist of a single unencapsidated molecule of dsRNA,

2.3–3.5 kbp in size, which encodes a single protein, the RNA-dependent RNA polymerase (RdRp), which is required to replicate the dsRNA (Hong *et al.* 1999). The second type of virus, exemplified by the d[1]-factor, replicates in the cytoplasm, and is not associated with mitochondrial defects. These cytoplasmic viruses generally cause milder symptoms than those associated with the mitoviruses (Sutherland & Brasier 1997) and belong to the *Partitivirus* genus of the Partitiviridae family. Partitiviruses consist of two molecules of dsRNA, 1.5–2.0 kbp in size, each separately encapsidated in isometric particles, 25–35 nm in diameter. One dsRNA encodes the capsid protein and the other dsRNA encodes the RdRp (Buck 1986, 1998).

Mixed infections with multiple (distinct) mitoviruses, and with mitoviruses and partitiviruses, occur commonly. Both mitoviruses and partitiviruses are associated with defective RNAs, derived from the virus genomic RNAs by deletions and rearrangements. Mixed infections and defective RNAs may play a role in inducing disease in the fungus by increasing the overall virus load on the cell, since isolates infected with single viruses are often symptomless. Mitoviruses and partitiviruses have also been found in *O. ulmi*. It has been suggested that virus infection contributed to the decline of *O. ulmi* in Europe in the 1940s (Brasier 1990b; Mitchell & Brasier 1994).

Viruses of *O. novo-ulmi* and *O. ulmi*, like those of other fungi, are transmitted by intracellular routes, through conidia and yeast-like cells, and via hyphal anastomosis (fusion of hyphae) (Brasier 1986b). Transmission into conidia and yeast-like cells is generally efficient, although some viruses are lost with a low, virus-specific frequency (Rogers *et al.* 1986; Cole *et al.* 1998). Virus transmission by hyphal anastomosis is dependent on the complement of vegetative incompatibility (*vic*) genes of the two intermingling fungal genotypes. Up to seven *vic* loci control the ability of different genotypes of *O. novo-ulmi* to form viable anastomoses (Brasier 1984; Milgroom & Brasier 1997). When all the *vic* loci of two genotypes are identical, the genotypes are of the same vegetative compatibility (vc) type and readily form viable anastomoses. Virus transmission by hyphal anastomoses between individuals of the same vc type is very efficient (approaching 100%) (Brasier 1986b).

When one or more *vic* loci between genotypes differ, the genotypes are of different vc types and hyphal fusion is followed by an incompatibility reaction resulting in death of the fused cell. The most severe type of incompatibility reaction is the wide or 'w' reaction, which occurs when genotypes differ at one or more *vic-w* loci. Differences at other *vic* loci can give rise to narrow (n), line (l) or line-gap (lg) reactions in decreasing order of incompatibility, but because 'w' reactions are dominant, the other types of vegetative incompatibility reactions are only observed between genotypes with the same *vic-w* alleles. In random pairings of wild isolates from European *O. novo-ulmi* populations, the 'w' reaction is by far the most common (Brasier 1984), probably reflecting both the dominance of the *vic-w* genes and the presence of multiple *vic-w* loci and/or alleles in the populations. Virus transmission is restricted by vegetative incompatibility reactions, particularly 'w' reactions (Brasier 1986b).

Viruses are generally excluded in *O. novo-ulmi* ascospores. *O. novo-ulmi* and *O. ulmi* are heterothallic with two mating types A and B. In sexual crosses between

O. novo-ulmi isolates of opposite mating types, no dsRNA was found in most single ascospore isolates even when both parents were virus infected (Brasier 1986b; Rogers *et al.* 1986).

Evidence for virus transmission from *O. ulmi* to *O. novo-ulmi* at *O. novo-ulmi* epidemic fronts

Populations of *O. novo-ulmi* at epidemic fronts are often clonal, consisting of a single vc type and mating type (see Figure 2.5). Isolates of *O. novo-ulmi* from the xylem of infected trees in the pathogenic phase of Dutch elm disease are generally free from deleterious viruses. However up to 90% of isolates taken from the bark in the ensuing saprotrophic phase are virus infected (Brasier 1986b, 1988b, 1990b). Until recently, the origin of these viruses was unknown. However evidence has now been

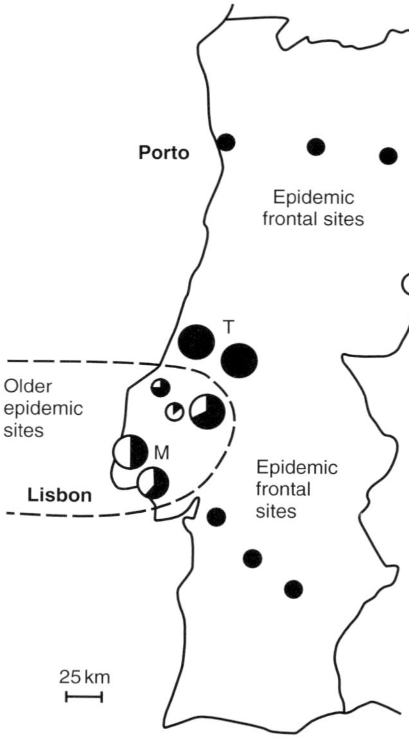

Figure 2.5 Distribution of the EASG vc type (black part of circles) and heterogeneous vc component genotypes (white part of circles) of *Ophiostoma novo-ulmi* subsp. *americana* in a survey in Portugal in 1985. The EASG vc type is the only genotype found at these epidemic front sites and the heterogeneous component is present only at the older epidemic sites. Smaller circles, 2–14 isolates; larger circles, 22–38 isolates. *O. ulmi* (not shown) represented about 74% of the overall *Ophiostoma* sample at the epidemic fronts and about 5% at the older epidemic sites. M, Mafra; T, Tomar. (Adapted from Brasier 1988b.)

obtained to suggest that they originate from the *O. ulmi* population present before the arrival of *O. novo-ulmi*. *O. ulmi* populations had declined by the 1940s and, since *O. ulmi* competes poorly with *O. novo-ulmi*, are soon completely replaced by the latter (Brasier 2001). Nevertheless when *O. novo-ulmi* first arrives at a new epidemic front, the resident *O. ulmi* population temporarily undergoes a considerable expansion. This is because the beetles which colonize the bark of the first trees to be killed by *O. novo-ulmi* will be carrying *O. ulmi* inoculum. Hence *O. novo-ulmi* which enters the bark from the xylem will initially be surrounded by a large mass of *O. ulmi*, resulting in physical contact between the two species (see Figure 2.3; Webber & Brasier 1984; Webber *et al.* 1988). Transmission of viruses by hyphal anastomosis would be expected to be largely confined to within a single species. However, the formation of transient hyphal bridges between the two species may be sufficient for virus transmission with low efficiency. Because the first arriving *O. novo-ulmi* population is clonal, consisting of a single vc type, once introduced, a virus would be expected to be transmitted rapidly through the network of interconnecting hyphae produced in the saprotrophic phase. Furthermore, the presence of only a single *O. novo-ulmi* mating type would preclude virus loss from the population by sexual crossing.

To obtain evidence for horizontal virus transmission from *O. ulmi* to *O. novo-ulmi*, it was necessary to compare virus sequences with those of their hosts. Because of their intracellular mode of replication, it has been suggested that mycovirus transmission may be largely vertical, i.e. from parent to progeny in mitosis or meiosis, and that the viruses have coevolved with their host species (Buck 1998; Ghabrial 1998). Differences in virus sequences would then be expected to reflect the divergence of *O. ulmi* and *O. novo-ulmi*. However, if recent horizontal virus transmission between *O. ulmi* and *O. novo-ulmi* had occurred as a result of transient hyphal fusions, sequences of viruses from *O. ulmi* and *O. novo-ulmi* isolates might be expected to be identical or nearly identical.

A commonly used method of determining horizontal transmission is to compare phylogenetic trees of host nuclear genes with trees of virus genes (Martin *et al.* 1999). If the order is the same in both species, it is assumed that the virus has coevolved with the host. If 'cross-overs' occur (character-state discordance), this may indicate horizontal transmission (Rosewich & Kistler 2000). However this method is not appropriate if the two host species between which the horizontal transmission is suspected are adjacent in host gene trees, as is the case for *O. ulmi* and *O. novo-ulmi* (Pipe *et al.* 2000a). An alternative approach is to compare the likely mutation rates of host and virus genes, and make predictions of the expected sequences of virus genes in the two species based on differences in the host genetic sequences between the two species, i.e. *O. ulmi* and *O. novo-ulmi*, assuming virus-host coevolution, i.e. no horizontal transmission.

In general, RNA viruses have a potential for more rapid evolution than nuclear DNA genes, because the RNA-dependent RNA polymerases (RdRps), which replicate RNA viruses, lack the proof-reading activities found in DNA polymerases, which replicate nuclear DNA. For example, the spontaneous mutation rate per replication has been estimated as $0.1–1.6 \times 10^{-4}$ for poliovirus (an RNA virus) and $2.0 \times$

Table 2.2 Comparisons of nucleotide sequences of OMV5 RdRp genes from *O. ulmi* and *O. novo-ulmi* isolates.

OMV5 RdRp sequences compared from pairs of isolates of:	% Nucleotide sequence identity
O. ulmi (Tomar)/*O. ulmi* (Tomar)	97.5–99.8
O. novo-ulmi (Tomar)/*O. novo-ulmi* (Tomar)	97.1–99.5
O. ulmi (Tomar)/*O. novo-ulmi* (Tomar)	96.8–100
O. ulmi (Tomar)/*O. ulmi* (Italy)	92.5–93.2
O. novo-ulmi (Tomar)/*O. ulmi* (Italy)	92.5–93.0
O. ulmi (Tomar)/*O. novo-ulmi* (England)	91.6–92.8
O. ulmi (Italy)/*O. novo-ulmi* (England)	92.2

The numbers OMV5 RdRp sequences from isolates from each geographical location, which were compared in all possible pairwise combinations, were as follows: *O. ulmi* (Tomar, Portugal), 7; *O. novo-ulmi* (Tomar, Portugal), 10; *O. ulmi* (Sicily, Italy), 1; *O. novo-ulmi* (England), 1.

10^{-11} for *Neurospora crassa*, which like *O. ulmi* and *O. novo-ulmi* is an ascomycete fungus (Drake 1991, 1993). It has been found that the nucleotide sequences of the cerato-ulmin (*cu*) gene of *O. ulmi* and *O. novo-ulmi* are 94.5% identical, with very little intraspecies differences, comparing isolates from widely separated geographical locations, including Britain, Canada, Iran, Italy, Netherlands, Portugal, the USA and Uzbekistan (Jeng *et al.* 1995, Pipe *et al.* 1997). The ribosomal internal transcribed spacer (ITS) sequences are more slowly evolving and are 99% identical between *O. ulmi* and *O. novo-ulmi* (Jeng *et al.* 1995). Given the potential for rapid evolution, it might be expected that endogenous *O. ulmi* and *O. novo-ulmi* viruses will differ in nucleotide sequences by more than the rapidly evolving *cu* genes of these species have done, i.e. have less than 94.5% sequence identity, with the proviso that evolution of virus genes may be subject to different structural and functional constraints (Simmonds & Smith 1999).

Comparisons of the nucleotide sequences of a 1.2 kbp region of the *Ophiostoma* mitovirus 5 (OMV5) RdRp gene of *O. ulmi* and *O. novo-ulmi* isolates from an epidemic site at Tomar, Portugal (Figure 2.5), together with those of an *O. ulmi* isolate from Italy and an *O. novo-ulmi* isolate from Britain, are shown in Table 2.2, and a phylogenetic tree of OMV5 RdRp amino acid sequences of some of the isolates is shown in Figure 2.6 (L.J. Crawford, K.W. Buck & C.M. Brasier, unpublished results).

Several deductions may be made from the data:

1 OMV5 has evolved at a higher rate than the *cu* gene. This is clear from the fact that OMV RdRp sequences within *O. ulmi* and *O. novo-ulmi* from different countries have diverged by up to 8% (92% identity), whereas *cu* sequences within *O. ulmi* or *O. novo-ulmi* isolates from many different countries were 99.8 to 100% identical.

2 OMV RdRp sequence similarities between Tomar *O. ulmi* and *O. novo-ulmi* isolates (96.8–100% identity) are greater than that between the *cu* genes of Tomar *O. ulmi* and *O. novo-ulmi* isolates (94.5% identity).

3 The OMV5 RdRp sequences cluster on a geographical basis rather than according to the fungal species (*O. ulmi* or O. *novo-ulmi*).

4 *O. ulmi* has been present in Portugal since the 1920s, whereas *O. novo-ulmi*

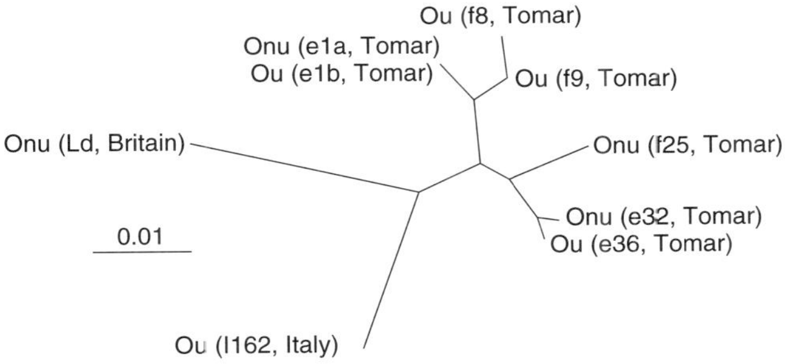

Figure 2.6 Neighbour-joining tree of OMV5 RdRp amino acid sequences from *Ophiostoma ulmi* (Ou) and *O. novo-ulmi* (Onu) isolates from a Dutch elm disease epidemic site in Tomar, Portugal, and isolates from Britain and Italy. The bar represents the number of amino acid substitutions per site.

subsp. *americana* was introduced into Europe from North America reaching Portugal in the 1980s. However, similarities between the OMV5 RdRp sequences of *O. ulmi* and *O. novo-ulmi* isolates from Tomar (96.8–100% identity) are greater than the similarities between OMV5 RdRp sequences of *O. ulmi* or *O. novo-ulmi* isolates from Tomar and those from *O. ulmi* or *O. novo-ulmi* isolates respectively from other countries (91.6–93.2%).

Taken together, the evidence is consistent with horizontal transmission of OMV5 between *O. ulmi* and *O. novo-ulmi*. Particularly persuasive is the identity of the OMV5 RdRp nucleotide sequences in an *O. ulmi* isolate (e1b) and an *O. novo-ulmi* isolate (e1a). Isolates e1a and e1b were obtained from the same small (4 mm^2) fragment of bark from a diseased elm tree in Tomar, Portugal, showing that physical contact between the colonies of the two species had actually occurred.

Circumstantial evidence suggests that the direction of virus transmission was from *O. ulmi* to *O. novo-ulmi* rather than *vice versa*. Deleterious viruses are scarce in North American populations of *O. novo-ulmi* subsp. *americana* and were initially absent from epidemic front populations of subsp. *americana* in Europe, which originated from the American populations (Figure 2.1b). By contrast, viruses are known to be common in European *O. ulmi* populations. The Italian isolate, from which the OMV5 RdRp region was sequenced, was obtained from a region of Italy (Sicily) at a time when only *O. ulmi* was present, confirming the presence of OMV5 in *O. ulmi* in Italy before the arrival of *O. novo-ulmi*.

It is noteworthy that the nucleotide sequence identities between OMV5 RdRp sequences of *O. ulmi* isolates from Portugal and Italy (92.5–93.2%), *O. ulmi* isolates from Portugal and an *O. novo-ulmi* isolate from England (91.6–92.8%), and an *O. ulmi* isolate from Italy and an *O. novo-ulmi* isolate from England (92.2%), are all similar. Given the divergence of OMV5 from *O. ulmi* isolates in Portugal and Italy, a greater divergence of OMV5 from *O. ulmi* and *O. novo-ulmi* from different countries might have been expected. A possible explanation is that OMV5 in the *O. novo-ulmi* isolate from England had also been transmitted from the resident *O. ulmi* population. In England, as in Portugal, *O. novo-ulmi* subsp. *americana* originated in North America and was probably initially virus free. Hence the comparisons of OMV5 sequences in isolates from England, Italy and Portugal may simply represent the degree of divergence of OMV5 in the *O. ulmi* population which spread across north-west Europe during the early part of the 20th century.

Evolution of *O. novo-ulmi* populations from clonality at epidemic fronts to heterogeneity in the postepidemic period

Although populations of *O. novo-ulmi* at Dutch elm disease epidemic fronts in Europe tend to be clonal, consisting of a single vc type and the B mating type, within a few years they become heterogeneous with the appearance of many new vc types and the A mating type (Brasier 1988b). The appearance of the A mating type allows sexual reproduction and the creation of yet more vc types as a result of recombination. There are at least three possibilities for the appearance of new vc types and the A mating type in *O. novo-ulmi* populations (Brasier *et al.* 1998): (i) Arrival of new *O. novo-ulmi* genotypes carrying different new *vic* alleles and the A mating type. (ii) Introgression of *vic* genes and the A mating type from *O. ulmi.* (iii) Mutation of *vic* genes to produce new alleles. Mutation of the B mating type to produce the A mating type is not possible, because, although the A and B mating type genes of ascomycete fungi have a common *mat* locus and behave as alleles in genetic crosses, they have little sequence similarity. Furthermore they encode different proteins which act as transcriptional regulators with different functions in the control of expression of genes required for sexual reproduction (Turgeon 1998). For this reason, the mating type genes are referred to as idiomorphs (Glass *et al.* 1988).

None of the above three possibilities is mutually exclusive, but observations favour a hybridization event over a direct mutation for the generation of new *vic* genotypes (Brasier *et al.* 1998). Such a hybridization event need not occur with high frequency or efficiency, since, because of its deleterious effects, mycovirus infection probably exerts a strong selection pressure for emergence of different virus-free vegetatively incompatible genotypes resistant to virus transmission. If new *vic* genotypes are generated as ascospore progeny arising from sexual hybridization, they would be virus free and, in the saprotrophic phase, would be expected to outgrow the virus-infected individuals of the original epidemic front *vic* genotype. This is consistent with the observation that in postepidemic *O. novo-ulmi* populations, the level of deleterious mycovirus infection drops to about 5% and the original epidemic front *vic* clone becomes a small proportion of the *O. novo-ulmi* population (Brasier 1988b). By contrast, since a high proportion of the epidemic front population in the saprotrophic phase is virus infected, if new *vic* genotypes were generated by direct mutation of *vic* genes, there would be a high probability that they would be virus infected. In this circumstance, the level of virus infection would be expected to remain high in the population, which is contrary to the above observations.

Additional evidence also supports the likelihood that the new *vic* genotypes and the introduction of the A mating type into *O. novo-ulmi* postepidemic populations was the result of hybridization of the original epidemic *vic* clone of *O. novo-ulmi* with resident *O. ulmi* genotypes, of which the A mating type constitutes about 50% of populations in Europe (Brasier 1984), rather than with a newly arrived *O. novo-ulmi* genotype. The increasing heterogeneity in vc types observed in postepidemic *O. novo-ulmi* populations is accompanied by the appearance of many novel morphological variants and, during the *O. ulmi/O. novo-ulmi* overlap period, a temporary but significant reduction in pathogenicity (Brasier 1988b). Based on their biological properties, gene sequencing and molecular fingerprinting. and gene mapping, some of the more unusual variants have been proved to be *O. ulmi/O. novo-ulmi* hybrids (Brasier *et al.* 1998). Such hybrids are of low pathogenicity and fitness and their existence appears to be transient. However, through natural backcrossing of these hybrids with *O. novo-ulmi*, it is possible for stable new *O. novo-ulmi* genotypes to be generated containing only beneficial 'introgressed' *O. ulmi* genes, such as *vic* genes and the A mating-type gene. Interestingly, evidence has been obtained that rare *O. novo-ulmi* isolates, which either have low pathogenicity or which do not produce the CU hydrophobin, have an introgressed *O. ulmi* pathogenity gene (Abdelali *et al.* 1999) or *cu* gene (Pipe *et al.* 2000b) respectively. In these cases, however, since the introgressed *O. ulmi* genes have been shown to have a negative effect on fitness, such introgressants are again unlikely to survive long in natural populations.

Introgression of *O. ulmi* vegetative incompatibility and mating type genes into *O. novo-ulmi*

To prove introgression of genes from *O. ulmi* to *O. novo-ulmi* during the early postepidemic phase of Dutch elm disease, *vic* genes which gave rise to the strongest (w) incompatibility reaction, and the mating type (*mat*) gene, were selected. *O. novo-ulmi* isolates were selected from epidemic sites at Tomar and Mafra in Portugal at which the transition of populations from clonality to heterogeneity is well documented, illustrated in Figure 2.5 (Brasier 1988b). One isolate belonged to the European subsp. *americana* vc supergroup (EASG) and B mating type, representing the dominant epidemic front clone found at these two sites. Two representative later epidemic isolates were chosen. These were of A mating type and produced 'w' vegetative incompatibility reactions with the EASG isolate and also with each other. Analysis of the progeny of genetic crosses conducted between pairs of these isolates established the presence of two independently segregating *vic* loci, and a single mating type locus. This is consistent with previous analyses of *vic* and *mat* loci in European and North American *O. novo-ulmi* populations (Brasier 1984; Milgroom & Brasier 1997). Interestingly, one of the *vic* loci and the *mat* locus were found to be linked (M. Paoletti, K. W. Buck & C. M. Brasier, unpublished results).

Bulked segregant analysis (Michelmore *et al.* 1991) was used to identify amplified fragment length polymorphic (AFLP) molecular markers (Vos *et al.* 1995) linked to *vic* and mating type genes. An AFLP genetic map of one of the *vic* loci was constructed with the program MAPMAKER (Lander *et al.* 1987). All the AFLP markers on either side of this *vic* locus in one of the later epidemic *O. novo-ulmi* isolates were found to be characteristic of *O. ulmi*, but not *O. novo-ulmi*, isolates. Confirmation that DNA flanking both sides of this *vic* gene in this *O. novo-ulmi* isolate was characteristic of *O. ulmi* was obtained by using cloned AFLP marker DNAs as probes in Southern blots in restriction endonuclease-digested DNA. Further work has shown that AFLP markers linked to the other *vic* locus and the *matA* locus in the second later epidemic *O. novo-ulmi* isolate were also characteristic of *O. ulmi*. Cloning and sequencing of the *matA* and *matB* regions of *O. ulmi* and *O. novo-ulmi* isolates has confirmed that the DNA sequence immediately flanking the *matA* gene in both of the two later epidemic *O. novo-ulmi* isolates is characteristic of *O. ulmi* (M. Paoletti, K.W. Buck & C.M. Brasier, unpublished results). Hence the evidence is consistent with the introgression of *O. ulmi vic* and mating type genes into later epidemic *O. novo-ulmi* populations. Interestingly it was found that more than 50% of the AFLP markers that distinguished the epidemic front EASG vc type, B mating type isolate from the two vegetatively incompatible later epidemic front isolates were linked to either the mating type or *vic* loci. This suggests that the two *vic* loci and the *matA* locus comprise the majority of *O. ulmi* DNA that has been introgressed into the later epidemic *O. novo-ulmi* isolates. The mechanism of introgression is unknown. It could have occurred via *O. ulmi/O. novo-ulmi* sexual hybridization, followed by backcrossing of the resulting hybrid with *O. novo-ulmi*. Alternatively, the mechanism might have involved somatic hybridization, following transient fusion of *O. ulmi* and *O. novo-ulmi* hyphae. Whatever the mechanism, only beneficial *O. ulmi* genes, such as those instrumental in reducing the load of deleterious viruses on the population, were stably retained, so increasing the overall fitness of the post-epidemic *O. novo-ulmi* population.

Sustained homogeneity of *O. novo-ulmi* populations in New Zealand in the absence of *O. ulmi* and deleterious virus infection

O. novo-ulmi subsp. *americana* was introduced into New Zealand in the late 1980s. The introduction consisted of a single vc type, corresponding to the major subsp. *americana* vc type (the EASG) found in Europe and the B mating type (Brasier & Gadgill 1992). *O. ulmi* is not present in New Zealand and therefore *vic* gene introgression from this species is not possible. Furthermore no deleterious viruses have been found which might act as selective agents for new *vic* gene alleles. Consistent with this environment, the *O. novo-ulmi* population in New Zealand has remained since its introduction as a single vc type and the B mating type (Brasier 2000b; C.M. Brasier, M. Dick & S. Kirk, unpublished results).

Impact of *vic* gene introgression on the prospects of developing mycoviruses as biological control agents for Dutch elm disease

Deleterious mycoviruses have been suggested as possible biological control agents for Dutch elm disease (reviewed by Brasier 2000b; Buck & Brasier 2002). The rootstock of diseased elm trees is generally not killed by *O. novo-ulmi* and suckers grow from the base of killed trees. There are millions of such suckers growing in hedgerows in the UK alone and these represent a considerable genetic resource for the eventual re-establishment of the elm. However when these reach a certain size, they are again colonized by bark beetles carrying *O. novo-ulmi* and become diseased. The heterogeneity of vc types which builds up in postepidemic *O. novo-ulmi* populations as a result of *vic* gene introgression from *O. ulmi* has resulted in the virtual elimination of deleterious virus from these populations and poses a serious impediment to the development of viruses for the control of *O. novo-ulmi*. A possible solution to this problem would be to introduce a cDNA copy of the virus genome into the genome of the fungus under the control of a fungal promoter, so that virus RNA is continuously synthesized from the cDNA insertion. Such a strategy has met with some success with another ascomycete fungal phytopathogen, *Cryphonectria parasitica* (see below). Application of this approach to the *Ophiostoma* mitoviruses may be difficult, as transformation of the mitochondrial DNA of filamentous ascomycete fungi has not yet been achieved. However the strategy should be readily applicable to the cytoplasmic partitiviruses. A range of *O. novo-ulmi vic* genotypes of A or B mating type with nuclear insertions of partitivirus cDNA could be released in controlled trials. Crossing with other *O. novo-ulmi* genotypes in the wild would produce new *vic* gene combinations, and loss of virus which normally occurs on ascospore formation would not be possible in the progeny containing the virus cDNA insertion. Hence the partitivirus would be introduced into a wide range of vc types, greatly increasing the chances of virus transmission by hyphal anastomosis in the saprotrophic stage in the bark.

Vegetative incompatibility and virus infections of *Cryphonectria parasitica*

Viruses also occur in populations of another invasive tree pathogen, *Cryphonectria parasitica*, the causative agent of chestnut blight, and virus transmission is again restricted by vegetative incompatibility sytems. It is of interest to examine some of the parallels between the *Ophiostoma*–virus–*vic* gene and *Cryphonectria*–virus–*vic* gene systems.

C. parasitica was introduced into North America about 1904 from the far east, probably Japan (reviewed by Anagnostakis 1982; Van Alfen 1986; Nuss 1992; Heiniger & Rigling 1994; Milgroom *et al.* 1996; Dawe & Nuss 2001). The disease gradually spread over most of the natural range of the American chestnut, *Castanea dentata*, on the eastern side of North America, eventually resulting in the destruction of billions of mature chestnut trees. As with Dutch elm disease, the roots of trees are not generally killed by the fungus, but suckers growing from the base of killed trees generally become infected again once they reach a certain size.

C. parasitica was also introduced into continental Europe in the late 1930s, again probably from Asia or from North America. Initially large numbers of the European chestnut, *Castanea sativa*, were destroyed by chestnut blight (Heiniger & Rigling 1994). However within 15 years, chestnuts in certain regions of Italy started to recover as a result of the spread of hypoviruses through the *C. parasitica* population. The hypoviruses (family Hypoviridae) consist of a single molecule of dsRNA ranging in size from 9.8 to 12.7 kbp and replicate in the cytoplasm. Unlike the *Ophiostoma* mitoviruses which reduce the ability of the fungus to cause disease by making the fungus respiratory deficient, *Cryphonectria* hypoviruses specifically attenuate the virulence of the fungus, at least partly by reducing the expression of heterotrimeric G proteins, key signalling molecules in *C. parasitica* pathogenesis (Dawe & Nuss 2001). The origin of the hypoviruses in Europe is not known, but since similar viruses have been found in *C. parasitica* isolates from China and Japan (Peever *et al.* 1998), they could have arrived in further *C. parasitica* introductions into Europe. Hypovirus spread in Italy, which as with other mycoviruses occurs by hyphal anastomosis, has led to a natural biological control of chestnut blight in that country. The spread of hypoviruses in Italy has been attributed in part to the relatively few vc types that occur in certain regions of the country (Heiniger & Rigling 1994; Cortesi *et al.* 1996, 1998).

Hypovirus-infected strains of *C. parasitica* have also been found in North America (Fulbright 1999), but have not had a major impact on chestnut blight in that country. This has often been attributed to the much greater diversity of vc types in North America than in Europe (Anagnostakis & Kranz 1987; Heiniger & Rigling 1994; Liu & Milgroom 1996; Bissegger *et al.* 1997; Liu *et al.* 1999). Consistent with this hypothesis, a negative correlation has been reported between hypovirus transmission and the number of *vic* genes that differed among isolates of a natural *C. parasitica* population (Liu & Milgroom 1996). Milgroom and Cortesi (1999) reported that European *C. parasitica* populations have lower vc diversity than North American populations because of a combination of lower *vic*-allele diversity, possibly resulting from founder effects, and limited recombination. Conversely the higher *vic*-allele diversity in North America could have resulted from a greater diversity in *C. parasitica* introductions from Japan combined with a greater sexual activity. However, in view of the introgression of *vic* genes from *O. ulmi* to *O. novo-ulmi*, the question arises whether *vic* genes may have been introduced into *C. parasitica* populations in North America from a related *Cryphonectria* species and, if so, whether virus-driven frequency-dependent selection of *vic* genes may have occurred. Hypoviruses could act as selective agents because infected individuals are female-sterile and asexual sporulation may be nearly completely inhibited (Van Alfen *et al.* 1975; Nuss 1992). Analysis of *vic*-allele frequencies in *C. parasitica* populations from Italy and North America has provided no support for frequency-dependent selection, except for the *vic2* gene which also exhibited the strongest overall inhibition of virus transmission (Milgroom & Cortesi 1999; Cortesi *et al.* 2001). However, in the relatively short time since *C. parasitica* was introduced from Asia, populations may not have reached equilibrium and it is still an open question whether virus introductions select for greater vc diversity, causing viruses to decline as a result of transmission blocks, or whether viruses fail to establish in North American *C. parasitica* populations with high vc diversity (Milgroom 1999).

Genetically modified hypoviruses as potential biological control agents for chestnut blight in North America

Attempts have been made to overcome the problem posed by high *vic*-allele diversity in *C. parasitica* populations in North America in using hypoviruses as biological control agents for chestnut blight by introducing a cDNA copy of a hypovirus genome into the nuclear DNA of *C. parasitica* (reviewed by Dawe & Nuss 2001). The cDNA was placed under the control of a strong *C. parasitica* promoter, so that hypovirus RNA was constitutively expressed from the *C. parasitica* nuclear DNA genome. Such transgenic strains showed the same range of hypovirulent properties as naturally infected isolates, including the ability to transmit the transcribed hypovirus RNA to other compatible isolates by hyphal anastomosis. Furthermore, the ability to produce hypovirus RNA was faithfully transmitted into ascospores through sexual reproduction and into asexual spores (conidia). Field trials with the genetically modified *C. parasitica* started in 1994. Initial surveys have shown that hypovirus transmission from transgenic hypovirulent strains to ascospore progeny occurred under field conditions and that the transgenic hypovirulent strains could be recovered from the test site two years after their introduction. However the frequency of released fungal genotypes fell rapidly during the first two years (Anagnostakis *et al.* 1998). To overcome this problem, more frequent introductions of the transgenic *C. parasitica* strains, and the development of new transgenic strains based on more moderate hypoviruses, are proposed (Dawe & Nuss 2001). The transgenic *C. parasitica* used in the field trials contained a hypovirus which resulted in small, superficial cankers that were generally devoid of spore-bearing stromal pustules (Choi & Nuss 1992). Other *C. parasitica* isolates infected with less aggressive hypoviruses have been described (Chen & Nuss 1999; Peever *et al.* 2000). Infection of chestnuts with these isolates produces cankers covered with stromal pustules containing large numbers of viable asexual spores. Hence a second generation of transgenic *C. parasitica*, based on the more moderate hypoviruses, may improve their dispersal and survival under field conditions (Cummings-Carlson *et al.* 1998; Dawe & Nuss 2001). *O. novo-ulmi* viruses which produce moderate effects on their hosts have also been proposed as the most effective biological control agents for Dutch elm disease (Sutherland & Brasier 1997).

Acknowledgements

Work in the authors' laboratories on virus transmission and gene introgression between *O. ulmi* and *O. novo-ulmi* was supported by the Natural Environment Research Council's thematic programme EDGE (Ecological Dynamics and Genes). We are grateful to Susan Kirk (FRA) for excellent technical assistance.

References

Abdelali, E., Brasier, C.M. & Bernier, L. (1999) Localization of a pathogenicity gene in *Ophiostoma novo-ulmi* and evidence that it may be introgressed from *O. ulmi*. *Molecular Plant–Microbe Interactions* **12**, 6–15.

Abu-Amero, S.N., Charter, N.W., Buck, K.W. & Brasier, C.M. (1995) Nucleotide sequence analysis indicates that a DNA plasmid in a diseased isolate of *Ophiostoma novo-ulmi* is derived by recombination between two long repeat sequences in the mitochondrial large subunit ribosomal RNA gene. *Current Genetics* **28**, 54–59.

Anagnostakis, S.L (1982) Biological control of chestnut blight. *Science* **215**, 466–471.

Anagnostakis, S.L. & Kranz, J. (1987) Population

dynamics of *Cryphonectria parasitica* in a mixed-hardwood forest. *Phytopathology* **77**, 751–754.
Anagnostakis, S.L., Chen, B., Geletka, L.M. & Nuss, D.L. (1998) Hypovirus transmission to ascospore progeny by field-released transgenic hypovirulent strains of *Cryphonectria parasitica. Phytopathology* **88**, 598–604.
Arnold, M.L. (1992) Natural hybridization as an evolutionary process. *Annual Review of Ecology and Systematics* **23**, 237–261.
Arnold, M.L. (1997) *Natural Hybridization and Evolution.* Oxford University Press, Oxford.
Arnold, M.L., Kentner, E.K., Johnston, J.A., Cornman, S. & Bouck, A.C. (2001) Natural hybridization and fitness. *Taxon* **50**, 93–104.
Bates, M.R., Buck, K.W. & Brasier, C.M. (1993a) Molecular relationships between *Ophiostoma ulmi* and the NAN and EAN races of *O. novo-ulmi* determined by restriction fragment length polymorphisms of nuclear DNA. *Mycological Research* **97**, 449–455.
Bates, M.R., Buck, K.W. & Brasier, C.M. (1993b) Molecular relationships of the mitochondrial DNA of *Ophiostoma ulmi* and the NAN and EAN races of *O. novo-ulmi* determined by restriction fragment length polymorphisms. *Mycological Research* **97**, 1093–1100.
Bisseger, M., Rigling, D. & Heiniger, U. (1997) Population structure and disease development of *Cryphonectria parasitica* in European Chestnut forests in the presence of natural hypovirulence. *Phytopathology* **87**, 51–59.
Brasier, C.M. (1983) A cytoplasmically transmitted disease of *Ceratocystis ulmi. Nature* **305**, 220–223.
Brasier, C.M. (1984) Intermycelial recognition systems in *Ceratocystis ulmi*: their physiological properties and ecological importance. In: *The Ecology and Physiology of the Fungal Mycelium* (eds D. H. Jennings & A. D. M. Rayner), pp. 451–497. Cambridge Univesity Press, Cambridge.
Brasier, C.M. (1986a) The population biology of Dutch elm disease: its principle features and some implications for other host–pathogen systems. *Advances in Plant Pathology* **5**, 55–118.
Brasier, C.M. (1986b) The d-factor in *Ceratocystis ulmi*—its biological characteristics and implications for Dutch elm disease. In: *Fungal Virology* (ed. K. W. Buck), pp. 177–208. CRC Press, Boca Raton, FL.
Brasier, C.M. (1988a) *Ophiostoma ulmi*, cause of Dutch elm disease. *Advances in Plant Pathology* **6**, 207–223.
Brasier, C.M. (1988b) Rapid changes in genetic structure of epidemic populations of *Ophiostoma ulmi. Nature* **332**, 538–541.
Brasier, C.M. (1990a) China and the origins of Dutch elm disease: an appraisal. *Plant Pathology* **39**, 5–16.
Brasier, C.M. (1990b) The unexpected element: mycovirus involvement in the outcome of two recent pandemics, Dutch elm disease and chestnut blight. In: *Pests, Pathogens and Communities* (eds J. J. Burdon & S. R. Leather), pp. 289–307. Blackwell Scientific Publications, London.
Brasier, C.M. (1991) *Ophiostoma novo-ulmi* sp. nov., causative agent of the current Dutch elm disease pandemics. *Mycopathologia* **115**, 151–161.
Brasier, C.M. (1995) Episodic selection as a force in fungal microevolution with special reference to clonal speciation and hybrid introgression. *Canadian Journal of Botany* **73**, S1213–S1221.
Brasier, C.M. (2000a) The rise of hybrid fungi. *Nature* **405**, 134–135.
Brasier, C.M. (2000b) Viruses as biological control agents of the Dutch elm disease fungus *Ophiostoma novo-ulmi.* In: *The Elms: Breeding, Conservation and Disease Management* (ed. C. P. Dunn), pp. 201–212. Kluwer Academic Publishers, Boston.
Brasier C.M. (2001) Rapid evolution of introduced plant pathogens via interspecific hybridization. *Bioscience* **51**, 123–133.
Brasier, C.M. & Buck, K.W. (2002) Rapid evolutionary changes in a globally invading fungal pathogen: the causative agent of Dutch elm disease. *Biological Invasions*, in press.
Brasier, C.M & Gadgill, P. (1992) Dutch elm disease outbreak in New Zealand. In: *Report on Forest Research 1996*, pp. 20–28 (ed. Forestry Commission). Her Majesty's Stationery Office, London.
Brasier C.M. & Kirk, S.A. (2001) Designation of the EAN and NAN races of *Ophiostoma novo-ulmi* as subspecies. *Mycological Research* **105**, 547–554.

Brasier, C.M., Kirk, S.A., Pipe, N.D. & Buck, K.W. (1998) Rare interspecific hybrids in natural populations of the Dutch elm disease pathogens *Ophiostoma ulmi* and *O. novo-ulmi*. *Mycological Research* **102,** 45–57.

Brasier, C.M., Cooke, D. & Duncan, J.M. (1999) Origin of a new *Phytophthora* pathogen through interspecific hybridization. *Proceedings of the National Academy of Sciences of the United States of America* **96,** 5878–5883.

Buck, K.W. (1986) Fungal virology—an overview. In: *Fungal Virology* (ed. K. W. Buck), pp. 1–84. Boca Raton, CRC Press, FL.

Buck, K.W. (1998) Molecular variability of viruses of fungi. In: *Molecular Variability of Fungal Pathogens* (eds P. D. Bridge, Y. Couteaudier & J. M. Clarkson), pp. 53–72. CAB International, Wallingford, UK.

Buck, K.W. & Brasier, C.M. (2002) Viruses of the Dutch elm disease fungi. In: *dsRNA Genetic Elements: Concepts and Applications in Agriculture, Forestry and Medicine* (ed. S. M. Tavantzis), pp. 165–190. CRC Press, Boca Raton, FL.

Burke, J.M. & Arnold, M.L. (2001) Genetics and fitness of hybrids. *Annual Review of Genetics* **35,** 31–52.

Charter, N.W., Buck, K.W. & Brasier, C.M. (1993) De-novo generation of mitochondrial DNA plasmids following transmission of a degenerative disease and associated double-stranded RNA in *Ophiostoma novo-ulmi*. *Current Genetics* **24,** 505–514.

Chen, B. & Nuss, D.L. (1999) Infectious cDNA clone of hypovirus CHV1-Euro7: a comparative virology approach to investigate virus-mediated hypovirulence of the chestnut blight fungus *Cryphonectria parasitica*. *Journal of Virology* **73,** 985–992.

Choi, G.H. & Nuss, D.L. (1992) Hypovirulence of chestnut blight fungus conferred by an infectious viral cDNA. *Science* **257,** 800–803.

Cole, T.E., Muller, B.M., Hong, Y., Brasier, C.M. & Buck, K.W. (1998) Complexity of virus-like double-stranded RNA elements in a diseased isolate of the Dutch elm disease fungus, *Ophiostoma novo-ulmi*. *Journal of Phytopathology,* **146,** 593–598.

Cortesi, P., Milgroom, M.G. & Bisiach, M. (1996) Distribution and diversity of vegetative compatibility types in subpopulations of *Cryphonectria parasitica*. *Mycological Research* **100,** 1087–1093.

Cortesi, P., Rigling, D. & Heiniger, U. (1998). Comparison of vegetative compatibility types in Italian and Swiss subpopulations of *Cryphonectria parasitica*. *European Journal of Forest Pathology* **28,** 167–176.

Cortesi, P., McCulloch, C.E., Song, H., Lin, H. & Milgroom, M.G. (2001) Genetic control of horizontal virus transmission in the chestnut blight fungus, *Cryphonectria parasitica*. *Genetics* **159,** 107–118.

Cummings-Carlson, J., Fulbright, D.W., MacDonald, W.L. & Milgroom, M.G. (1998) West Salem: a research update. *Journal of the American Chestnut Foundation* **12,** 24–26.

Dawe, A.L. & Nuss, D.L. (2001) Hypoviruses and chestnut blight: exploiting viruses to understand and modulate fungal pathogenesis. *Annual Review of Genetics* **35,** 1–29.

Drake, J.W. (1991) A constant rate of spontaneous mutation in DNA-based microbes. *Proceedings of the National Academy of Sciences of the United States of America* **88,** 7160–7164.

Drake, J.W. (1993) Rates of spontaneous mutation among RNA viruses. *Proceedings of the National Academy of Sciences of the United States of America* **90,** 4171–4175.

Fulbright, D.W. (1999) Hypovirulence to control fungal pathogenesis. In: *Handbook of Biological Control* (eds T. W. Fisher & T. S. Bellows), pp. 691–698. Academic Press, San Diego, CA.

Ghabrial, S. A. (1998) Origin, adaptation and evolutionary pathways of fungal viruses. *Virus Genes* **16,** 119–131.

Glass, N.L., Vollmer, S.J., Staben, C., Grotelueschen, J. & Metzenberg, R.L. (1988) DNAs of the two mating type alleles of *Neurospora crassa* are highly dissimilar. *Science* **241,** 570–573.

Gonthier P., Garbelotto M., Varese G.C. & Nicolotti G. (2001) Relative abundance and potential dispersal range of intersterility groups of *Heterobasidion annosum* in pure and mixed forests. *Canadian Journal of Botany* **79,** 1057–1065.

Heiniger, U. & Rigling, D. (1994) Biological control of chestnut blight in Europe. *Annual Review of Phytopathology* **32**, 681–599.

Hong, Y., Dover, S.L., Cole, T.E., Brasier, C.M. & Buck, K.W. (1999) Multiple mitochondrial viruses in an isolate of the Dutch elm disease fungus *Ophiostoma novo-ulmi. Virology*, **258**, 118–127.

Jeng, R.S., Hintz, W.E., Barden, C.G., Horgen, P.A. & Hubbes, M.M. (1996) A comparison of the nucleotide sequence of the cerato-ulmin gene and the rDNA ITS between aggressive and non-aggressive isolates of *Ophiostoma ulmi sensu lato*, the causal agent of Dutch elm disease. *Current Genetics* **29**, 168–173.

Lander, E., Abrahamson, J., Barlow, A. *et al.* (1987) MAPMAKER, a computer package for constructing genetic linkage maps. *Cytogenetics and Cell Genetics* **46**, 642–647.

Liu, Y.-C. & Milgroom, M.G. (1996) Correlation between hypovirus transmission and the number of vegetative incompatibility (*vic*) genes different among isolates of natural populations of *Cryphonectria parasitica. Phytopathology* **86**, 79–86.

Liu, Y.-C., Durrett, R. & Milgroom, M.G. (1999) A spatially-structured stochastic model to stimulate heterogenous transmission of viruses in fungal populations. *Ecological Modeling* **127**, 291–308.

Man in't Veldt, W.A., Veenbaas-Rijks, W.J., Illieva, E. *et al.* (1998) Natural hybrids of *Phytophthora nicotianae* and *P. cactorum* demonstrated by isozyme analysis and random amplified polymorphic DNA. *Phytopathology* **88**, 922–929.

Martin, J., Herniou, E., Cook, J., O'Neill, R. W. & Tristem, M. (1999) Interclass transmission and phyletic host tracking in murine leukaemia virus-related retroviruses. *Journal of Virology* **73**, 2442–2449.

Michelmore, R.W., Paran, I. & Kesseli, R.V. (1991) Identification of markers linked to disease-restance genes by bulked segregant analysis—a rapid method to detect markers in specific genomic regions by using segregating populations. *Proceedings of the National Academy of Sciences of the United States of America* **88**, 9828–9832.

Milgroom, M.G. (1999) Viruses in fungal populations. In: *Structure and Dynamics of Fungal Populations* (ed. J. Worrall), pp. 283–305. Chapman & Hall, London.

Milgroom, M.G. & Brasier, C.M. (1997) Potential of vegetative compatibility types of *Ophiostoma novo-ulmi* in North America. *Mycologia* **89,** 722–726.

Milgroom M.G. & Cortesi P. (1999) Analysis of population structure of the chestnut blight fungus based on vegetative incompatibility genotypes. *Proceedings of the National Academy of Sciences of the United States of America* **96**, 10518–10523.

Milgroom, M.G., Wang, K., Zhou, Y., Lipari, S. E. & Kaneko, S. (1996) Intercontinental population structure of the chestnut blight fungus. *Mycologia* **88**, 179–190.

Mitchell, A.G. & Brasier, C.M. (1994) Contrasting structure of European and North American populations of *Ophiostoma ulmi. Mycological Research* **98**, 576–582.

Neuffer, B., Auge, H., Mesch, H., Amarell, U. & Bradll, R. (1999) Spread of violets in polluted pine forests: morphological and molecular evidence for the ecological importance of interspecific hybridization. *Molecular Ecology* **8**, 365–377.

Newcombe, G., Stirling, B., McDonald, S. & Bradshaw, J.R. (2000) *Melampsora* × *columbiana*, a natural hybrid of *M. medusae* and *M. occidentalis. Mycological Research* **104**, 261–274.

Nuss, D.L. (1992) Biological control of chestnut blight: an example of virus mediated attenuation of fungal pathogenesis. *Microbiological Reviews* **56**, 561–576.

Peever, T.L., Liu, Y.-C., Wang, K. *et al.* (1998) Incidence and diversity of double-stranded RNAs occurring in the chestnut blight fungus, *Cryphonectria parasitica*, in China and Japan. *Phytopathology* **88**, 811–817.

Peever, T.L., Liu, Y.-C., Cortesi, P. & Milgroom, M.G. (2000) Variation in tolerance and virulence in the chestnut blight fungus-hypovirus interaction. *Applied and Environmental Microbiology* **66**, 4863–4869.

Pipe, N.D., Buck, K.W. & Brasier, C.M. (1995). Molecular relationships between *Ophiostoma ulmi* and the NAN and EAN races of *O. novo-ulmi* determined by RAPD markers. *Mycological Research* **99**, 653–658.

Pipe, N.D., Buck, K.W. & Brasier, C.M. (1997) Comparison of the cerato-ulmin (*cu*) gene sequences of the recently identified Himalayan Dutch elm disease fungus *Ophiostoma himal-ulmi* with those of *O. ulmi* and *O. novo-ulmi* suggests that the *cu* gene of *O. novo-ulmi* is unlikely to have been acquired recently from *O. novo-ulmi*. *Mycological Research* **101**, 415–421.

Pipe, N. D., Brasier, C. M. & Buck, K. W. (2000a) Evolutionary relationships of the Dutch elm disease fungus *Ophiostoma novo-ulmi* to other *Ophiostoma* species investigated by RFLP analysis of the rDNA region. *Journal of Phytopathology* **148**, 533–539.

Pipe, N.D., Brasier, C.M. & Buck, K.W. (2000b) Naturally occurring cerato-ulmin (CU)-deficient mutants of *Ophiostoma novo-ulmi* have either an introgressed *O. ulmi cu* gene or an *O. novo-ulmi cu* gene with a mutation in an intron splice consensus sequence. *Molecular Plant Pathology* **1**, 373–376.

Riesberg, L.H. & Wendel, J.F. (1993) Introgression and its consequences in plants. In: *Hybrid Zones and the Evolutionary Process* (ed. R. G. Harrison), pp. 70–109. Oxford University Press, New York.

Rogers, H.J., Buck, K.W. & Brasier, C.M. (1986) Transmission of double-stranded RNA and a disease-factor in *Ceratocystis ulmi*. *Plant Pathology* **35**, 277–287.

Rogers, H.J., Buck, K.W. & Brasier, C.M. (1987) A mitochondrial target for double-stranded RNA in diseased isolates of the fungus that causes Dutch elm disease. *Nature* **329**, 558–560.

Rosewich, U.L. & Kistler, H.C. (2000) Role of horizontal gene transfer in the evolution of fungi. *Annual Review of Phytopathology* **38**, 325–363.

Simmonds, P. & Smith, D.B. (1999) Stuctural constraints on RNA virus evolution. *Journal of Virology* **73**, 5787–5794.

Sutherland, M.L. & Brasier, C.M. (1997) A comparison of thirteen d-factors as potential biological control agents of *Ophiostoma novo-ulmi*. *Plant Pathology* **46**, 680–693.

Temple B. & Horgen, P.A. (2000) Biological roles for cerato-ulmin, a hydrophobin secreted by the elm pathogens, *Ophiostoma ulmi* and *O. novo-ulmi*. *Mycologia* **92**, 1–9.

Turgeon, B.G. (1998) Application of mating type technology to problems in fungal biology. *Annual Review of Phytopathology*, **36**, 115–137.

Van Alfen, N.K. (1986) Hypovirulence of *Endothia* (*Cryphonectria*) *parasitica* and *Rhizoctonia solani*. In: *Fungal Virology* (ed. K. W. Buck), pp. 143–162. CRC Press, Boca Raton, FL.

Van Alfen, N.K., Jaynes, R.A., Anagnostakis, S.L. & Day, P.R. (1975) Chestnut blight: biological control by transmissible hypovirulence in *Endothia parasitica*. *Phytopathology* **82**, 1206–1211.

Vos, P., Hogers, R., Bleeker, M. *et al.* (1995) AFLP: a new technique for DNA fingerprinting. *Nucleic Acids Research* **23**, 4407–4414.

Webber, J.F. & Brasier, C.M. (1984) The transmission of Dutch elm disease: a study of the processes involved. In: *Invertebrate–microbial Interactions* (eds J. M. Anderson, A. D. M. Raynor & D. Walton), pp. 271–306. Cambridge University Press, Cambridge.

Webber, J.F., Brasier, C.M. & Mitchell, A.G. (1988) The role of the saprophytic phase in Dutch elm disease. In: *Plant Infecting Fungi* (eds G. F. Pegg & P. G. Ayres), pp. 298–313. Cambridge University Press, Cambridge.

Wosten H.A.B. (2001) Hydrophobins: multipurpose proteins. *Annual Review of Microbiology* **55**, 625–646.

Chapter 3

The population genetic consequences of range expansion: oak gallwasps as a model system

G. N. Stone, R. J. Atkinson, G. Brown, A. Rokas and G. Csóka*

Introduction

The distributions of all organisms are not static in time and space, and range expansion on some temporal and spatial scale is probably a feature of the history of most species. Human introductions of alien species, whether deliberate or accidental, are a feature of many ecosystems, and represent major threats to natural patterns of biodiversity. This significance has resulted in continuing efforts to understand the consequences of invasion processes for the invading species and the communities they invade (for recent reviews see Ruiz *et al.* 2000; Sakai *et al.* 2001). A large body of theoretical and simulation-based literature shows that the processes underlying range expansion have important consequences for the genetic diversity and spatial structure of populations, and this has resulted in turn in the development of methods allowing a range of aspects of the invasion process to be inferred using population genetic data. As long as suitable genetic markers are available, it is possible to identify the origin of invaders, their pattern and sometimes mode of dispersal, and the direction and timescale of major changes in population size. This chapter highlights pattern and process during the colonization of northern Europe by a group of insect species—oak-feeding gallwasps (Hymenoptera: Cynipidae). We summarize existing work on patterns of genetic diversity and spatial structure arising from common models of range expansion, and illustrate application of analytical techniques designed to detect such patterns using our gallwasp data.

Oak gallwasps as a model system in invasion biology

Many invasions have been investigated using a diversity of population genetic methods. This diversity makes it difficult, however, to identify which factors are correlated empirically with the direction and strength of genetic patterns associated with range expansion: differences in variables of interest (such as the temporal or spatial scale of the invasion) are often confounded with variation in the taxa studied

* Correspondence author: graham.stone@ed.ac.uk

or methods applied. Our aim in the gallwasp work has been to use a consistent set of methods to generate population genetic data sets for a group of closely related species representing replicates of the invasion process, and whose histories differ in variables of interest to our understanding of range expansion.

Oak gallwasps (Hymenoptera: Cynipidae, Cynipini) develop in galls induced on oaks and closely related Fagaceae (Stone *et al.* 2002). Almost all oak gallwasps have cyclically parthenogenetic life cycles, with a single spring sexual generation and a single summer/autumn asexual generation each year. Individual gallwasp species can only induce galls on specific host oak taxa, and so have distributions intimately tied to those of the oaks. During the last ice age, European oaks and oak gallwasps were restricted to three glacial refugia—the Iberian peninsular, southern Italy and the Balkans (Stone & Sunnucks 1993; Dumolin-Lapègue 1997; Csóka *et al.* 1998; Hewitt 1999; Rokas *et al.* 2001; Stone *et al.* 2001). With the retreat of the ice, oak and gallwasp species richness remained highest in these refugia, and only two oak species (*Quercus robur* and *Q. petraea*) successfully expanded their ranges far to the north. Fossil pollen data suggest that these oak species reached Britain and much of northern Europe at least 6000–7000 years ago (Huntley & Birks 1983; Bennett 1986). Most of the gallwasp species whose life cycle requires only these oaks also expanded their ranges northwards, and have probably also been present in northern Europe for thousands of years.

Many other European gallwasp species have natural distributions restricted to southern Europe. This group includes a large clade of species in the genus *Andricus* that (unusually for gallwasps) have host-alternating life cycles (Stone *et al.* 2002). All of these life cycles involve obligate alternation between an asexual generation on oaks in the section Quercus (which includes *Q. robur* and *Q. petraea*) and a sexual generation on oaks in the section Cerris. In Europe, the section Cerris includes *Q. cerris* (Turkey oak), which is native to Italy and the Balkans, and *Q. suber* (Cork oak), which has disjunct races in Italy, the Iberian Peninsula and northwestern Africa (Figure 3.1). No section Cerris oaks were able to migrate northwards from their glacial refugia following retreat of the ice, and hence host-alternating gallwasps are also naturally restricted to these refugia.

From the 15th century onwards, however, *Q. cerris* became a popular ornamental tree, and has been dispersed far beyond its native range by human trade. *Q. cerris* reached Britain by the early 18th century, and is now self-seeding through much of its introduced range (see Figure 3.1). Introduction of *Q. cerris* has created a patchy landscape in which *Q. cerris* grows within a natural matrix of native *Q. petraea* and *Q. robur*. As a result, seven host-alternating gallwasp species have invaded northern Europe, without known direct human assistance (Stone *et al.* 2002). Three *Andricus* species native to the Balkans and Italy reached Britain, apparently independently, between 1950 and 1970 (*A. corruptrix*, *A. lignicola* and *A. quercuscalicis*; Stone *et al.* 2002). *A. quercuscalicis* caused considerable public concern through spectacular increase in population size over the decades following its arrival. Because the asexual generation of this species develops in conspicuous acorn galls, and destroys acorn crops (a valuable animal forage in the recent past), range expansion in this species is

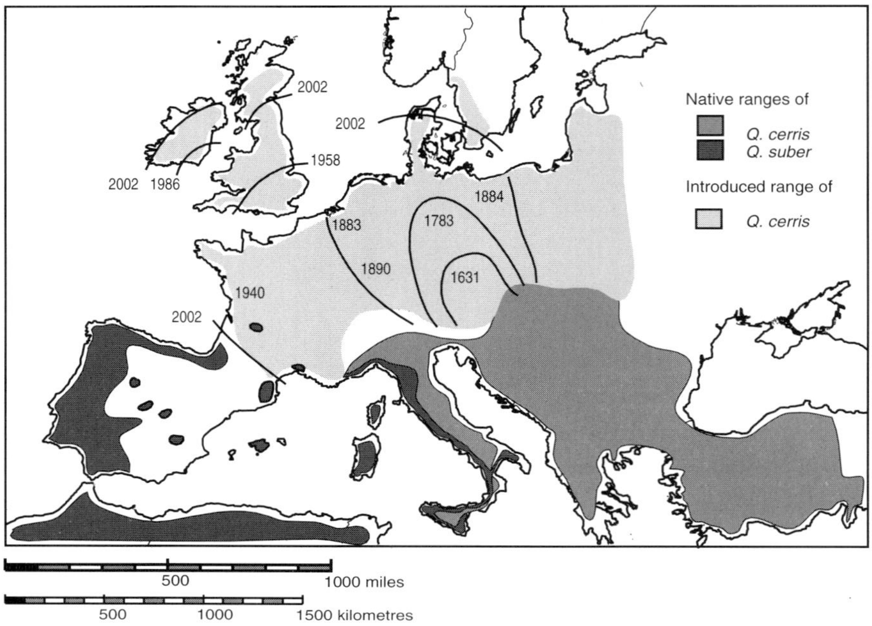

Figure 3.1 The natural distributions of *Quercus cerris* and *Q. suber*, and the introduced range of *Q. cerris* (redrawn from Stone *et al.* 2001). The dated lines indicate estimates of the progress of invasion by *Andricus quercuscalicis* based on literature sources (Stone & Sunnucks 1993).

also well documented (Stone & Sunnucks 1993; Stone *et al.* 2002) (see Figure 3.1). Three more species native to central Europe reached Britain at the end of the 20th century (*Andricus aries, Andricus lucidus, Aphelonyx cerricola*) and are currently expanding their distribution northwards from southeastern England. An eighth species, *Andricus kollari*, is the most widespread host-alternating invader in northern Europe. In addition to unassisted range expansion, this species was introduced to southwestern Britain in large numbers in the 1830s as a source of tannins for the manufacture of dyes. Within 40 years it had dispersed as far as Scotland and undergone dramatic population growth. The population dynamics and community impacts of these invasions have been the subject of detailed study (Csóka *et al.* 1998; Schönrogge & Crawley 2000; reviewed in Stone *et al.* 2002).

European oak gallwasps allow us to address a range of population genetic questions associated with range expansion:

1 Can population genetic data identify the origin of gallwasp invasions, where alternatives occur? Can we infer why one potential source population has been more successful in range expansion than another?

2 What are the consequences of recent, rapid range expansion for population genetic diversity? Do invasions over a similar timescale show similar patterns? Are there consistent differences between patterns seen in recent invasions and patterns generated by ancient postglacial range expansion? This comparison gives an indication of the significance of glacial refugia as long-term centres of genetic diversity.

3 Can we infer (i) which of a range of colonization models is most appropriate to gallwasp invasions, and (ii) whether patterns of genetic variation seen in gallwasps are more compatible with predicted effects of selection or genetic drift?

Invasion models and patterns in genetic diversity

Invasion models

A range of approaches have been used to model range expansion processes (Williamson 1996; Shigesada & Kawasaki 1997; Clobert *et al.* 2001). The vast majority of population genetic work on spatial patterns of genetic diversity has contrasted the predictions arising from two patch occupancy models of range expansion: island models and stepping stone models (Goodisman *et al.* 1998; Le Corre & Kremer 1998; Gandon & Rousset 1999; Irwin & Taylor 2000). In island models, unoccupied patches are colonized independently by migrants from a single specified source population, or are taken at random from the entire set of populations. In stepping stone models, colonists arriving at a given patch are drawn from a limited neighbourhood around the patch. The probability that a population will contribute colonists to another patch is modelled as a function of distance between patches through the stepping stone network. Models vary from colonization only from neighbouring patches, to those in which dispersal distances fit a given distribution.

Island and stepping stone models are two ends of a continuum, and for each the proportion of source genetic diversity that is present in an invaded patch depends on: (i) the founding population size; (ii) the number of colonists per generation; (iii) the generation time of the species; and (iv) the spatial distribution of patches that contribute those colonists (Austerlitz *et al.* 1997; Le Corre & Kremer 1998; Comps *et al.* 2001). The smaller the number of founders (or of migrants between existing populations), the greater the potential for changes in gene frequencies through genetic drift (Nei *et al.* 1975; Barton & Charlesworth 1984). Simulation studies have shown the loss of diversity along an invasion route to be most severe in stepping stone models (Ibrahim *et al.* 1996; Le Corre & Kremer 1998). This is because stepping stone colonization involves repeated founder effects as the invasion front moves through an array of patches, and colonists are derived principally from neighbouring, recently founded populations which have themselves experienced genetic bottlenecks in the recent past (Le Corre & Kremer 1998).

Invaded range populations need not always have lower genetic diversity than source populations. Invasions may have multiple origins and occur over a range of timescales—such that the patterning seen in a given invaded area represents the summation of these multiple processes. Where alternative origins have diverged genetically, and both contribute adequate numbers of colonists to the same invaded range, it is possible for newly colonized areas to have higher genetic diversity than either origin (Slatkin 1977; Comps *et al.* 2001).

Selective sweeps of neutral variation

Even for truly neutral markers patterns of diversity may not be devoid of the effects of selection. When a selectively advantageous mutation spreads through a population to fixation (a 'selective sweep'), a particular selectively neutral variant at a closely linked locus may also be fixed (Maynard Smith & Haigh 1974). A similar effect can be generated where two genetic entities are not physically linked, but share the same mode of inheritance. Mitochondrial DNA can be exposed to sweeps by selection acting on other maternally inherited entities, such as the endoparasitic bacterium *Wolbachia*. In some host species, infection by *Wolbachia* results in cytoplasmic incompatibility, such that crosses between hosts infected with different *Wolbachia* strains are infertile (Turelli *et al.* 1992; Werren 1997). Cytoplasmic incompatibility prevents any recombination between host mtDNA haplotypes and the *Wolbachia* strain present in the host germline, and spread of a single dominant *Wolbachia* strain through the host population may lead to fixation of a single mitochondrial haplotype (e.g. Turelli *et al.* 1992). The high frequency of *Wolbachia* infection in arthropods means that for these organisms there is a real risk of confusing population bottlenecks and selective sweeps as causes of reduced haplotype diversity. These potential causes can be distinguished by comparing patterns in nuclear and mitochondrial markers: a *Wolbachia* selective sweep should affect only mitochondrial diversity, while a bottleneck should also reduce diversity at nuclear markers.

Relatively few gallwasp species are infected with *Wolbachia*. One species that is infected is *Biorhiza pallida*, the oak apple gallwasp (Rokas *et al.* 2001). As introduced above, this species shows the negative correlation between *Wolbachia* infection rates and mitochondrial haplotype diversity characteristic of a selective sweep. However, in this case low haplotype diversity was also associated with low diversity at nuclear loci, implying loss of diversity through population bottlenecks rather than a selective sweep.

Discrimination among alternative origins for gallwasp invasions

The utility of genetic markers in discriminating among alternative invasion origins depends on a range of factors, including the invasion timescale, the evolutionary rate of the marker in question, the information content of the marker and the effective population size of the marker (Roderick 1996). As illustrated below for the gallwasp *A. kollari*, glacial refuges that have been separated without migrants for 10 000 years may be in the process of sorting ancestral polymorphism for allozyme loci (high population size, low mutation rate), but have mutually discrete clades of mitochondrial haplotypes (a quarter of the population size for nuclear loci, higher mutation rate).

A. kollari is the only one of the invading, host-alternating gallwasps to be native to all three southern European glacial refugia, providing multiple potential origins for

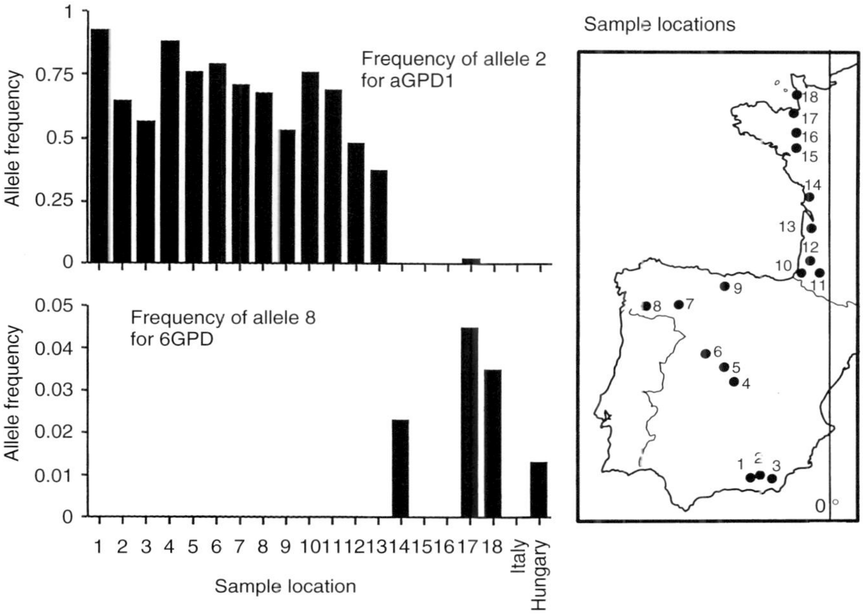

Figure 3.2 The genetic contact zone between Iberian and Balkan populations of *Andricus kollari*, as revealed by changes in the frequency of refuge-specific private allozyme alleles. The left–right arrangement of sites in the bar chart corresponds to the south–north ordering of samples sites in the map. The transition in *A. kollari* gene frequencies occurs at the northern limits of the natural distribution of the cork oak, *Quercus suber*, which is the sexual generation host for the Iberian race of *A. kollari*. (Data from Stone *et al.* 2001.)

colonization of northern Europe. In the Balkans and Italy, the sexual generation host of this species is Turkey oak, *Q. cerris*. In Spain, Portugal and the Atlas Mountains of Morocco, however, the sexual generation host is the cork oak, *Q. suber*. Colonization of introduced *Q. cerris* from Iberia would thus require a host shift. Screening of polymorphic allozyme loci showed these two life-cycle groups to share many ancestral polymorphisms, but also to possess several refuge- and life-cycle-specific alleles, suggesting limited gene flow between them. As predicted from its smaller effective population size and higher mutation rate, far better resolution was achieved using mitochondrial haplotype data. Sequence data for a fragment of the mitochondrial cytochrome b gene showed the Iberian populations to form one monophyletic group, and the individuals sampled from Italy and the Balkans another. Such refuge-specific polymorphism allows the origin of invaders in northern Europe to be assigned unambiguously. Sampling through France, Germany and the Low Countries showed populations in the continental European part of the invaded range of *A. kollari* to be derived entirely from the Balkan refuge (Figure 3.2) (Stone *et al.* 2001).

The inability of Iberian *A. kollari* to escape their refuge suggests that populations that usually exploit *Q. suber* may be unable to exploit *Q. cerris*. Exploitation of

Q. cerris by Iberian populations would require: (i) that asexual-generation females adapted to location of *Q. suber* can locate *Q. cerris*, and receive suitable oviposition stimuli; (ii) that host tissues are located in the narrow time window during which they are developmentally susceptible to gall induction; and (iii) that their sexual generation offspring can successfully induce galls on the novel host (Stone *et al.* 2001). Inability of western *A. kollari* to exploit *Q. suber* could result from failure in any one of these requirements. Host races have been demonstrated in two gallwasps—one on oaks (*Andricus mukaigawae*, Abe 1991) and one on roses (Stille 1985)—and in both systems, host races are characterized by demonstrable divergence in the cues which elicit oviposition. Rearing experiments suggest that *A. kollari* adapted to one sexual generation host may indeed not be able to recognize the alternative host as an oviposition site (Stone *et al.* 2001).

The question of origins for *A. kollari* in Britain is potentially more complex. The large introduction of asexual generation galls from the eastern mediterranean in the 1830s means that current British populations could be derived from these imports, or from more recent invaders arriving without direct human assistance (as for *Andricus quercuscalicis*) or from a combination of the two. Analyses of allozyme allele frequency and mitochondrial cytochrome b haplotype sequence data show: (i) that Britain contains allozyme alleles absent from all sampled European populations; (ii) that all British haplotype sequences form a clade that is discrete from all European sequences; and (iii) that the British haplotypes show greater sequence diversity than seen in equivalent numbers of sequences sampled over a much wider geographical range in continental Europe. These data strongly argue that British *A. kollari* are derived from a source outside continental Europe, with the origin of the 1830s imports (to date unsampled in Kurdistan) as the most probable source.

Loss of genetic diversity in invading gallwasps

Genetic diversity at polymorphic allozyme or microsatellite loci is commonly summarized by two statistics—heterozygosity (the mean probability of a locus being heterozygous) and allelic diversity (the mean number of alleles per locus). Allelic diversity is more sensitive to population bottlenecks than heterozygosity, and heterozygosity is only seriously reduced for very low population sizes (Nei *et al.* 1975; Leberg 1992). Comparison of patterns in these two measures with expectations based on mutation drift equilibrium forms the basis of a test for detection of bottlenecks that has been applied to data for a range of invasion processes (Cornuet & Luikart 1996; Jimenez *et al.* 1999; Comps *et al.* 2001). A drawback of this test is that comparison of observed and expected genetic diversity requires polymorphic loci—and the bottlenecks associated with range expansion may reduce polymorphism to a level where the test has inadequate statistical power. This is indeed the case for our current gallwasp data set. Nevertheless, comparison of patterns in these two measures provides an indication of the severity of bottlenecks associated with the colonization process.

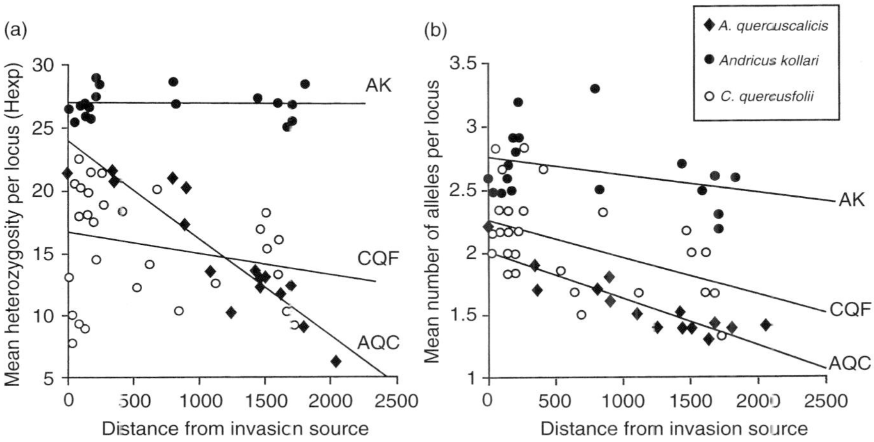

Figure 3.3 Changes in genetic diversity with distance from invasion origins for the oak gallwasps *Andricus quercuscalicis* (AQC), *A. kollari* (AK) and *Cynips quercusfolii* (CQF). (a) Mean heterozygosity per locus, calculated from observed allele frequences assuming Hardy–Weinberg equilibrium (Hexp). (b) Mean number of alleles per locus (allelic diversity). (Data from Stone & Sunnucks 1993; Stone *et al.* 2001.)

Recent invaders: *Andricus quercuscalicis* and *A. kollari*

Taking the invasion origin as the site with the greatest number of alleles, both mean heterozygosity and allelic diversity per locus show a dramatic decline along the invasion route in *A. quercuscalicis* (Figure 3.3), compatible with severe founder effects during colonization (Stone & Sunnucks 1993). Similar patterns, though with lower sampling resolution, were found for *A. corruptrix* and *A. lignicola*. In all three species, examination of individual loci shows that rare alleles are lost sequentially along the invasion route, a pattern suggesting sequential bottlenecks, and hence of stepping stone colonization. The strong patterns in *A. quercuscalicis* raise the question of whether such trends in genetic diversity are a general feature of recent invaders dependent on *Quercus cerris*, or are instead the product of a specific aspect of the biology of this gallwasp species.

The deliberate introduction of *A. kollari* to Britain means that patterns in *A. kollari* and *A. quercuscalicis* should only be compared for the continental European invaded range. *A. kollari* was sampled throughout continental Europe, and genetic variation screened at the same loci scored for *A. quercuscalicis*. The contrast between the two gallwasp species is dramatic. Along its invasion route, *A. kollari* shows little or no loss of heterozygosity, and only a slight loss of allelic diversity (see Figure 3.3). There is no evidence of the severe founder effects characteristic of *A. quercuscalicis*. Possible reasons behind this contrast between two recent invaders are discussed below.

Ancient invaders: *Cynips quercusfolii*

C. quercusfolii has a life cycle in which both generations are able to develop on both *Q. petraea* and *Q. robur*. This species is a widespread native in northern Europe, and probably tracked its hosts during their northwards range expansion 8000–10 000 years ago. This species is also native only to the Italian and Balkan refugia, allowing us to compare changes in genetic diversity along the probable northward invasion route with those seen in *A. quercuscalicis* and *A. kollari* (see Figure 3.3). Data for the same allozyme loci in *C. quercusfolii* show a decrease in both measures of genetic diversity, with slopes intermediate between the two recent invaders. These data show: (i) that for this small sample of species there is no clear divide in trends in recent and ancient invaders; and (ii) that glacial refugia remain centres of genetic diversity even for winged insects over postglacial timescales. At first sight, such loss of diversity seems surprising given that postglacial colonization by the oaks was relatively slow (of the order of 1 km a year (Bennet 1986), compared to the 15–20 km a year achieved by *A. kollari* after its introduction into Britain). In principal, large numbers of gallwasp colonists should have been able to keep up with the northwards advance of their hosts. However, *C. quercusfolii* is rare and patchily distributed through the northernmost current distribution of its oak hosts, and it is quite possible that population densities were similarly low (and the potential for founder effects associated with leptokurtic dispersal suitably high) during postglacial recolonization.

Patterns in genetic diversity observed in other species

The vast majority of empirical studies of apparently selectively neutral genetic variation in invading organisms show reduced diversity in the invaded range, particularly in cases of rare long-range anthropogenic dispersal (e.g. Tarr *et al.* 1998; Davies *et al.* 1999). Postglacial range expansion is also associated with lower genetic diversity at higher latitudes (Figure 3.4a) in a wide range of taxa (e.g. Ball *et al.* 1988; Dumolin-Lapègue *et al.* 1997; Merilä *et al.* 1997; Armbruster *et al.* 1998; Broyles 1998). Exceptions to this general rule can be divided into three types: (i) species in which colonization of lower latitude distributions occurred from high latitude refugia (Leonard *et al.* 2000); (ii) species in which higher latitude invaded ranges have received genetic contributions from multiple refugia (e.g. Figure 3.4b; Green *et al.* 1996; Fry & Zink 1998); and (iii) cases in which the colonization process has generated conflicting signals in allelic diversity and heterozygosity (Comps *et al.* 2001). In this last case, populations in the postglacial invaded range of European beech show loss of alleles (due to founder effects) but an increase in heterozygosity (due to contributions from multiple sources) relative to refuge populations. Despite such variation, the general persistence of postglacial clines in genetic diversity underlines the value of low latitude refugia as centres of genetic diversity.

Inference of invasion processes from spatial patterns in gene frequencies

Many studies have examined the consequences of range expansion for patterns of genetic subdivision between populations, and on the persistence of those patterns

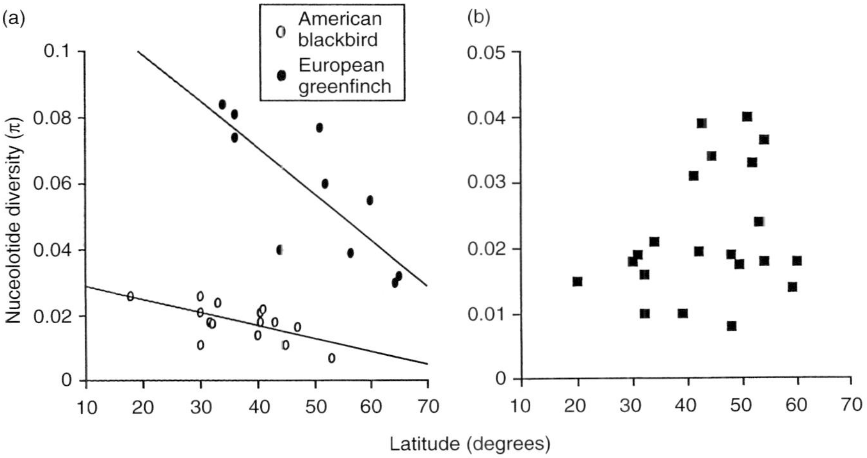

Figure 3.4 Patterns in genetic diversity associated with postglacial range expansion. (a) Decline in nucleotide diversity with increasing latitude for two North American bird species, the red-winged blackbird (redrawn from data in Ball *et al.* 1988) and the greenfinch (redrawn from data in Merilä *et al.* 1997). (b) A non-significant increase in genetic diversity with increasing latitude in the song sparrow (redrawn from data in Fry & Zink 1998).

through time (Wade & McCauley 1988; Comps *et al.* 2001). As with genetic diversity, patterning in spatial structure depends on the origin and frequency of migrants (Slatkin 1977; Wade & McCauley 1988; Nichols & Hewitt 1994; Ibrahim *et al.* 1996). Rare, long-range dispersal, as encompassed in leptokurtic dispersal models, predicts both the establishment of stronger genetic subdivision in the invaded range than other models, and greater persistence of this substructuring through time (Nichols & Hewitt 1994; Ibrahim *et al.* 1996; Le Corre *et al.* 1997). The persistent cline in genetic diversity seen along postglacial invasion routes in many species is taken as qualitative support for the significance of leptokurtic dispersal.

The modelling of invasion processes has led to the development of techniques designed to reveal range expansion from spatial patterning in genetic data. Two widely applied approaches are spatial analyses of allele or haplotype frequency data (using Mantel tests or spatial autocorrelation analysis) and nested clade analysis of the distribution of mitochondrial haplotypes (Templeton 1998). The former reveals patterns by comparing signatures across multiple loci, while the latter can reveal multiple effects through spatial patterning for a single marker. In both approaches, very high levels of gene flow among patches (effective panmixia) make distinction among alternate models of dispersal either difficult or impossible. However, under a range of circumstances, island and stepping stone models predict different spatial structures for selectively neutral markers. Island models predict neighbouring occupied patches to be no more similar genetically than expected by chance, because each patch is colonized independently from the source or from a spatially random set of sites. By contrast, stepping stone models predict neighbouring patches to be more similar than expected by chance. The relationship between similarity and the spatial separation between sites is also sensitive to the extent and direction of gene flow, producing differences that allow the identification of directional invasion processes. If gene flow occurs without directional bias across an array of patches, then stepping stone models predict the genetic similarity between patches to decline until it approaches that expected by chance. This signature is characteristic of isolation by distance, and is predicted for arrays of long-established populations with limited gene flow. However, if dispersal has a strong directional bias (as in invasions), distant sites are also predicted to be significantly more different than expected by chance (Sokal *et al.* 1997).

Here we illustrate the application of one of these approaches—spatial autocorrelation—to analysis of patterns along the entire invasion route, and of contrasts between the native and invaded range, in *A. kollari*, *A. quercuscalicis* and *C. quercusfolii*. Spatial autocorrelation quantifies the similarity in allele frequency between two sites, relative to the similarity expected for the same trait were any two sites compared at random. The correlation is expressed as a normalized autocorrelation coefficient (such as Moran's I), which ranges from 1 (maximum similarity) through 0 (no more similar than expected at random) to −1 (maximum difference). The pairwise comparisons for each allele are sorted into sets (distance classes) on the basis of the spatial separation between them, and the spatial structure in the data visualized by plotting the autocorrelation coefficient as a function of distance class. The resulting diagram is termed a correlogram. Extensive simulation studies show that a range of microevolutionary processes can be inferred by examining the shape of correlograms for individual alleles, and the congruence in pattern among loci (Sokal & Jacquez 1991; Sokal *et al.* 1997). For unlinked, selectively neutral loci, the stepping stone scenarios above predict qualitatively similar patterns across loci, and consistent differences between models involving panmixia, isolation by distance, and directional migration (Sokal & Jacquez 1991; Sokal *et al.* 1997) (Figure 3.5a–c). Spatial autocorrelation also allows the identification of patterns in allele frequency resulting from selection: an allele showing clinal variation across a series of patches will show high similarity in neighbouring patches but differ more significantly in frequency than predicted by chance when the two ends of the cline are compared (Figure 3.5d). This is a similar pattern to that expected for directional migration, except that here the pattern is expected only for the locus (or set of linked loci) under selection, and other unlinked and non-selected loci are expected to indicate processes affecting the genome as a whole.

Spatial autocorrelation analysis of invading gallwasps

Andricus quercuscalicis

Across the whole invaded range of *A. quercuscalicis*, spatial autocorrelation shows a strong signature of directional stepping stone migration followed by limited gene flow between populations (compare Figure 3.6 with Figure 3.5c), and there is no evidence that one among the sample of loci is responding to selection. The possibility

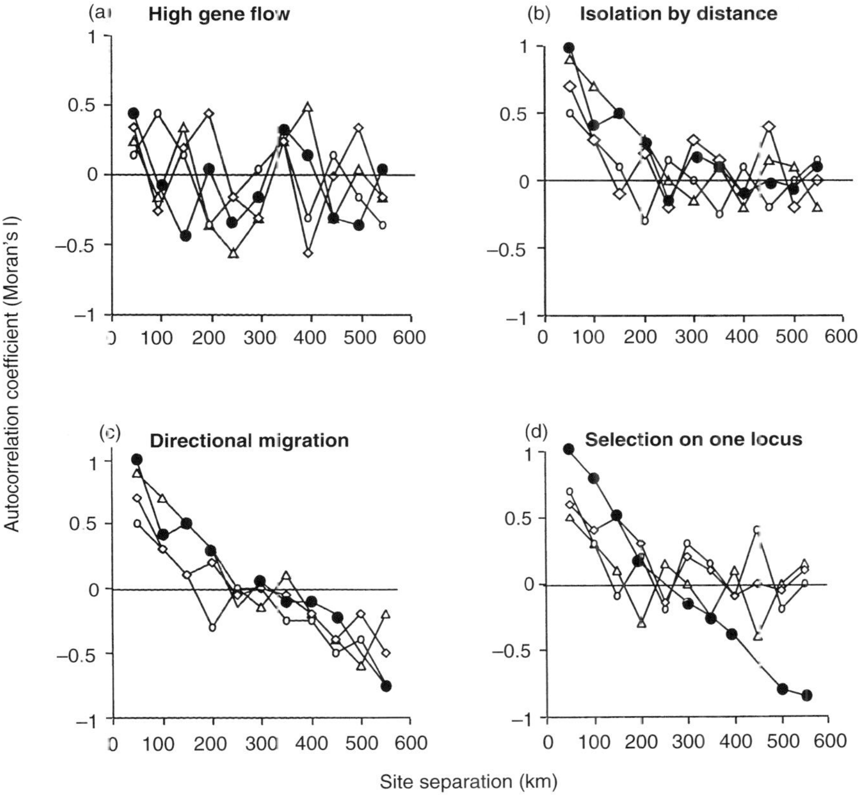

Figure 3.5 Spatial autocorrelation patterns generated by different microevolutionary patterns: (a) panmixia; (b) isolation by distance; (c) directional stepping stone migration with limited gene flow between populations; (d) clinal selection on a single locus. (Figures based on Sokal & Jacquez 1991; Sokal *et al.* 1997.)

remains that all of the selected loci are responding to selection acting on all of them at the same spatial scale, although this is extremely unlikely.

Patterns over the same spatial scale in the native range (Hungary and Slovenia) and the edge of the invaded range (Britain) are compared in Figure 3.7. The native range shows no evidence of spatial structure (compare with Figure 3.5a), indicating high levels of gene flow across populations. By contrast, data for the British populations reveal an even stronger signature of directional stepping stone migration than the invaded range as a whole. One obvious potential contributing factor to this contrast is the lower density and greater patchiness of *Q. cerris* in the invaded range. Oak gallwasp populations are prone to local extinction through a range of causes (Stone *et al.* 2002), and observed genetic structure is likely to represent the combined effects of any directional processes and local population turnover. If the greater distance

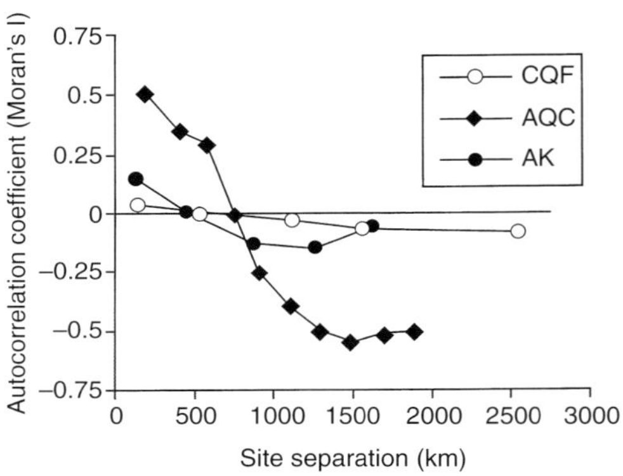

Figure 3.6 Spatial autocorrelation of patterns in allele frequency through the naturally invaded ranges of *Andricus quercuscalicis* (AQC), *A. kollari* (AK) and *Cynips quercusfolii* (CQF). The plots shown are mean values across correlograms generated for the same set of loci (GOT-s, GOT-m, PEP-b and PGM) for each species. (Data from Stone & Sunnucks 1993; Stone *et al.* 2001.)

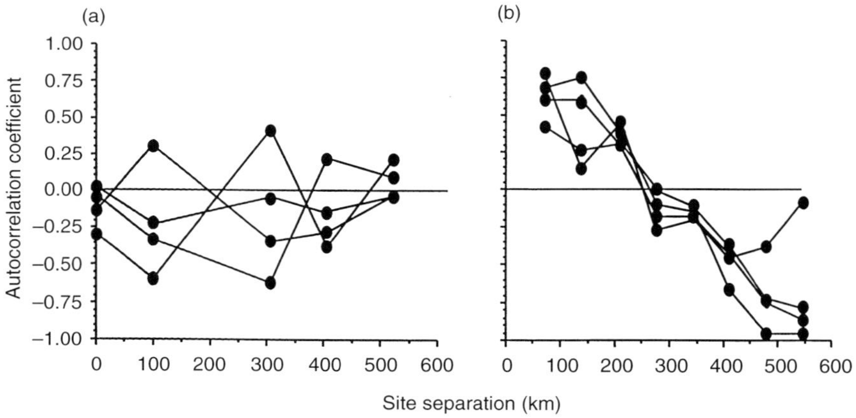

Figure 3.7 A comparison of patterns of spatial genetic structure in *Andricus quercuscalicis* between (a) a set of Hungarian populations in the natural, refuge distribution of this species, and (b) a set of populations over the same spatial scale in Britain and Ireland. Data are for the alleles using data for the loci GOT-s, GOT-m, PEP-b and PGM from Stone and Sunnucks (1993).

between *Q. cerris* patches in the invaded range results in a higher frequency of founder effects in the colonization of vacant patches, then populations here will be more prone to divergence through genetic drift. If this is a general feature of gallwasps dependent on *Q. cerris*, we expect to see similar patterns in other host-alternating species with similar invasion histories. Such comparisons must await more intense sampling of *A. corruptrix* and *A. lignicola.*

Andricus kollari

The large-scale introduction of *A. kollari* into Britain means that any comparison of patterns in Britain between this species and *A. quercuscalicis* are confounded by differences in invading population size and date of arrival. However, comparisons can be made between *A. quercuscalicis* and *A. kollari* in continental Europe. Spatial autocorrelation of allele frequencies for *A. kollari* in continental Europe gives a signature of directional stepping stone migration followed by limited gene flow between populations (see Figure 3.6), but the magnitude of the effect is far weaker than for *A. quercuscalicis*.

Why are the patterns in genetic diversity and spatial structure for these two species so different, given close similarities in life cycle and invasion timescale? The differences suggest that an underlying dependence on the distribution of *Q. cerris* is not enough to produce similar spatial patterns in genetic diversity, and that other aspects of the ecology of these two species are important. Stronger spatial structuring and loss of diversity for *A. quercuscalicis* could result from higher rates of local extinction or population bottlenecks. There is no reason to believe that the dispersal abilities of the two species differ, and we believe that the answer is more likely to lie in factors limiting local population size. Though little is known in general about bottom-up effects in gallwasp population dynamics (Hails & Crawley 1991; Schönrogge & Crawley 2000, Stone *et al.* 2002), it is possible that the differences between these species may stem from differences in the host plants attacked, and in the sites they require for gall induction. First, the asexual-generation galls of *A. kollari* can develop on either *Q. petraea* or *Q. robur*, while the same generation of *A. quercuscalicis* is almost entirely restricted to *Q. robur*. Though these two oaks and their hybrids are found in close proximity through much of their distribution, restriction to *Q. robur* probably excludes *A. quercuscalicis* from regions with higher rainfall where *Q. petraea* tends to dominate (Jalas & Suominen 1987). Second, galls of the asexual generation of *A. kollari* develop on buds, while those of *A. quercuscalicis* develop on acorns. While buds are not known to vary periodically in abundance, acorns certainly do, and cycling in acorn availability has been shown to drive the population size of *A. quercuscalicis* galls (Crawley & Long 1995). Although the effects of the dynamics of one generation on the effective population size of a gallwasp species remain unclear, it is possible that the patterns in *A. quercuscalicis* stem from a cycle in abundance driven by resource availability.

Cynips quercusfolii

Along its entire invasion route, *C. quercusfolii* shows a signature of directional migration, but it is the weakest of the three species (see Figure 3.6). This relative lack of structure, despite persistent loss of genetic diversity along the invasion route, is compatible with a lower incidence of local founder effects following colonization. Several factors could contribute to such a pattern. The most obvious is that the oak hosts for this species are abundant and broadly distributed in comparison to *Q. cerris*. There has also been far longer in this species for subsequent modification of any patterns established during the original range expansion process.

Future directions

The dramatic difference between patterns in *A. kollari* and *A. quercuscalicis*, which are similar in almost all aspects of their biology, highlights the need for detailed data sets for further species before general patterns can be identified. The possible impact of resource availability on genetic diversity and population structure in gallwasps highlights the need to link population genetics with population dynamics in future work. More broadly, work on oak gallwasps has shown how the population genetics of an invading insect may be influenced by aspects of the biology of its host plants. An exciting step forward is examination of the consequences of gallwasp population structure for the rich communities of natural enemies and inquilines centred around oak galls (Stone *et al.* 2002). Most of the members of oak gallwasp communities are specialists in oak galls, and an analysis of their population structure would reveal the extent to which the population structures of their gallwasp hosts structure those of species in higher and dependent trophic levels.

Acknowledgements

We would like to thank all those who have helped with comments and ideas throughout the work leading to this chapter. We thank Rich Bailey, Mick Crawley, George Melika, Jose-Luis Nieves-Aldrey, Juli Pujade-Villar and Karsten Schönrogge for their many thoughts and discussions on the subjects covered in this chapter. This work was supported by the following grants to GNS: a NERC EDGE programme grant (Gst/03/2035) and NERC grant (GR12847), a Small Ecological Research grant from the British Ecological Society, and a Royal Society Research grant.

References

Abe, Y (1991) Host race formation in the gallwasp *Andricus mukaigawae*. *Entomologia Experimentalis et Applicata* **58**, 15–20.

Armbruster, P., Bradshaw, W.E. & Holzapfel, C.M. (1998) Effects of postglacial range expansion on allozyme and quantitative genetic variation of the pitcher plant mosquito, *Wyeomyia smithii*. *Evolution* **52**, 1697–1704.

Austerlitz, F., Jung-Muller, B., Godelle, B. & Gouyon, P-H. (1997) Evolution of coalescence times, genetic diversity and structure during colonization. *Theoretical Population Biology* **51**, 148–164.

Ball, R.M., Freeman, S., James, F.C., Bermingham, E. & Avise, J.C. (1988) Phylogeographic population structure of red-winged blackbirds assessed by mitochondrial DNA. *Proceedings of the National Academy of Sciences of the United States of America* **85**, 1558–1562.

Barton, N.H. & Charlesworth, B. (1984) Genetic revolutions, founder effects and speciation. *Annual Review of Ecology and Systematics*, **5**, 133–164.

Bennet, K.D. (1986) The rate of spread and population increase of forest trees during the postglacial. *Philosophical Transactions of the Royal Society of London Series B* **314**, 523–531.

Broyles, S.B. (1998) Postglacial migration and the loss of allozyme variation in northern populations of *Asclepias exaltata* (Asclepiadaceae). *American Journal of Botany* **85**, 1091–1097.

Clobert, J., Danchin, E., Dhondt, A.A. & Nichols, J.D. (2001) *Dispersal*. Oxford University Press, Oxford.

Comps, B., Gömöry, D., Letouzey, J., Thiébault, B. & Petit, R.J. (2001) Diverging trends between heterozygosity and allelic richness during

postglacial colonization in the European beech. *Genetics* **157**, 389–397.
Cornuet, J.M. & Luikart, G. (1996) Description and power analysis of two tests for detecting recent population bottlenecks from allele frequency data. *Genetics* **144**, 2001–2014.
Crawley, M.J. & Long, C.R. (1995) Alternate bearing, predator satiation and seedling recruitment in *Quercus robur*. *Journal of Ecology* **83**, 683–696.
Csóka, G., Stone, G.N., Atkinson, R. & Schönrogge, K. (1998). The population genetics of postglacial invasions of northern Europe by cynipid gallwasps (Hymenoptera: Cynipidae). In: *The Biology of Gall-inducing Arthropods* (eds G. Csóka, W. Mattson, G. N. Stone & P. W. Price). General Technical Report NC 199. USDA. North Central Research Station, St Paul, MN.
Davies, N., Villablanca, F.X. & Roderick, G.K. (1999) Bioinvasions of the medfly *Ceratitis capitata*: source estimation using DNA sequences at multiple intron loci. *Genetics* **153**, 351–360.
Dumolin-Lapègue, S., Demesure, B., Fineschi, S., Le Corre, V. & Petit, R.J. (1997) Phylogeographic structure of White oaks throughout the European continent. *Genetics* **146**, 1475–1487.
Fry, A.J. & Zink, R.M. (1998) Geographic analysis of nucleotide diversity and song sparrow (Aves: Emberizidae) population history. *Molecular Ecology* **7**, 1303–1313.
Gandon, S. & Rousset, F. (1999) Evolution of stepping stone dispersal rates. *Proceedings of the Royal Society of London Series B* **266**, 2507–2513.
Goodisman, M.A.D., Shoemaker, D.D. & Asmussen, M.A. (1998). Cytonuclear theory for haplodiploid species and X-linked genes. 2. Stepping stone models of gene flow and application to a fire ant hybrid zone. *Evolution* **52**, 1423–1440.
Green, D.M., Sharbel, T.F., Kearsley, J. & Kaiser, H. (1996) Postglacial range fluctuation, genetic subdivision and speciation in the western North American spotted frog complex, *Rana pretiosa*. *Evolution* **50**, 374–390.
Hails, R.S. & Crawley, M.J. (1991) The population dynamics of an alien insect: *Andricus quercuscalicis* (Hymenoptera: Cynipidae). *Journal of Animal Ecology* **60**, 545–562.
Hewitt, G.M. (1999) Postglacial re-colonization of European biota. *Biological Journal of the Linnean Society* **68**, 87–112.
Huntley, B.H. & Birks, J.B. (1983) *An Atlas of Past and Present Pollen Maps for Europe: 0–13,000 Years Ago*. Cambridge University Press, Cambridge.
Ibrahim, K.N., Nichols, R.A. & Hewitt, G.M. (1996) Spatial patterns of genetic variation generated by different forms of dispersal during range expansion. *Heredity* **77**, 282–291.
Irwin, A.J. & Taylor, P.D. (2000) Evolution of dispersal in a stepping stone population with overlapping generations. *Theoretical Population Biology* **58**, 321–328.
Jalas, J. & Suominen, J. (1987) *Atlas Florae Europeae. Distribution of Vascular Plants in Europe*, Vol.2. Cambridge University Press, Cambridge.
Jimenez, P., Agundez, D., Alia, R. & Gil, L. (1999) Genetic variation in central and marginal populations of *Quercus suber* L. *Silvae Genetica* **48**, 278–284.
Leberg, P.L. (1992) Effects of population bottlenecks on genetic diversity as measured by allozyme electrophoresis. *Evolution* **46**, 477–494.
Le Corre, V. & Kremer, A. (1998) Cumulative effects of founding events during colonization on genetic diversity and differentiation in an island and stepping stone model. *Journal of Evolutionary Biology* **11**, 495–512.
Le Corre, V., Machon, N., Petit, R.J. & Kremer, A. (1997) Colonization with long distance seed dispersal and genetic structure of maternally inherited genes in forest trees: a simulation study. *Genetical Research* **69**, 117–125.
Leonard, J.A., Wayne, R.K. & Cooper A. (2000) Population genetics of ice age brown bears. *Proceedings of the National Academy of Sciences of the United States of America* **97**, 1651–1654.
Maynard Smith, J. & Haigh, J. (1974) The hitch-hiking effect of a favorable gene. *Genetical Research* **23**, 23–35.
Merilä, J., Björklund, M. & Baker, A.J. (1997) Historical demography and present day population structure of the Greenfinch, *Carduelis chloris*—an analysis of mtDNA control-region sequences. *Evolution* **51**, 946–956.
Nei, M., Maruyama, T. & Chakraborty, R. (1975)

The bottleneck effect and genetic variability in populations. *Evolution* **29**, 1–10.

Nichols, R.A. & Hewitt, G.M. (1994) The genetic consequences of long distance dispersal during colonization. *Heredity* **72**, 312–317.

Roderick, G.K. (1996) Geographic structure of insect populations: gene flow, phylogeography, and their uses. *Annual Review of Entomology* **41**, 325–352.

Rokas, A., Atkinson, R.J., Brown, G.S., West, S.A. & Stone, G.N. (2001) Understanding patterns of genetic diversity in the oak gallwasp *Biorhiza pallida*: demographic history or a *Wolbachia* selective sweep? *Heredity* **87**, 294–304.

Ruiz, G.M., Fofonoff, P.W., Carlton, J.T., Wonham, M.J. & Hines, A.H. (2000) Invasion of coastal marine communities in North America: apparent patterns, processes and biases. *Annual Review of Ecology and Systematics* **31**, 481–531.

Sakai, A.K., Allendorf, F.W., Holt, J.S. *et al.* (2001) The population biology of invasive species. *Annual Review of Ecology and Systematics* **32**, 305–332.

Schönrogge, K. & Crawley, M.J. (2000) Quantitative webs as a means of assessing the impact of alien insects. *Journal of Animal Ecology* **69**, 841–868.

Shigesada, N. & Kawasaki, K. (1997). *Biological Invasions: Theory and Practice. Oxford Series in Ecology and Evolution.* Oxford University Press, Oxford.

Slatkin, M. (1977) Gene flow and genetic drift in a species subject to frequent local extinctions. *Theoretical Population Biology* **12**, 253–262.

Sokal, R.R. & Jacquez, G.M. (1991) Testing inferences about microevolutionary processes by means of spatial autocorrelation analysis. *Evolution* **5**, 152–168.

Sokal, R.R., Oden, N.L. & Thomson, B.A. (1997) A simulation study of microevolutionary inferences by spatial autocorrelation analysis. *Biological Journal of the Linnean Society* **60**, 73–93.

Stille, B. (1985) Host plant specificity and allozyme variation in the parthenogenetic gallwasp *Diplolepis mayri* and its relatedness to *D. rosae* (Hymenoptera: Cynipidae). *Entomologia Generalis* **10**, 87–96.

Stone, G.N. & Sunnucks, P.J. (1993) The population genetics of an invasion through a patchy environment: the cynipid gallwasp *Andricus quercuscalicis. Molecular Ecology* **2**, 251–268.

Stone, G.N., Atkinson, R., Rokas, A., Csóka, G. & Nieves-Aldrey, J-L. (2001) Differential success in northwards range expansion between ecotypes of the marble gallwasp *Andricus kollari*: a tale of two lifecycles. *Molecular Ecology* **10**, 761–778.

Stone, G.N., Schönrogge, K., Atkinson, R.J., Bellido, D. & Pujade-Villar, J. (2002) The population biology of oak gallwasps (Hymenoptera: Cynipidae). *Annual Review of Entomology* **47**, 633–668.

Sunnucks, P.J. & Stone, G.N. (1996) Genetic structure of invading insects and the case of the knopper gallwasp. In: *Frontiers of Population Ecology*, pp. 485–496. CSIRO.

Tarr, C.L., Conant, S. & Fleischer, R.C. (1998) Founder events and variation at microsatellite loci in an insular passerine bird, the Laysan finch (*Telespiza cantans*). *Molecular Ecology* **7**, 719–731.

Templeton, A.R. (1998) Nested clade analyses of phylogeographic data: testing hypotheses about gene flow and population history. *Molecular Ecology* **7**, 381–397.

Turelli, M., Hoffmann, A.A. & McKechnie, W.S. (1992) Dynamics of cytoplasmic incompatibility and mtDNA variation in natural *Drosophila simulans* populations. *Genetics* **132**, 713–723.

Wade, J.W. & McCauley, D.E. (1988) Extinction and recolonization: their effects on the genetic differentiation of local populations. *Evolution* **42**, 995–1005.

Werren, J.H. (1997). Biology of *Wolbachia. Annual Review of Entomology* **42**, 587–609.

Williamson, M.H. (1996). *Biological Invasions.* Chapman and Hall, London.

Zink, R.M., Barrowclough, G.F., Atwood, J.L & Blackwell-Rago, R.C. (2000) Genetics, taxonomy, and conservation of the threatened California gnatcatcher. *Conservation Biology* **14**, 1394–1405.

Chapter 4

A genetic perspective of *Drosophila melanogaster* latitudinal body size clines

*W. J. Kennington, J. Gockel, D. B. Goldstein and L. Partridge**

Introduction

Body size has profound consequences for animal life history and ecology (Roff 1992; Stearns 1992). It is therefore likely to be influenced by a diverse array of potentially conflicting selection pressures, and for this reason, consistent spatial correlations with particular ecological variables are not expected. Nevertheless, many species exhibit latitudinal variation in body size, with larger individuals found in populations derived from higher latitudes. For endotherms, these geographical clines conform to the predictions of Bergmann's (1847) rule that states that animals tend to evolve larger body size in colder conditions (Mayr 1963). The physiological explanation for this pattern is that the surface area and volume of body size are important in determining rates of heat production and dissipation, and hence in maintaining standard body temperature.

Latitudinal clines in body size are not restricted to endotherms, however, as several species of ectotherms show clinal patterns, even when different populations are reared under standard conditions. Such clines have been demonstrated in the house fly *Musca domestica* (Bryant 1977), the honey bee *Apis melifera* (Alpatov 1929), an ant lion *Myrmeleon immaculatus* (Arnett & Gotelli 1999), a copepod *Scottolana canadensis* (Lonsdale & Levinton 1985) and several species of fish (the Atlantic silverside *Menidia menidia* (Conover & Present 1990; Billerbeck *et al.* 2000), striped bass *Morone saxatilis* (Brown *et al.* 1998) and the Mummichog *Fundulus heteroclitus* (Schultz *et al.* 1996)). Latitudinal size clines have also been found in several species of *Drosophila* (for review see Partridge & French 1996).

Here, we review recent and ongoing genetic studies on latitudinal body size clines in *Drosophila melanogaster*. These studies fall into two main categories 'quantitative trait locus' (QTL) mapping and analyses of spatial genetic structure. Our aims are to report on current efforts to find the genes responsible for latitudinal variation in body size and to understand the role of demographic processes in shaping these clines.

* Correspondence author: l.partridge@ucl.ac.uk

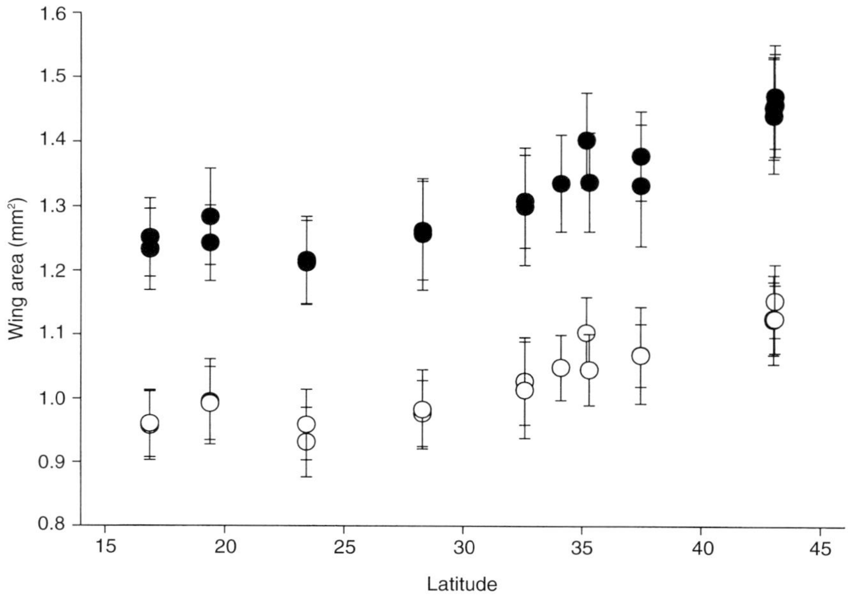

Figure 4.1 Body size in relation to latitude for flies raised under standard laboratory conditions from populations of *Drosophila melanogaster* along the east coast of Australia. Solid circles, females; open circles, males. (Reproduced with permission from Gockel *et al.* 2001.)

Background

D. melanogaster has a global distribution, and latitudinal clines in body size have been found in all the main continents (e.g. Middle East, Europe, Africa (Tantawy 1961; David & Bocquet 1975), Japan (Watada *et al.* 1986), North America (Coyne & Beecham 1987, Long & Singh 1995), Eastern Europe, Central Asia (Imasheva *et al.* 1994), Australia (James *et al.* 1995, 1997; Gockel *et al.* 2001, Figure 4.1) and South America (van't Land *et al.* 1995, 1999)). These clines are evident no matter the temperature of the standard rearing conditions (Coyne & Beecham 1987; James *et al.* 1997; van't Land *et al.* 1999). This is despite temperature having a direct environmental effect on body size and development time in *Drosophila,* and other ectotherms, with both traits increasing as temperature decreases (Atkinson 1994). Latitudinal clines are also evident in flies collected directly from the field. Wild-caught flies tend to be smaller and have steeper clines than laboratory-reared flies. The reduced size of wild flies is thought to be due to reduced nutrition, whereas the steeper slopes have been attributed to an environmental response to latitudinal variation in temperature (James *et al.* 1997).

The repeatability of genetically based latitudinal clines in body size in the same species in different geographical regions of the world, and the establishment of such clines soon after species arrive in new continents (Huey *et al.* 2000), demonstrates that they are the products of natural selection. In most cases the selective agent(s) producing the cline has not been identified. Many ecological variables, such as day length, population density and temperature, change with latitude, and could therefore be implicated. However, several lines of evidence point to temperature. First, body size has been shown to increase not only with latitude, but also with altitude in *Drosophila* (Stalker & Carson 1948) and a frog *Rana sylvatica* (Berven 1982), and in the cooler periods of the breeding season in *Drosophila* (Stalker & Carson 1949; Tantawy 1964). Second, replicated populations of *Drosophila* have been maintained at different temperatures in the laboratory for many generations, and the resulting cold-adapted lines are larger than the warm-adapted ones when reared at a common temperature, irrespective of whether the comparison is conducted in the warm or the cold environment (Anderson 1966, 1973; Cavicchi *et al.* 1985, 1989; Partridge *et al.* 1994a).

Body size in *Drosophila* is only one of a suite of traits that show latitudinal clines and that evolve in response to thermal selection in the laboratory. For example, high-latitude and low-temperature populations show increased pre-adult growth efficiency and growth rate (Partridge *et al.* 1994b; Neat *et al.* 1995; Robinson & Partridge 2001), more rapid pre-adult development (James & Partridge 1995; James *et al.* 1995; van't Land *et al.* 1999), increased egg size (Azevedo *et al.*1996) and greater ovariole number (Capy *et al.* 1993; Watada *et al.* 1986; Azevedo *et al.* 1996) relative to low-latitude and warm-temperature populations. All of these characters may be targets of natural selection. Alternatively, some may evolve as correlated responses to selection on others, if there are positive genetic correlations between them.

One explanation why body size evolves in response to temperature is that persistently higher nutrient availability during growth at lower temperature may have driven an evolutionary reallocation of resources to growth. This reallocation would in turn lead to evolutionary increased growth efficiency, growth rate and adult body size. This idea is plausible since it would be expected to lead to increased growth rate and to reduce development time at low-temperature and high-latitude populations, all of which are observed. Increased efficiency of food could also explain increased ovariole number and egg size in the adult (Partridge & French 1996). Rates of heat exchange and equilibrium body temperature are unlikely to be important, because organisms such as *Drosophila* that show thermal evolution of body size are too small to retain or exclude heat from their environment (Stevenson 1985).

QTL mapping

The mapping of genomic regions or QTL can provide valuable information on the ecological and evolutionary processes influencing latitudinal clines. It is the first step towards identifying the genes producing a response to selection, and in determining the environmental factors(s) that change the fitness of body size at different positions along the cline. It can also provide answers to other evolutionary points of interest. For example, once genomic regions responsible for phenotypic differences between extreme ends of a cline are mapped, it will be possible to assess if the same regions contribute to the variation along the entire cline by comparing the degree of clinality of neutral markers that are closely and distantly linked to these locations, i.e. we will be able to determine how far these selected alleles spread along the cline. On a larger scale it will be possible to determine if the same genomic regions consistently respond to selection on different continents. Finally, QTL mapping will be able to determine whether genomic regions responsible for body size variation also control other clinal traits, and hence whether they are physiological determinants of body size (pleiotropy) or co-selected traits.

QTL mapping also provides an opportunity to test the concept of using spatial distributions of neutral molecular markers to map genomic regions under clinal selection. This concept is of broad, general interest, because it could provide a method for detecting QTL in genetically less tractable organisms (e.g. humans) in which traditional mapping approaches are not possible. A framework for testing this concept follows. First, QTL mapping identifies genomic regions responsible for phenotypic differences between cline ends. Second, the spatial distributions of densely spaced neutral markers inside and outside the regions of the QTL are examined. If neutral markers inside the region of the QTL tend to be more clinally distributed than neutral markers outside this region, then this would provide support for the concept. These alleles could have been spread along the cline due to gene flow or be a historical contingency. If clinal patterns were not evident it would suggest that the method was not a practical way of mapping genomic regions under clinal selection.

To date, studies aimed at detecting QTL for body size variation have only identified broad genomic regions responsible for differences between cline ends. However, these studies have shown a remarkable consistency in the genomic regions implicated in controlling body size, both within and between continents. These data have also provided a means of carrying out preliminary investigations on the association between the positions of QTL and molecular markers with clinal distributions.

Two different genetic approaches have been used to map genomic regions responsible for differences in body size between extreme cline ends, chromosome substitution and classical QTL analysis. These studies have been carried out on populations collected from two latitudinal clines in the southern hemisphere, one on the east coast of Australia, the other on the west coast of South America. Results from both experiments are reported in Gockel *et al.* (2002).

The chromosome substitution method (for a detailed description of methods see Kearsey & Pooni 1996) uses balancer chromosomes, which contain multiple overlapping inversions, a dominant phenotypic marker (e.g. 'curly wings') and a recessive lethal, to substitute chromosomes between different parental lines collected from extreme cline ends. True-breeding lineages of all possible substitution lines can be generated using standard crossing procedures. These lines can then be measured for wing area under standard laboratory conditions. Because *D. melanogaster* has three main chromosomes, the total number of substitution lines available is six (3 chromosome combinations × 2 parental genetic backgrounds). Results obtained using this method revealed that on both the South American and Australian continents, genomic regions on the third chromosome control most of the variation in body size between cline ends (Figure 4.2). On average, chromosome 3 accounted for

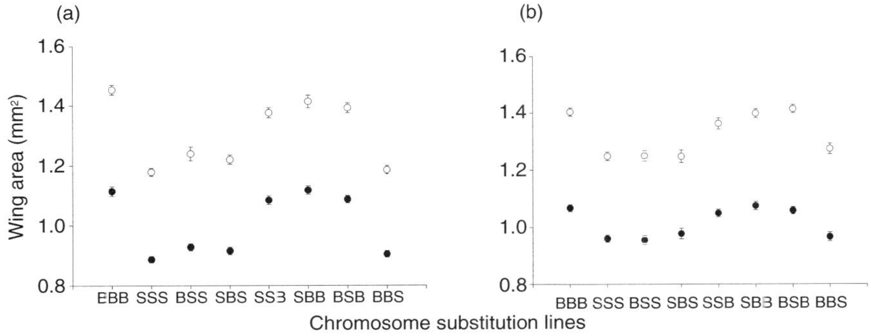

Figure 4.2 Wing areas of (a) South American and (b) Australian chromosome substitution lines. B represents big (high latitude) chromosome, S represents small (low latitude) chromosome; position of the letter represents chromosome number, e.g. BBS = big X chromosome, big chromosome 2 and small chromosome 3. Solid circles, females; open circles, males; error bars, 95% confidence intervals. The figure shows that whenever chromosome 3 is small, flies are small and that whenever chromosome 3 is big the flies are big. (Reproduced with permission from Gockel *et al.* 2002.)

77% of the size difference between the parental populations, while the second and X chromosomes both accounted for 13% of the total variation. Analysis of variance indicated that all three chromosomes had significant effects on body size, and also that there was a significant chromosome 3 by continent interaction. Chromosome 3 also had a larger effect on the body size of South American flies than Australian flies. This result supports the view that there may be some underlying differences in the genetic basis of phenotypic differences in body size on the two continents, as suggested by the line-cross study by Gilchrist and Partridge (1999), and the comparisons of cellular basis of body size by Zwaan *et al.* (2000). This is an important result, as it lends further support to the view that it is body size itself, rather than an associated trait, which is the target of selection.

Analysis of chromosome substitution data also revealed there were significant interaction effects between chromosomes. Evidence of interactions between genes or epistasis contributing to differences in body size has also been demonstrated by Gilchrist and Partridge (1999). These results are highly relevant because epistasis can create adaptive peaks and valleys that can prevent deterministic evolution (see Whitlock *et al.* 1995; Fenster *et al.* 1997).

If epistatic contributions to phenotypic differences in body size are common between clinal populations, evolutionary responses to environmental changes may be limited unless circumvented by a 'shifting balance process' (Wright 1931, 1977), involving selection among populations carrying different gene combinations.

Classical QTL mapping methods use marker-trait associations in crosses between diverged lines to identify genomic positions responsible for phenotypic differences. Using microsatellites as molecular markers, Gockel *et al.* (2002) used this approach to map genomic regions responsible for differences between cline ends in the

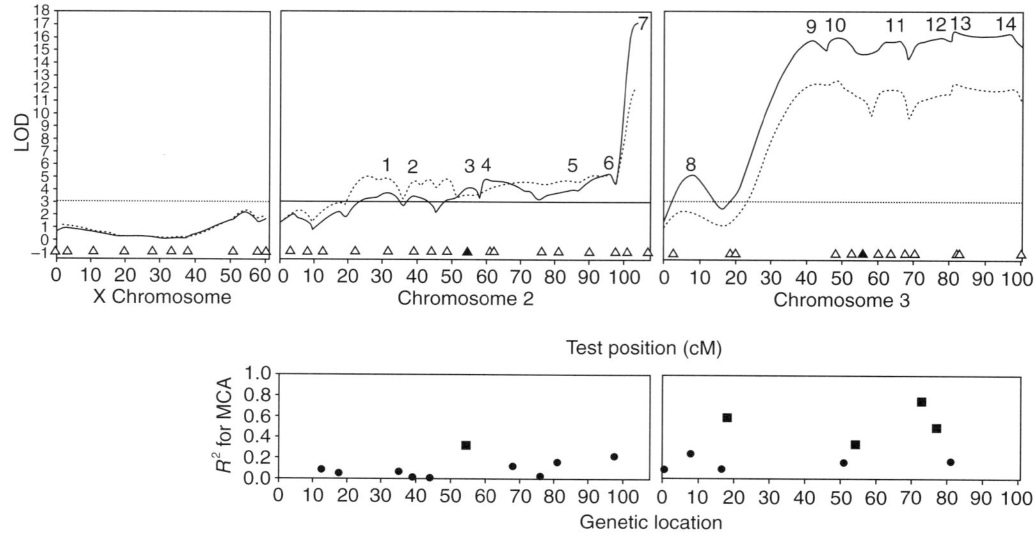

Figure 4.3 The likelihood map for wing area QTL positions in a cross of two flies from different ends of a body size cline. Evidence of a QTL is given when the likelihood function exceeds the significance threshold (3.1, indicated by horizontal line). The dashed curve gives the LOD (likelihood of odds) scores based on interval mapping (IM), and the solid curve the LOD scores based on composite interval mapping (CIM). Numbers refer to significant QTL peaks for CIM identified with the mapping program QTL CARTOGRAPHER. Triangles on the x-axis correspond to marker positions used in the analysis. Two markers that previously showed significant clinal variation in Gockel *et al.* (2001) are indicated as black triangles. Underneath the likelihood maps are the locations of 19 microsatellite loci tested for clinal variation in Gockel *et al.* (2001) with estimates of the proportion of variation explained by latitude (depicted by R^2 values). Loci for which latitude explained a significant proportion of the variation are shown as squares. MCA, most common allele. (Reproduced with permission from Gockel *et al.* 2002.)

Australian cline. As with the chromosome substitution analysis, we found that genomic regions with the largest effect on body size were on the third chromosome. Genomic regions with significant effects on body size were also found on the second chromosome, but none was found on the X chromosome. In total, 14–17 significant QTL peaks were identified using the QTL CARTOGRAPHER mapping program (Basten *et al.* 1997), depending on the mapping method employed (Figure 4.3). However, given the shallow valleys between many of these peaks, especially on chromosome 3, a more cautious assessment would suggest there are as few as five different QTL.

Comparison of the locations of these QTL with the locations of microsatellite loci with clinal distributions in the frequency of their most common allele in the Australian cline (see Gockel *et al.* 2001) revealed that there was a tendency for these markers to fall within the range of significant QTL peaks. Of the five clinal loci examined, four fell within the range of the identified QTL (see Figure 4.3). However, resampling tests revealed that these associations were not significantly different to those expected at random (Gockel *et al.* 2002).

The lack of association between the positions of clinal markers and QTL is consistent with the idea that different QTLs are responsible for phenotypic variation along the length of the cline. If this is the case, it follows that examining spatial distributions of clinal markers is not a practical way of detecting genomic regions under clinal selection. However, the non-significant association found in this study may, in part, be due to the poor resolution of QTL peaks. With fine-scale QTL maps, it is entirely possible that significant associations between the positions of QTL and clinal markers may be found. These issues, therefore, remain unresolved.

Spatial genetic structure

Studies investigating spatial genetic structure also improve our understanding of latitudinal clines. These studies use selectively neutral molecular markers to assess the spatial distribution of genetic variation. In contrast to phenotypic variation, which may be influenced by selection, neutral genetic variation is largely determined by demographic factors such as migration and colonization history. These data provide a means to assess the relative influence of different evolutionary processes in shaping the cline. For example, Gockel *et al.* (2001) used regression analysis to compare the magnitude of latitudinal effects for microsatellite and body size variation in the Australian cline. The results revealed that approximately 80% of body size variation was explained by latitude, whereas the average amount of microsatellite variation explained by latitude was 21%. Since the 95% confidence intervals of these estimates were non-overlapping, the null hypothesis that the morphological variation was exclusively shaped by demographic processes was rejected, and we were able to infer that natural selection was responsible for the cline. Differences in the apportionment of molecular (allozyme) and morphological variation has also implicated natural selection as the factor responsible for a *D. melanogaster* body size cline in North America (Long & Singh 1995) and differences in body size among the populations of the cactophilic species, *D. buzzatii* (Prout & Barker 1993).

An important point to note about these results is that even though a common demographic explanation for the molecular and morphological variation can be rejected, the presence of some underlying demographic processes cannot necessarily be ruled out. We investigated this possibility using spatial autocorrelation, which assesses the correlation between samples in reference to their spatial location. Spatial autocorrelation is present if samples are interrelated, and there is spatial pattern to the correlation. It can be quantified with several different measures, each with their own strengths and weaknesses. One of the most popular is the Moran's *I* measure, which ranges from –1 to +1, where zero reflects a completely random distribution (i.e. no spatial trend). Significant positive values indicate overall resemblance between samples, and significant negative values indicate dissimilarity.

To characterize the pattern of genotypic variation along the Australian cline, Moran's *I* was independently calculated for 10 microsatellite loci, by comparing all possible pairs of individuals (559 individuals from 32 sites distributed along the cline) using the SGS program (Degen 2000). These comparisons were carried out in distance classes of 200 km intervals, ranging from 0–200 km to 1601–1800 km. The level of significance for each measure was based on a permutation procedure using 500 random redistributions of individual genotypes over the spatial coordinates of the sampled points. Spatially random distributions results in a series of non-significant Moran's *I* coefficients at all distances. A decreasing set of Moran's *I* coefficients, from positive significant to negative significant, describes a cline, whereas decreasing coefficients from positive significant to non-significant at large distances is expected for allele frequencies under isolation by distance, i.e. when genetic variability reflects genetic drift and short-range gene flow.

Our analyses revealed that all 10 microsatellite loci considered depart significantly from randomness, and that although some differences exist between loci, they share common features (Table 4.1). First, although Moran's *I* coefficients are small, they are highly significant. Second, spatial autocorrelation is significantly positive for all loci at the first distance class, 0–200 km and these tend to decline with distance to insignificant values beyond 200 or 400 km. Finally, all loci exhibit negative values beyond the 1001–1200 km distance class, most of which are significant. This overall pattern of variation is clearly clinal, with individuals at close proximity showing the highest similarity, and individuals at extremes of the geographical range showing the highest divergence. It differs to the pattern expected with isolation by distance, because distant populations are not expected to be significantly different under this process.

Parallel patterns of geographical variation at several loci are regarded as evidence for evolutionary processes acting on the entire genome, rather than single genes. It is therefore unlikely that selection has played a major role in shaping this gradient in microsatellite variation. A more likely scenario is that the gradient has resulted from a directional range expansion. This idea is supported by a negative relationship between the average expected heterozygosity (gene diversity) within populations and latitude ($R^2 = 0.43$; $P < 0.001$), a pattern that is consistent with a southward range expansion accompanied by repeated founder effects.

Alternative explanations for the cline in genetic diversity are a declining population density gradient from low to high latitudes or a latitudinal gradient in habitat heterogeneity. A population density gradient is likely to have strong influences on patterns of migration along a latitudinal cline and this, in turn, will influence the ability of populations to respond to selection (see Slatkin 1985, 1987). For example, a declining population density gradient from low to high latitudes in the Australian cline may cause higher emigration rates from populations at low latitudes, and this could reduce the magnitude of genetic differentiation between them. James *et al.* (1995) have suggested that this as an explanation for the non-linear shape of the Australian body size cline, which is characterized by small changes in body size at low latitudes, and significant changes at higher latitudes.

Table 4.1 Spatial autocorrelation of microsatellite variation at 10 loci.

	Distance class limits (km)								
Locus	0–200	201–400	401–600	601–800	801–1000	1001–1200	1201–1400	1401–1600	1601–1800
AC005555	0.023***	0.001	0.014***	−0.005	0.004	0.003	−0.014**	−0.027**	−0.040***
AC005889	0.023***	0.015***	0.011***	−0.008*	−0.013**	−0.013**	−0.012**	−0.010	−0.020*
AC004343	0.030***	0.014***	0.011***	0.001	−0.018**	−0.013***	−0.030***	−0.029**	0.011
DROPROSA	0.017***	−0.006	−0.005	0.007**	−0.008	−0.013**	−0.000	−0.013*	−0.009
AC004759	0.092***	0.040***	0.011*	0.036***	0.018**	−0.056***	−0.033***	−0.098***	−0.131***
DMMP20	0.017**	0.030***	−0.011*	−0.021**	0.003	0.000	−0.011*	−0.034***	−0.016
AC005115	0.046***	0.016*	0.004	0.005	0.005	−0.032***	−0.036***	−0.021*	−0.003
AC004658	0.031***	0.004	0.003	−0.004	−0.020*	−0.001	−0.004	−0.024*	−0.037**
DMU1951	0.023***	0.002	0.006**	0.011***	−0.002	−0.003	−0.014***	−0.035***	−0.062***
AC004641	0.020***	−0.003	0.008**	0.003*	−0.011**	−0.011**	−0.001	−0.024***	−0.020***

* $P<0.05$, ** $P<0.01$, *** $P<0.001$.

Conclusions and future work

In this review we have outlined various ways that genetic approaches can be used to gather information on the ecological and evolutionary processes influencing latitudinal clines. This information can be substantial and conclusive. For example, by simply comparing the amounts of molecular and morphological variation explained by latitude, Gockel *et al.* (2001) were able to reject a historical explanation for a body size cline on the east coast of Australia, and hence provide strong evidence that this cline was due to natural selection.

However, by far the most direct and potentially rewarding genetic approach employed to date has been QTL mapping. Using this approach the genomic regions influencing phenotypic variation can be identified, and this will ultimately lead to the identification of the genes producing the response to selection. The prospects of achieving this goal are greatly increased for model organisms such as *D. melanogaster* for which the complete genome sequence is now available. This will make it more straightforward to go from quantitative trait locus mapping to gene, using microarray technology. Using this information it will be possible to understand at the most fundamental level how environmental factors change the fitness of body size at different positions along the cline.

The resolutions of body size QTL maps produced to date are still too low to warrant application of microarray technology, but they do suggest there may be a number of common genomic regions responding to selection on different continents. Whether or not this reflects a common genetic basis for independent clines remains unresolved until more detailed QTL mapping has been carried out. Elucidation of whether genomic regions responsible for differences between cline ends contribute to the phenotypic cline, and the related issue of whether clinally varying molecular markers such as microsatellites can be used to indirectly identify QTLs of clinal traits, also remain unresolved until higher resolution QTL maps become available. Definitive answers to these issues are of general interest because they will provide information on the spatial scale over which alleles are selected, and test the applicability of using spatial distributions of molecular markers to detect clinal polymorphisms.

Finally, the role of migration and its interplay with selection in shaping the latitudinal clines remains poorly understood and is a topic requiring further work. Of special interest is the extent to which asymmetrical migration from warm, low latitudes to cool, high latitudes may constrain the evolutionary potential of populations. The cline on the east coast of Australia, characterized by non-linear changes in body size and indirect evidence of a population density gradient, represents an ideal system for the investigation of these processes further.

Acknowledgements

This work was funded by the Natural Environment Research Council's thematic programme EDGE (Ecological Dynamics and Genes).

References

Alpatov, W.W. (1929) Biometrical studies on variation and races of the honey bee (*Apis melifera*). *Quarterly Review of Biology* **4**, 1–58.

Anderson, W.W. (1966) Genetic divergence in M. Vetukhiv's experimental populations of *Drosophila pseudoobscura*. *Genetical Research Cambridge* **7**, 255–266.

Anderson, W.W. (1973) Genetic divergence in body size among experimental populations of *Drosophila pseudoobscura* kept at different temperatures. *Evolution* **27**, 278–284.

Arnett, A.E. & Gotelli, N.J. (1999) Geographic variation in life-history traits of the ant lion, *Myrmeleon immaculatus*: evolutionary implication of Bergmann's rule. *Evolution* **53**, 1180–1188.

Atkinson, D. (1994) Temperature and the organism size: a biological law for ectotherms? *Advances in Ecological Research* **25**, 1–58.

Azevedo, R.B.R., French, V. & Partridge, L. (1996) Thermal evolution of egg size in *Drosophila melanogaster*. *Evolution* **50**, 2338–2345.

Basten, C.J., Weir, B.S. & Zeng, Z.B. (1997) *QTL Cartographer: A Reference Manual and Tutorial for QTL Mapping*. Department of Statistics, North Carolina State University, Raleigh, NC.

Bergmann, C. (1847) Uber die verhältnisse der Wärmeökonomie der Thiere zu ihrer Grösse. *Göttinger Studien* **1**, 595–708.

Berven, K. A. (1982) The genetic basis of altitudinal variation in the wood frog *Rana sylvatica*. I. An experimental analysis of life history traits. *Evolution* **36**, 962–983.

Billerbeck, J.M., Schultz, E.T. & Conover, D.O. (2000) Adaptive variation in energy acquisition and allocation among latitudinal populations of the Atlantic silverside. *Oecologia* **122**, 210–219.

Brown, J.J., Ehtisham, A. & Conover, D.O. (1998) Variation in larval growth rate among striped bass stocks from different latitudes. *Transactions of the American Fisheries Society* **127**, 598–610.

Bryant, E. H. (1977) Morphometric adaptation of the housefly, *Musca domestica* L., in the United States. *Evolution* **31**, 580–596.

Capy, P., Pla, E. & David, J.R. (1993) Phenotypic and genetic variability of morphometrical traits in natural populations of *Drosophila melanogaster* and *Drosophila simulans*. I. Geographic variations. *Genetics, Selection and Evolution* **25**, 517–536.

Cavicchi, S., Guerra, D., Giorgi, G. & Pezzoli, C. (1985) Temperature-related divergence in experimental populations of *Drosophila melanogaster*. I. Genetic and developmental basis of wing size and shape variation. *Genetics* **109**, 665–689.

Cavicchi, S., Guerra, D., Natali, V., Pezzoli, C. & Giorgi, G. (1989) Temperature-related divergence in experimental populations of *Drosophila melanogaster*. II. Correlation between fitness and body dimensions. *Journal of Evolutionary Biology* **2**, 235–251.

Conover, D. O. & Present, T.M.C. (1990) Countergradient variation in growth rate: compensation for length of the growing season among Atlantic silversides from different latitudes. *Oecologia* **83:** 316–324.

Coyne, J. A. & Beecham, E. (1987) Heritability of two morphological characters within and among natural populations of *Drosophila melanogaster*. *Genetics* **117**, 727–737.

David, J. R. & Bocquet, C. (1975) Evolution in a cosmopolitan species: genetic latitudinal clines in *Drosophila melanogaster* wild populations. *Experientia* **31**, 164–166.

Degen, B. (2000) *Spatial Genetics Software*. Silvolab Guyane, INRA Station de Recherches Forestières, France DOM, Guyane.

Fenster, C.B., Galloway, L.F. & Chao, L. (1997) Epistasis and its consequences for the evolution of natural populations. *Trends Ecology and Evolution* **12**, 282–286.

Gilchrist, A.S. & Partridge, L. (1999) A comparison of the genetic basis of wing size divergence in three parallel body size clines of *Drosophila melanogaster*. *Genetics* **153**, 1775–1787.

Gockel, J., Kennington, W.J., Hoffmann, A.A., Goldstein, D.B. & Partridge, L. (2001) Nonclinality of molecular variation implicates selection in maintaining a morphological cline of *Drosophila melanogaster*. *Genetics* **158**, 319–323.

Gockel, J., Robinson, S.J.W., Kennington, W.J., Goldstein, D.B. & Partridge, L. (2002) Quantitative genetic analysis of natural variation in body size in *Drosophila melanogaster. Heredity*, **89**, 145–153.

Huey, R.B., Gilchrist, G.W., Carlson, M.L., Berrigan, D. & Serra, L. (2000) Rapid evolution of a geographic cline in size in an introduced fly. *Science* **287**, 308–309.

Imasheva, A.G., Bubli, O.A. & Lazebny, O.E. (1994) Variation in wing length in Eurasian natural populations of *Drosophila melanogaster. Heredity* **72:** 508–514.

James, A.C. & Partridge, L. (1995) Thermal evolution of rate of larval development in *Drosophila melanogaster* in laboratory and field populations. *Journal of Evolutionary Biology* **8**, 315–330.

James, A. C., Azevedo, R.B.R. & Partridge, L. (1995) Genetic and environmental responses to temperature of *Drosophila melanogaster* from a latitudinal cline. *Genetics* **146**, 881–890.

James, A. C., Azevedo, R.B.R. & Partridge, L. (1997) Cellular basis and developmental timing in a size cline of *Drosophila melanogaster. Genetics* **140**, 659–666.

Kearsey, M. J. & Pooni, H. S. (1996) *The Genetical Analysis of Quantitative Traits.* Chapman & Hall, London.

van't Land, J., van Putten, P., Villarroel, H., Kamping, A. & van Delden, W. (1995) Latitudinal variation in wing length and allele frequencies for *Adh* and *Gpdh* in populations of *Drosophila melanogaster* from Ecuador and Chile. Drosophila *Information Service* **76:** 156.

van't Land, J., van Putten, P., Zwaan, B, Kamping, A. & van Delden, W. (1999) Latitudinal variation in wild populations of *Drosophila melanogaster*: heritabilities and reaction norms. *Journal of Evolutionary Biology* **12**, 222–232.

Long, A. D. & Singh, R.S. (1995) Molecules versus morphology: the detection of selection acting on morphological characters along a cline in *Drosophila melanogaster. Heredity* **74**, 569–581.

Lonsdale, D.J. & Levinton, J.S. (1985) Latitudinal differentiation in copepod growth: an adaptation to temperature. *Ecology* **66**, 1397–1407.

Mayr, E. (1963) *Animal Species and Evolution.* Harvard University Press, Cambridge, MA.

Neat, F., Fowler, K., French, V. & Partridge, L. (1995) Thermal evolution of growth efficiency in *Drosophila melanogaster. Proceedings of the Royal Society of London Series B* **260**, 73–78.

Partridge, L. & French, V. (1996) Why get big in the cold? In: *Animals and Temperature* (eds I. A. Johnston & A. B. Bennett), pp. 265–292. Cambridge University Press, Cambridge.

Partridge, L., Barrie, B., Fowler, K. & French, V. (1994a) Evolution and development of body-size and cell-size in *Drosophila melanogaster* in response to temperature. *Evolution* **48**, 1269–1276.

Partridge, L., Barrie, B., Fowler, K. & French, V. (1994b) Thermal evolution of pre-adult life history traits in *Drosophila melanogaster. Journal of Evolutionary Biology* **7**, 645–663.

Prout, T. & Baker, J. S. (1993) *F* statistics in *Drosophila buzzatii*: selection, population size and inbreeding. *Genetics* **134**, 369–375.

Robinson, S.J.W. & Partridge, L. (2001) Temperature and clinal variation in larval growth efficiency in *Drosophila melanogaster. Evolutionary Biology* **14**, 14–21.

Roff, D.A. (1992) *The Evolution of Life Histories: Theory and Analysis.* Chapman & Hall, New York.

Schultz, E.T., Reynolds, K.E. & Conover, D.O. (1996) Countergradient variation in growth among newly hatched *Fundulus heteroclitus*: geographic differences revealed by common-environment experiments. *Functional Ecology* **10**, 366–374.

Slatkin, M. (1985) Gene flow in natural populations. *Annual Review of Ecology and Systematics* **16**, 393–430.

Slatkin, M. (1987) Gene flow and the geographical structure of natural populations. *Science* **263**, 787–792.

Stalker, H.D. & Carson, H.L. (1948) An altitudinal transect of *Drosophila robusta* Sturtevant. *Evolution* **2**, 295–305.

Stalker, H.D. & Carson, H.L. (1949) Seasonal variation in the morphology *of Drosophila robusta* Sturtevant. *Evolution* **3**, 330–343.

Stearns, S.C. (1992) *The Evolution of Life Histories.* Oxford University Press, Oxford.

Stevenson, R.D. (1985) Body size and limits to daily range of body temperature in terrestrial ectotherms. *American Naturalist* **125**, 102–117.

Tantawy, A. O. (1961) Effects of temperature on productivity and genetic variance of body size in *Drosophila pseudoobscura*. *Genetics* **46**, 227–238.
Tantawy, A. O. (1964) Studies on natural populations of *Drosophila*. III. Morphological and genetic differences of wing length in *Drosophila melanogaster* and *D. simulans* in relation to season. *Evolution* **18**, 560–570.
Watada, M., Ohba, S. & Tobari, Y.N. (1986) Genetic differentiation in Japanese populations of *Drosophila simulans* and *Drosophila melanogaster*. II. Morphological variation. *Japan Journal of Genetics* **61**, 469–480.
Whitlock, M.C., Philips, P.C., Moore, F.B–G. & Tonsor, S.J. (1995) Multiple fitness peaks and epistasis. *Annual Review of Ecology and Systematics* **26**, 601–629.
Wright, S. (1931) Evolution in mendelian populations. *Genetics* **16**, 97–159.
Wright, S. (1977) *Evolution and the Genetics of Populations*. Vol 3. *Experimental Results and Evolutionary Deductions*. University of Chicago Press, Chicago.
Zwaan, B., Azevedo, R.B.R., James, A.C., van't Land, J. & Partridge, L. (2000) Cellular basis of wing size variation in *Drosophila melanogaster*: a comparison of latitudinal clines on two continents. *Heredity* **84**, 338–347.

Chapter 5

Spatial geometry determines gene flow in plant populations

T. R. Meagher and C. Vassiliadis*

Introduction

If pollen falls in the forest and there is no recipient, does it form a seed? Obviously, the answer is no. Successful reproduction requires equal contributions of maternal as well as paternal gametes. Although this point may seem self-evident, most studies of pollen flow in plant populations primarily emphasize dispersal profiles from paternal source plants. The presence of maternal receptors in sufficient proximity to paternal donors is a critical factor underlying reproduction and ultimately fitness.

The introduction of genetic marker-based paternity analysis into plant population biology has greatly increased the level of insight we have into the precise genetic dynamics of natural populations. Prior to the mid 1980s, reproductive success in plants was typically measured in terms of seed set, which emphasizes maternal contributions. With the ongoing development of ever more sophisticated methods of paternity analysis (e.g. Meagher 1986; Thompson 1986; Roeder *et al.* 1989; Adams & Birkes 1991; Adams *et al.* 1992; Smouse & Meagher 1994; Smouse *et al.* 1999), paternal contributions to reproductive success have become a major focus of study, encompassing characterization of variation in paternal fitness (Devlin & Ellstrand 1990b; Smouse & Meagher 1994; Elle & Meagher 2000; Vassiliadis *et al.* 2002), the relationship between phenotype and fitness (Devlin *et al.* 1992; Snow & Lewis 1993; Smouse *et al.* 1999; Elle & Meagher 2000; Morgan & Conner 2001) and pollen dispersal patterns (see paragraph below). Comparisons of variation in maternal and paternal reproductive success have provided more insight into sex-specific selection pressures (e.g. Devlin & Ellstrand 1990b; Conner *et al.* 1996a, 1996b; Elle & Meagher 2000; T. R. Meagher *et al.* personal observation).

Pollen dispersal profiles associated with paternal reproduction have been a particular focus of many paternity analyses. An excess of near-neighbour mating as compared to random expectation has been reported in insect-pollinated plants (Meagher 1986; Devlin & Ellstrand 1990a; Godt & Hamrick 1993; Schnabel & Hamrick 1995; Smouse *et al.* 1999; Elle & Meagher 2000). Such effects have also

* Correspondence author: trm3@st-and.ac.uk

been found in the seaweed *Gracilaria gracilis* (Engel *et al.* 1999). Results for wind-pollinated plants are mixed, with distance effects found in some wind-pollinated trees (Burczyk *et al.* 1996; Burczyk & Prat 1997; Kaufman *et al.* 1998; Streiff *et al.* 1999; Vassiliadis *et al.* 2002) but not others (Adams & Birkes 1991; Latta *et al.* 1998; Schuster & Mitton 2000). Recent landscape-level approaches to gene flow in metapopulation systems have been directed towards identifying the composition of available paternal gametes at different locations (Sork *et al.* 1998, 1999; Dyer & Sork 2001; Smouse *et al.* 2001). Another context in which pollen dispersal as a function of distance has been attracting attention is risk assessment of transgenic crop species (Lavigne *et al.* 1998; Amand *et al.* 2000; Giddings 2000; Belanger *et al.* 2002).

For the most part, previous work on distance effects on reproductive success has focused on pollen-mediated gene flow. What are now needed are studies that address the interaction between maternal and paternal components of reproductive success, which can be viewed on several levels, including distance between mating individuals, phenotypic similarities and assortative or disassortative mating, and genetic relatedness and compatibility. Overall, one can consider individual patterns of success as either a paternal or maternal parent (in the case of hermaphroditic or co-sexual plants, each individual plays both roles), evaluating success as either pollination (paternal success) or seed set (maternal success). However, reproductive success is actually a population-level process, because it involves successful combination of paternal and maternal gametes to form the next generation in that population. Thus, it is unrealistic to consider reproductive success in isolation from the interaction that occurs between paternal and maternal parents. There must be a maternal receptor as well as a paternal source for reproduction and genetic transmission to the next generation to occur. In plants, a major feature of this interaction is the spatial proximity of individuals, so that distance between potential mates is likely to play an important role in pollen gene flow and concomitant reproductive success. Effects of distance on reproductive success are the primary focus of this chapter.

The frequency distribution of potential intermate distances within a population is defined by the spatial geometry or pattern of that population. Three types of spatial geometry are typically associated with plant populations (Pielou 1977; Grieg-Smith 1983; Li & Reynolds 1994): a regular distribution in which plants are systematically and evenly distributed; a random distribution in which every potential location on a map of the population has an equal probability of being occupied by a plant (based on a Poisson process); and an aggregated distribution in which plants occur in one or more clusters. The precise spatial geometry manifested by a population may be due to historical factors, such as colonization history (seed dispersal and seedling establishment) or ecological heterogeneity in which some patches are suitable for establishment and others are not (e.g. Harper 1977; Travis & Dytham 1999). In the case of crop species or experimental populations, plants may be intentionally regularly spaced.

The main goal of this chapter is to explore the impact of spatial geometry on reproductive success by computer simulation. We begin by outlining a negative exponential model of pollen dispersal, based on maximum-likelihood (ML) paternity analysis. Then a range of parameter values associated with the negative exponential model in selected empirical studies is explored. Finally, we simulate reproductive success in populations showing contrasting spatial geometries and varying negative exponential parameters in order to evaluate the combined impacts of spatial geometry and pollen dispersal on reproductive success.

Modelling paternal reproductive success

Recent progress in our understanding of the spatial interaction between maternal and paternal reproductive success has been based upon application of ever more sophisticated methods of paternity analysis. In plant populations, the high number of potential paternal parents typically requires a ML approach to determine the distribution of paternal reproductive success, except in a very few cases where molecular markers allow complete exclusion (e.g. Krauss 1999; Streiff *et al.* 1999). The most recent iteration of ML paternity analysis makes use of genotype profiles of candidate paternal parents and progeny arrays to assign ML estimates of fertility to each candidate paternal parent (Roeder *et al.* 1989; Smouse & Meagher 1994). The reproductive success of all males in a population can be represented by a set of relative fertility values, $\{\gamma_k\}$, representing each male's proportionate genetic contribution to progeny, i.e. their relative fitnesses. Using ML methods, a ML distribution of such paternal fertilities over all of the $k = 1, \ldots, K$ potential paternal parents, $\Lambda = \{\gamma_k\}$, can be determined by iterative maximization of the likelihood function:

$$L(\Lambda \mid \text{genetic data}) \propto \prod_{i=1}^{N} (\lambda_1 x_{i1} + \lambda_2 x_{i2} + \ldots + \lambda_K x_{iK}) \quad (5.1)$$

where x_{ik} represents the Mendelian segregation probability of the kth putative paternal parent on the ith maternal parent : progeny combination based on observed genetic marker data (Smouse & Meagher 1994). Mendelian segregation probabilities are obtained by determining the probability that a given offspring would segregate from a cross involving its known maternal parent and each potential paternal parent. In many cases, the segregation probability will be zero, representing genetic exclusion, but there will also be many instances in most data sets where multiple paternal parents could have given rise to a given progeny, in which case their segregation probabilities will provide information about their relative likelihoods of paternal contribution. The fertility profile obtained by this method is an estimate of realized male reproductive success based on the representation of genes that may have come from that male in the progeny array. Thus, relative male fertilities $\{\gamma_k\}$ can be interpreted as either a predictive measure of the reproductive potential of males or of the realized fitness of males.

An example of a paternal fertility profile obtained by this method in the dioecious, insect-pollinated plant *Chamaelirium luteum* (Smouse & Meagher 1994) is shown in Figure 5.1. This pattern is typical for plant species under a range of pollination

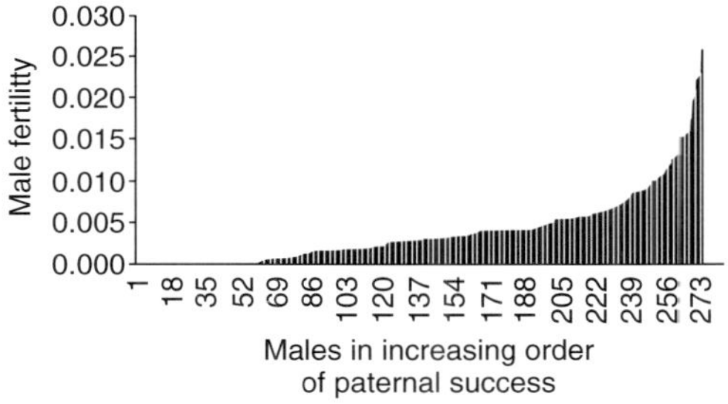

Figure 5.1 Individual male fertilities $\{\gamma_k\}$ of individuals in a population of *Chamaelirium luteum* estimated by ML paternity analysis.

mechanisms, including insects, wind and water (Streiff 1998; Engel *et al.* 1999; Elle & Meagher 2000; Vassiliadis *et al.* 2002): there is considerable variation in paternal fertility, with some candidate paternal parents making little or no contribution to the progeny array, and others making substantial contributions.

One of the major advantages of the ML model of paternal fertility outlined above is that it provides an analytical framework for evaluating impacts of population spatial structure. This chapter will explore impacts of intermate distance distributions on reproductive success. In order to do this, the set of fertilities of K candidate paternal parents on I potential maternal parents, $\{\gamma_{ik}\}$, is modelled as follows:

$$\lambda_{ik} = e^{\gamma\delta_{ik}} / \sum e^{\gamma\delta_{ik}} \qquad (5.2)$$

where δ_{ik} is a function of intermate distance and γ is a 'distance effect' parameter. The distance effect refers to the rate at which paternal genetic contributions are expected to decline with increasing distance to maternal recipients. This model follows Adams and Birkes (1991), and details of the estimation of γ under ML paternity analysis are outlined in Smouse *et al.* (1999). Empirical applications of this model have found that defining δ_{ik} as the natural logarithm of Euclidean distance between putative paternal parent k and target maternal parent i provides a good fit in natural and experimental populations (see next section). Thus, we will use this definition of δ_{ik} in the simulations outlined below. In addition to distance effects, this model has also been adapted to address the impact of plant features, such as overall size, floral characteristics, etc., on male success (Smouse *et al.* 1999), but the present paper is limited to a consideration of spatial effects.

Under this model, negative values of γ reflect an exponential decay of pollen dispersal from pollen sources (e.g. Smouse *et al.* 1999; Austerlitz & Smouse 2001), which, as noted in the following section of this paper, shows a good fit to observed pollen dispersal in natural plant populations; and $\gamma = 0$ reflects a panmictic population in which all potential paternal parents have equal probability of success irrespective of distance to potential maternal parents (Figure 5.2). Mathematically, this model would also allow for $\gamma > 0$, but we consider this to be biologically unrealistic and thus only consider $\gamma \leq 0$.

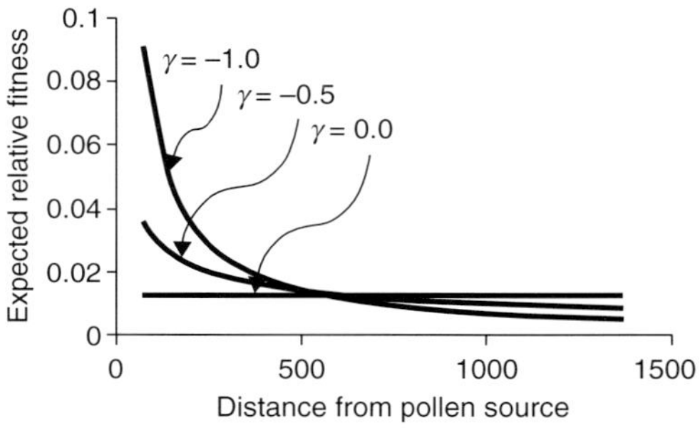

Figure 5.2 Expected relative male reproductive success $\{\gamma_k\}$ according to a negative exponential model of pollen dispersal (equation 5.2) with different values for the distance effect parameter γ.

Using the negative exponential model of pollen dispersal, one can enumerate the cumulative reproductive success expected at increasing intermate distances. Considering all of the pollen sources in a population, reproductive success at a particular intermate distance depends not only on the amount of pollen available, but also on whether or not there are recipients present at that distance. Cumulative reproductive success as a function of intermate distance will show a diminishing return at increasing distances as pollen sources reach saturation and pollen availability from more distant sources declines in its representation in the pollen pool (e.g. see Figures 5.3 and 5.6). If there are no recipient plants in close proximity to pollen sources, then reproductive success at nearby distances may be low even though that is where the majority of the pollen is available, and thus cumulative reproductive success as a function of distance would increase slowly. Alternatively, if there are many nearby recipients, then cumulative reproductive success may saturate very quickly. Saturation refers to the extent to which the progeny of a given maternal parent can be genetically accounted for by potential male parents within a given distance radius. As this distance radius increases, including more potential paternal parents, the probability of including all pollen donors for the progeny array increases to a maximum that approaches one. Thus, the shape of this diminishing return of paternal gamete capture depends on the precise spatial geometry of the population and the corresponding distribution of potential intermate distances. Reproductive success will accumulate at a rate that is determined by how often and at what distances potential pollen recipients are encountered by pollen dispersed from source plants.

Distance effects in natural and experimental populations

In order to provide a biological context for our simulation models, we will briefly review several empirical assessments of distance effects. The first example involves the paternity analysis of a natural population of *Chamaelirium luteum* (Meagher 1986, 1991; Smouse *et al.* 1999). This species is insect pollinated. Plants in this population

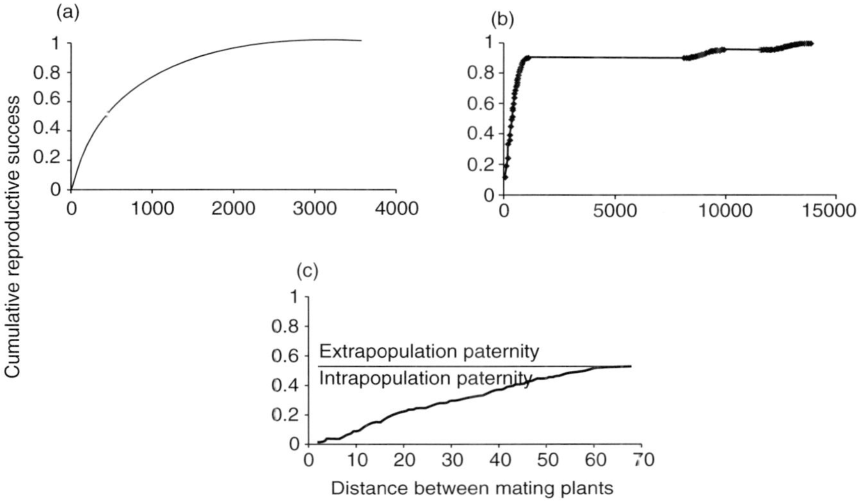

Figure 5.3 Cumulative reproductive success as a function of distances between potentially mating pairs based on the negative exponential model of pollen dispersal and the spatial geometry of the population for three empirical studies: (a) *Chamaelirium luteum* (Smouse *et al.* 1999). $\gamma = -0.90$, distance in cm; (b) *Silene latifolia* (T. R. Meagher *et al.*, personal observation), $\gamma = -0.88$, distance in cm; (c) *Phillyrea angustifolia* (Vassiliadis *et al.* 2002), $\gamma = -0.34$, distance in m.

are more or less randomly distributed, with a few major clusters of plants. The ML estimate of γ obtained for this population is -0.90, which results in a negative exponential model of pollen dispersal (e.g. Figure 5.2; $\gamma = -1.0$). Under this model, pollination success, and hence paternal reproductive success, is expected to drop off rapidly, to near zero values at distances greater than 100 cm. However, if one considers potential paternal success taking into account the proximity of available maternal parents (Figure 5.3a), a different pattern emerges in which realized paternal success is expected to occur at relatively high levels to a distance of approximately 100 cm before dropping off.

Our second example is a paternity analysis on an experimental population of *Silene latifolia* that follows a regular distribution (T. R. Meagher *et al.*, unpublished observations). This species is also insect pollinated. We consider here one of two sets of four 1000 cm × 1000 cm experimental plots established at the corners of one hectare fields. Within each 1000 cm × 1000 cm plot, plants are arranged at 100 cm spacing, with alternative male and female plants of this dioecious species. This experiment was designed to assess impacts of male phenotype on reproductive performance, and the plots were separated by 8000 cm (adjacent corners) or 11 300 cm (diagonal corners) so that each plot could be considered independently. Thus, the goal of assessing spatial effects in this population was primarily to ensure that there was limited gene flow between plots. The estimate for γ obtained for one set of four plots was −0.88, ensuring reproductive isolation whether one is looking at source- or recipient-based representations of reproductive success (Figure 5.3b).

Our final example of spatial impacts on reproductive success involves a naturally occurring population of the androdioecious (both hermaphrodite and male individuals in the same population), wind-pollinated shrub *Phillyrea angustifolia* (Vassiliadis *et al.* 2002). This study involved a paternity analysis of a local population of *P. angustifolia* in which only 52.7% of the progeny array were found to have resulted from pollen sources within the sampled population (of 0.49 ha). In this population, an estimate of $\gamma = -0.34$ was obtained, reflecting the fact that spatial scale is an important feature of distance effects analyses (Figure 5.3c). In this case, there were a number of potential paternal parents that were not accounted for in the spatial model. However, even on the local scale under consideration, a negative exponential model of pollen dispersal was obtained.

These various examples show a range of features to be expected under natural conditions. One feature that has come out of these examples, and similar results presented elsewhere for other wind- (e.g. Burczyk *et al.* 1996) and insect-pollinated (e.g. Devlin & Ellstrand 1990a) species, is that the exponential decay model of pollen dispersal is applicable to both pollination syndromes. Thus, given a population sample encompassing the majority of the potential paternal parents, a negative exponential model provides a reasonably accurate representation of how pollen dispersal is expected to occur. The second feature is that spatial geometry can impact on the rate of saturation of paternal reproductive success. For example, in the *S. latifolia* example, where most if not all of the potential pollen sources were considered, saturation occurs very locally, with more distant males contributing little (see Figure 5.3b). By contrast, in the *P. angustifolia* example, where there were many potential pollen sources that were not taken into account, the observed rate of saturation of paternal success is much slower (see Figure 5.3c).

The remainder of this paper will explore how the spatial distribution of the population impacts on saturation of reproductive success with distance under negative exponential decay of pollen dispersal.

Impacts of distance and spatial geometry in simulated populations

In order to assess impacts of spatial geometry on reproductive success, populations were simulated representing different extreme spatial geometries. For all simulations, a constant square map size of 1000×1000 units and a constant population size of $N = 195$ were used. This ensured that all spatial geometries occurred at the same average plant density. For purposes of simulating reproductive success, we considered these populations to be composed of self-incompatible hermaphroditic plants, the prevalent breeding system found in flowering plants (Renner & Ricklefs 1995). Each plant was then both a pollen source and pollen receiver.

Four spatial geometries were used:

1 Regular spatial distribution was simulated using a hexagonal pattern in order to maximize the number of intermate pairs that could occur at each distance.

2 Random spatial distribution was simulated by assigning plants to x and y coordinates using a pseudorandom number generator function (see below).

Aggregated spatial distributions, in which plants were arranged in clusters of varying sizes, were simulated in two ways:

3 Simultaneous aggregation in which a subset (1/20) of plants was simultaneously assigned random locations, and then all subsequent plants were placed within a set distance (a 100 unit radius) of the initial colonizers. This is meant to represent a population history where a population that is initially at a low density gradually fills in.

4 Sequential aggregation in which a first plant is assigned to a random location, and subsequent plants are assigned either to a random location or within a set distance (a 100 unit radius) from preceding plants. This was meant to simulate a history of population establishment based on a limited number of initial colonizers.

Patterns of reproductive success with distance were evaluated for each simulation over three values of γ, representing either negative exponential pollen dispersal ($\gamma = -1.0$ or -0.5) or panmictic pollen dispersal ($\gamma = 0.0$).

The regular spatial geometry was fixed, so there could only be one such geometry generated. However, the random, simultaneous aggregated, and sequential aggregated geometries were simulated over 100 iterations. The purpose of this replication was to provide a means of distinguishing the impacts of distance effects under the different spatial geometries. In order to evaluate the impacts of the different spatial geometries and distance effect γ parameter values ($-1.0, -0.5, 0.0$) on the profile of realized reproductive success, values of the cumulative reproductive success were determined at quarterly distance increments (distance quartiles), calculated up to the maximum distance that occurred in a given simulation. Once cumulative reproductive successes by distance quartiles were calculated, a two-way ANOVA involving spatial geometry model and distance effect γ values effects was performed on the first three distance quartiles using PROC GLM within the SAS software package (Release 8.00, 1999, SAS Institute, Cary, NC, USA). Because all of these cumulative paternal success profiles were approaching a maximum value of 1.0, our expectation is that there would be strong impacts of spatial geometry and distance effects in the earlier quartiles with a gradual diminution in the higher quartiles. By definition, cumulative reproductive success at the fourth distance quartile was 1.0 under all simulations, and therefore this distance quartile was not included in the ANOVA.

Simulation results and discussion

Examples of population maps under the four spatial geometries under consideration are shown in Figure 5.4. Although all maps have the same number of potential mating pairs (37 830), each of these spatial geometries results in a different type of intermate distance distribution (Figure 5.5). Random and regular spatial geometries yield similar intermate distance profiles (Figure 5.5a,b), with the main differ-

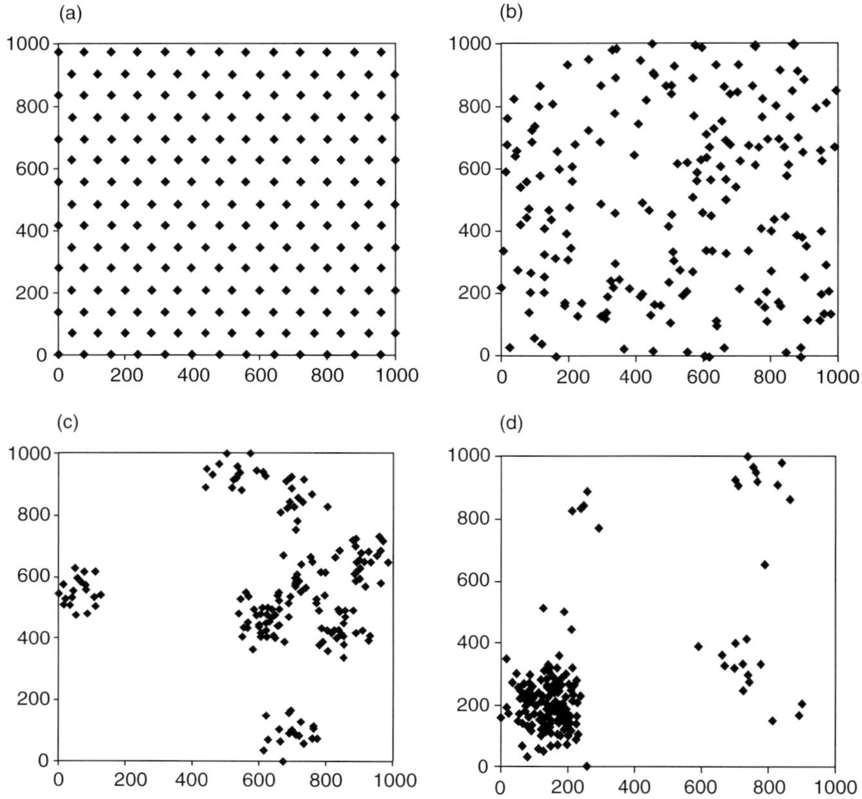

Figure 5.4 Examples of simulated maps for populations ($N = 195$) showing four different spatial geometries: (a) regular (hexagonal); (b) random; (c) simultaneous aggregated; (d) sequential aggregated.

ence being a more jagged intermate distance profile in the regular distribution than in the random distribution, due to fewer discrete classes of intermate distances in the former (81 vs. 17 616). The aggregated intermate distance profiles yield multimodal distributions, with the clustering of intraclump (shorter) versus interclump (longer) distances (Figure 5.5c,d). The sequential aggregated simulation gave the more extremely multimodal distribution with one major peak of short intermate distances corresponding to the main clump (Figure 5.5d).

The different spatial geometries and γ values also yielded different patterns of cumulative reproductive success with increasing intermate distance (Figure 5.6). Random spatial geometries showed a smooth and gradual approach to saturation of reproductive success (Figure 5.6b). The regular spatial geometry gave very similar results to the random spatial geometry, with a slightly more jagged approach to saturation (Figure 5.6a). However, the aggregated spatial geometries showed an accelerated approach to saturation, reflecting the effects of the high number of

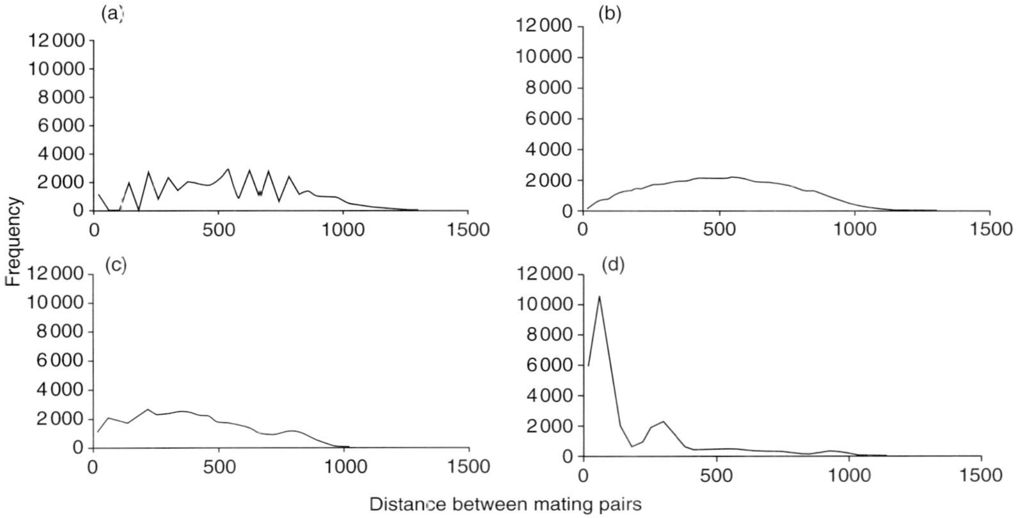

Figure 5.5 Frequency distribution of distances between potentially mating pairs (195 different hermaphroditic self-incompatible individuals = 18 915 potential pairs, thus leading to a maximum of 18 915 different distances) for the four population maps shown in Figure 5.4: (a) regular (hexagonal); (b) random; (c) simultaneous aggregated; (d) sequential aggregated.

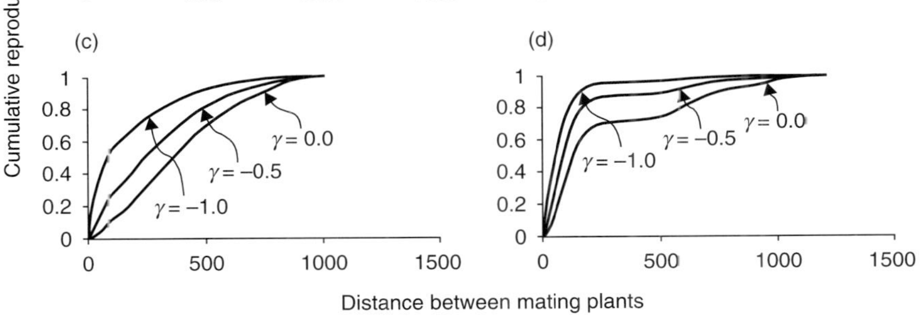

Figure 5.6 Cumulative reproductive success under the negative exponential pollen dispersal model for different values of γ (−1.0, −0.5, 0.0) for the four population maps shown in Figure 5.4: (a) regular (hexagonal); (b) random; (c) simultaneous aggregated; (d) sequential aggregated.

Table 5.1 Mean distance quartile values for cumulative reproductive success over 100 simulations of each of the spatial geometry models with corresponding standard deviation.

γ	First quartile	Second quartile	Third quartile
Random			
−1.0	0.555 ± 0.015	0.869 ± 0.015	0.990 ± 0.003
−0.5	0.379 ± 0.015	0.782 ± 0.022	0.981 ± 0.006
0.0	0.252 ± 0.013	0.696 ± 0.028	0.971 ± 0.009
Aggregated (simultaneous)			
−1.0	0.751 ± 0.041	0.905 ± 0.027	0.982 ± 0.009
−0.5	0.494 ± 0.058	0.772 ± 0.054	0.953 ± 0.022
0.0	0.280 ± 0.054	0.626 ± 0.076	0.906 ± 0.044
Aggregated (sequential)			
−1.0	0.976 ± 0.012	0.985 ± 0.008	0.994 ± 0.004
−0.5	0.917 ± 0.033	0.945 ± 0.028	0.975 ± 0.019
0.0	0.788 ± 0.075	0.842 ± 0.076	0.926 ± 0.051

Table 5.2 ANOVA of distance quartile data represented in Table 5.1.

			First quartile		Second quartile		Third quartile	
Source	MS_{error}	df	*F*	*P*	*F*	*P*	*F*	*P*
1. Spatial model (M)	(3)	2	37.35	0.0026	16.40	0.0118	3.42	0.1364
2. Distance effect (γ)	(3)	2	14.04	<0.0155	21.91	<0.0071	9.19	<0.0320
3. M by γ interaction	(4)	4	320.24	<0.0001	70.09	<0.0001	39.76	<0.0001
4. Error		891						

intraclump (short) versus interclump (long) distances (Figure 5.6c,d). Saturation occurred more rapidly under the sequential aggregated simulation. Across all models, the negative exponential model of pollen dispersal ($\gamma < 0$) led to more rapid saturation than did random mating ($\gamma = 0$).

Quartile measures of cumulative reproductive success over the replicated simulations show that the impacts of spatial geometry model and distance are general (Table 5.1). The two-way ANOVA of the quartile measures (Table 5.2) clearly showed significant effects of both spatial geometry model and pollen dispersal profile (γ), with aggregated spatial geometries and more extreme negative exponential pollen dispersal ($\gamma = 1$) both leading to more rapid saturation of cumulative reproductive success. Although this phenomenon was clearly indicated in the examples illustrated in Figure 5.6, the ANOVA confirms that differences between spatial geometries in patterns of cumulative male reproductive success with increasing distance are indeed consistent for the 100 replicates of each spatial geometry/distance effect combination. Under the aggregate spatial geometry, there will be an excess of nearby mating opportunities, thus explaining the more rapid saturation. The sequential aggregation geometry would be expected to be even more extreme, as we observed, because there is essentially one large clump with several increasingly smaller clumps being established over time, so that the frequency distribution pattern of intermate distances will emphasize the nearby intermate distances in the first clump to become established. The association of more rapid saturation with increasingly negative parameter values in the negative exponential pollen dispersal model is also intuitive; more negative values of γ result in a higher percentage of the pollen being presented to nearby recipients. Finally, there is a strong interaction between spatial geometry and distance effect, indicating that the impact of a given magnitude of the distance effect parameter γ will depend on the spatial geometry of the population.

Conclusion and future directions

We have demonstrated that spatial geometry can have a substantial impact on patterns of pollination success. Pollen dispersal is only part of the story; the maternal dimension of reproductive success has to be an explicit part of population models of gene flow.

Another important factor is spatial scale. When considering a local population, at some level plants are typically randomly distributed. However, as the extent of the spatial scale is increased, the distribution may gradually take the form of an aggregated distribution, as the population under consideration encompasses a series of more local subpopulations. Indeed, the focus of gene flow studies has been gradually shifting from exploration of intrapopulation processes to gene flow within and among spatially distinct local populations (Sork *et al.* 1998, 1999; Smouse *et al.* 2001). Our sequential aggregation simulation approximates such a spatially structured landscape. Most natural populations show a patchy spatial distribution, with local aggregations of plants serving collectively as pollen sources and recipients but with gene flow among as well as within the units. In this situation, the dynamics of gene flow and realized patterns of reproductive success could have a bearing not only on interpretation of the historical population process, but also on the prediction of conservation and genetic implications of habitat fragmentation. Our findings clearly demonstrate the importance of taking into account the distribution of maternal recipients in terms of evaluating the overall reproductive dynamics of a metapopulation system.

Having established the basic nature of impacts of spatial geometry on reproductive success in natural populations, there are a number of ways in which this modelling effort might be extended to take into account other details of plant population biology. First, one could incorporate seed dispersal models (Nathan & Muller-Landau 2000; Travis & French 2000) in order to reconstruct historical processes that give rise to different spatial geometries. Second, one could incorporate variance in maternal reproductive capacity, since the availability of ovules is a critical factor in ultimate reproductive output of a population. Third, one could include population genetic processes in the simulations to enable a temporal analysis of genetic turnover at the population level. Finally, there is a need to integrate our demonstrated consequences of spatial geometry with empirical data, so that we can better interpret the impacts of spatial geometry on the reproductive dynamics of natural populations.

Moving beyond basic plant population biology, models of the sort discussed here have potential practical application in plant breeding, seed production and risk assessment in connection with genetically modified organisms (GMOs). For horticultural and silvicultural varieties that are propagated by seed, added value is attached to seed production by the certification of either seed purity or genetic enhancement. Since seed production often occurs in open plantations, isolation of seed nurseries from potential sources of pollen contamination is a major factor in determining seed certification. In the case of GMOs, there is the opposite concern that transgenic pollen from cultivars may hybridize with surrounding natural populations, releasing genetic traits such as herbicide resistance into related weed species (Lavigne *et al.* 1998; Amand *et al.* 2000; Giddings 2000; Belanger *et al.* 2002). In either case, the precise spatial configuration of surrounding source or recipient populations for gene flow to or from cultivars needs to be assessed. Our results show that it is not only distance per se, but also the pattern exhibited by those distances, e.g. spatial structure at the landscape level, that needs to be considered in assessing potential seed quality or risk assessment of transgenic hybridization.

Acknowledgements

We thank the organizers of the 2001 BES/NERC EDGE symposium at St Catherine's College, Oxford for inviting this contribution. We also thank Rosie Hails and four anonymous reviewers for their constructive comments on an earlier draft of this paper. Finally, C.V. was supported during this study by US National Science Foundation grant DEB00–96216.

References

Adams, W.T. & Birkes, D.S. (1991) Estimating mating patterns in forest tree populations. In: *Biochemical Markers in the Population Genetics of Forest Trees* (eds S. Fineschi, M. E. Malvolti, F. Cannata & H. H. Hattemer), pp. 157–172. SPB Academic Publishing, The Hague.

Adams, W.T., Griffin, A.R. & Moran, G.F. (1992) Using paternity analysis to measure effective pollen dispersal in plant population. *American Naturalist* **140**, 762–780.

Amand, P.C.S., Skinner, D.Z. & Peaden, R.N. (2000) Risk of alfalfa transgene dissemination and scale-dependent effects. *Theoretical and Applied Genetics* **101**, 107–114.

Austerlitz, F. & Smouse, P.E. (2001) Two-generation analysis of pollen flow across a landscape. II. Relation between Phi(ft), pollen dispersal and interfemale distance. *Genetics* **157**, 851–857.

Belanger, F.C., Meagher, T.R., Day, P.R., Plumley, K. & Meyer, W. (2002) Interspecific hybridization between *Agrostis stolonifera* (creeping bentgrass) and related *Agrostis* species under field conditions. *Crop Science*, in press.

Burczyk, J. & Prat, D. (1997) Male reproductive success in *Pseudotsuga menziesii* (Mirb.) Franco: the effects of spatial structure and flowering characteristics. *Heredity* **79**, 638–647.

Burczyk, J., Adams, W.T. & Shimizu, J.Y. (1996) Mating patterns and pollen dispersal in a natural knobcone pine (*Pinus attenuata* Lemmon) stand. *Heredity* **77**, 251–260.

Conner, J. K., Rush, S. & Jennetten P. (1996a) Measurements of natural selection of floral traits in wild radish (*Raphanus raphanistrum*). I. Selection through lifetime female fitness. *Evolution* **50**, 1127–1136.

Conner, J. K., Rush, S., Kercher, S. & Jennetten, P. (1996b) Measurements of natural selection of floral traits in wild radish (*Raphanus raphanistrum*). II. Selection through lifetime male and total fitness. *Evolution* **50**, 1137–1146.

Devlin, B. & Ellstrand, N.C. (1990a) The development and application of a refined method for estimating gene flow from angiosperm paternity analysis. *Evolution* **44**, 248–259.

Devlin, B. & Ellstrand, N.C. (1990b) Male and female fertility variation in wild radish, a hermaphrodite. *American Naturalist*, **136**, 87–107.

Devlin, B., Clegg, J. & Ellstrand, N.C. (1992) The effect of flower production on male reproductive success in wild radish populations. *Evolution* **46**, 1030–1042.

Dyer, R.J. & Sork, V.L. (2001) Pollen pool heterogeneity in shortleaf pine, *Pinus echinata* Mill. *Molecular Ecology* **10**, 859–866.

Elle, E. & Meagher, T.R. (2000) Sex allocation and reproductive success in the andromonoecious perennial *Solanum carolinense* (Solanaceae). II. Paternity and functional gender. *American Naturalist* **156**, 622–636.

Engel, C.R., Wattier, R., Destombe, C. & Valero, M. (1999) Performance of non-motile male gametes in the sea: analysis of paternity and fertilization success in a natural population of a red seaweed, *Gracilaria gracilis*. *Proceedings of the Royal Society of London Series B* **266**, 1879–1886.

Giddings, G. (2000) Modelling the spread of pollen from *Lolium perenne*. The implications for the release of wind-pollinated transgenics. *Theoretical and Applied Genetics* **100**, 971–974.

Godt, M.J. & Hamrick, J.L. (1993) Patterns and levels of pollen-mediated gene flow in *Lathyrus latifolius*. *Evolution* **47**, 98–110.

Grieg-Smith, P. (1983) *Quantitative Plant Ecology*, 3rd edn. Blackwell Scientific, Oxford.

Harper, J.L. (1977) *Population Biology of Plants*. Academic Press, London.

Kaufman, S.R., Smouse, P.E. & Alvarez-Buylla, E.R. (1998) Pollen-mediated gene flow and differential male reproductive success in a tropical pioneer tree, *Cecropia obtusifolia* Bertol. (Moraceae): a paternity analysis. *Heredity* **81**, 164–173.

Krauss, S.L. (1999) Complete exclusion of nonsires in an analysis of paternity in a natural plant population using amplified fragment length polymorphism (AFLP). *Molecular Ecology* **8**, 217–226.

Latta, R.G., Linhart, Y.B., Fleck, D. & Elliot, M. (1998) Direct and indirect estimates of seed versus pollen movement within a population of ponderosa pine. *Evolution* **52**, 61–67.

Lavigne, C., Klein, E.K., Vallee, P. *et al.* (1998) A pollen-dispersal experiment with transgenic oilseed rape. Estimation of the average pollen dispersal of an individual plant within a field. *Theoretical and Applied Genetics* **96**, 886–896.

Li, H. & Reynolds, J.F. (1994) A simulation experiment to quantify spatial heterogeneity in categorical maps. *Ecology* **75**, 2446–2455.

Meagher, T.R. (1986) Analysis of paternity within a natural population of *Chamaelirium luteum*. I. Identification of most-likely male parents. *American Naturalist* **128**, 199–215.

Meagher, T.R. (1991) Analysis of paternity within a natural population of *Chamaelirium luteum*. II. Patterns of male reproductive success. *American Naturalist* **137**, 738–752.

Morgan, M.T. & Conner, J.K. (2001) Using genetic markers to directly estimate male selection gradients. *Evolution* **55**, 272–281.

Nathan, R. & Muller-Landau, H.C. (2000) Spatial patterns of seed dispersal, their determinants and consequences for recruitment. *Trends in Ecology and Evolution* **15**, 278–285.

Pielou, E.C. (1977) *Mathematical Ecology*. Wiley-Interscience, New York.

Renner, S.S. & Ricklefs, R.E. (1995) Dioecy and its correlates in the flowering plants. *American Journal of Botany* **82**, 596–606.

Roeder, K., Devlin, B. & Lindsay, B.G. (1989) Application of maximum likelihood methods to population genetic data for the estimation of individual fertilities. *Biometrics* **45**, 363–379.

Schnabel, A. & Hamrick, J.L. (1995) Understanding the population genetic structure of *Gleditsia triacanthos* L.: the scale and pattern of pollen gene flow. *Evolution* **49**, 921–931.

Schuster, W.S.F. & Mitton, J.B. (2000) Paternity and gene dispersal in limber pine (*Pinus flexilis* James). *Heredity* **84**, 348–361.

Smouse, P.E. & Meagher, T.R. (1994) Genetic analysis of male reproductive contributions in *Chamaelirium luteum* (L.) Gray (Liliaceae). *Genetics* **22**, 369–370.

Smouse, T.E., Meagher, T.R. & Kobak, C.J. (1999) Parentage analysis in *Chamaelirium luteum* (L.) Gray (Liliaceae): why do some males have higher reproductive contributions? *Journal of Evolutionary Biology* **12**, 1069–1077.

Smouse, P.E., Dyer, R.J., Westfall, R.D. & Sork, V.L. (2001) Two-generation analysis of pollen flow across a landscape. I. Male gamete heterogeneity among females. *Evolution* **55**, 260–271.

Snow, A.A. & Lewis, P.O. (1993) Reproductive traits and male fertility in plants: empirical approaches. *Annual Review of Ecology and Systematics* **24**, 331–351.

Sork, V.L., Campbell, D.R., Dyer, R. *et al.* (1998) Proceedings from a Workshop on Gene Flow in Fragmented, Managed, and Continuous Populations. Research Paper No. 3. National Center for Ecological Analysis and Synthesis, University of California, Santa Barbara. Available at: http://www.nceas.ucsb.edu/nceas-web/projects/2057/nceas-paper3/

Sork, V.L., Nason, J., Campbell, D.R. & Fernandez, J.F. (1999) Landscape approaches to historical and contemporary gene flow in plants. *Trends in Ecology and Evolution* **14**, 219–224.

Streiff, R. (1998) *Structure génétique spatiale et flux polliniques dans une chênaie mixte* Quercus petraea *(Matt.) Liebl. et* Quercus robur *L.* PhD thesis, Université des Sciences et Techniques du Languedoc, Montpellier.

Streiff, R., Ducousso, A., Lexer, C., Steinkellner, H., Gloessl, J. & Kremer, A. (1999) Pollen dispersal inferred from paternity analysis in a mixed oak stand of *Quercus robur* L. and *Quercus petreae* (Matt.) Liebl. *Molecular Ecology* **8**, 831–841.

Thompson, E.A. (1986) *Pedigree Analysis in Human Genetics.* John Hopkins University Press, Baltimore.

Travis, J.M.J. & Dytham, C. (1999) Habitat persistence, habitat availability and the evolution of dispersal. *Proceedings of the Royal Society of London Series B* **266**, 723–728.

Travis, J.M.J. & French, D.R. (2000) Dispersal functions and spatial models: expanding our dispersal toolbox. *Ecology Letters* **3**, 163–165.

Vassiliadis, C., Saumiou-Laprade, P., Lepart, J. & Viard, F. (2002) High male reproductive success of hermaphrodites in the androdioecious *Phillyrea angustifolia.* Evolution **56**, 1362–1373.

Chapter 6

Measuring genetic variation in wild populations: from molecular markers to adaptive traits

G. R. Carvalho, C. van Oosterhout, L. Hauser and A. E. Magurran*

Introduction: population dynamics and adaptive change

Traditionally, population ecologists study changes in the numbers of organisms, whereas population geneticists focus on gene frequency changes. A central problem in evolutionary ecology is to determine how changes in gene frequency influence, and are influenced by, population characteristics such as birth and death rates, age distribution, competitive processes and population size and density. Perhaps the most forthright early recognition that population dynamics and genetic structure are inextricably linked was by Charles Elton (1924), who considered how factors such as phenotype, deleterious mutations and genetic variation may drive changes in birth and death rates, which in turn influence population dynamics. His classical consideration of population cycles and selection represented a key stage in the integration of evolutionary thinking into ecological studies (Lindström *et al.* 2001), though translating such theoretical principles into empirical demonstrations continues to be a major challenge. Since anthropogenic influences are predicted to result in rapid environmental change in the next decades (e.g. elevated CO_2, global warming, increased UV, latitudinal shifts in climate), ecological assessment of population and community response will presumably be in urgent demand. Such information requires a sound knowledge of the genetic basis of traits under selection, as well as reliable indicators of assaying the nature and speed of adaptive shifts.

To investigate how genetic structure and phenotypic variation interact requires an analysis of the constraints and opportunities afforded by the evolution of adaptive variation. Adaptive genetic variation, hereafter referred to simply as 'adaptive variation', can be defined as heritable phenotypic variation that is influenced predominantly by natural selection, so enhancing fitness in specific environments. Three aspects arise from this definition: first, 'heritable phenotypic variation' indicates the need to understand the genetic basis and mode of inheritance of trait variation; second, tools must recognize and quantify the extent and dynamics of the

* Correspondence author: g.r.carvalho@hull.ac.uk

differences among phenotypes; finally, a study of fitness variation in traits must focus on how phenotypes respond to environmental variation, and how such interactions translate into viability and reproductive benefits. Perhaps the reason why we find few convincing cases of adaptive dynamics in the wild (e.g. Grant & Grant 2000; Merilä & Sheldon 2000) is the requirement for effective integration of data from disparate sources, often involving molecular genetic markers, laboratory breeding studies or inferences on relatedness and extensive field work. In particular, the need to estimate the proportion of phenotypic variance in a trait that is passed on to the next generation, or narrow-sense heritability (hereafter referred to simply as 'heritability'), is crucial to predicting how a trait will respond to selection. Although heritability studies on organisms other than *Drosophila* and mice have increased recently (Mousseau *et al.* 2000), the logistical problems of estimating heritabilities in ecologically well-documented taxa, especially vertebrates, remain a major disincentive.

Here we consider ecologically tractable approaches to studying adaptive variation in the wild, with particular focus on the extent to which molecular genetic markers provide informative indicators of fitness variation. We then examine the underlying genetic basis of trait variation, followed by a consideration of the relationship between molecular diversity and quantitative traits. The variety of approaches used to investigate adaptive variation in the laboratory and field are examined. Emphasis is placed on the choice of molecular tools and experimental approaches likely to prove most effective in vertebrates, where inferences on genetic relatedness may prove especially challenging. Logistical considerations in breeding experiments, particularly the appropriate sample sizes to detect heritable variation, are then discussed. Comprehensive reviews on the determinants of adaptive variation and its detection in the wild are presented elsewhere (Endler 1986, 2000; Milligan *et al.* 1994; Hard 1995; Lynch & Walsh 1998; Hoffmann 2000; Hedrick 2001).

The genetic basis of adaptive variation

In outbreeding populations, individuals differ generally in their morphology, behaviour and life history, traits that may be influenced by many hundreds or thousands of loci (Lynch & Walsh 1998). An important objective in quantitative genetics is to estimate the heritability of such traits by assessing the proportional contribution of environmental and genetic variation. In sexual species, new genetic variation is generated in every generation by mutation and recombination, though migration may also introduce novel variants into the local gene pool. In equilibrium populations, the input of such variation is balanced by a loss due to genetic drift and selection (Lynch & Walsh 1998). Although the mutation rate of genes is very low (typically reported estimates in the order of 10^{-5} to 10^{-6} mutations per gene per generation for allozyme loci (Lynch & Walsh 1998)), the typically large number of loci determining quantitative traits means that the cumulative effect of mutations on a trait may be considerable. Indeed, the character mutation rate of an estimated 10^{-1} indicates that the input of novel genetic variation per generation will contribute significantly to heritable genetic variation. For instance, long-term selection lines of *Drosophila melanogaster*, initially with homozygous individuals, show that the accumulating effects of mutations can lead to appreciable response to selection equivalent to 7 to 8 phenotypic standard deviations in bristle number per 125 generations (Mackay *et al.* 1994).

Although mutations may generate novel genetic and phenotypic variation necessary for a continued adaptive response to selection, the majority of mutations have deleterious effects on fitness (Mackay *et al.* 1994). Nevertheless, deleterious mutations play a major role in the biology and evolution of species because they tend to accumulate as mutational load (Lande 1994; Lynch *et al.* 1995; Charlesworth & Charlesworth 1987, 1998). Perhaps counterintuitively, the cumulative fitness effects of deleterious mutations over generations is independent of the severity of the mutation or selection coefficient (Haldane 1937), and the mutation load in populations is largely dependent on the dominance coefficient of mutations and the mutation rate (Charlesworth & Charlesworth 1998).

The inheritance of polygenic traits, in particular their phenotypic expression and potential gene interactions, continue to constitute a major challenge to understanding the dynamics of adaptive shifts (Hedrick 2001). The astronomical number of potential gene interactions and their resulting effects on phenotype may be categorized into three classes: additive, dominance and epistatic effects.

With additive effects, the total genetic variance in the population is the sum of paternally and maternally inherited genes (Falconer 1989). The mean phenotypic value of selected parents yields a predicted phenotypic response in the offspring, and this fundamental relation is known as the breeders' equation. Additive genetic variation and heritability of traits are commonly estimated using offspring–parent regressions, though more advanced breeding designs reveal other sources of genetic variation (Lynch & Walsh 1998).

Non-additive variation derives from interactions between alleles within a locus (dominance) or between loci (epistasis). Dominance variation arises when a recessive allele is not expressed when accompanied by a dominant allele on an homologous chromosome. The phenotypic effects of such recessive alleles remain hidden except in homozygous individuals, which may be produced by a switch from clonal reproduction to bisexuality (Lynch & Gabriel 1983) or by inbreeding. Under such circumstances, dominance effects may increase phenotypic variation.

Linking molecular diversity to quantitative genetic variation

Foremost among the priorities of evolutionary and conservation biologists is the exploration of factors governing the levels and distribution of genetic variation. A major focus is to identify linkages between the dynamics of temporal and spatial patterns of genetic structure, revealed usually by molecular genetic markers, and fitness variation in ecologically significant traits. The assumption is often made that genetic parameters that quantify genetic diversity (e.g. heterozygosity, single-locus polymorphism) and population subdivision (Wright's F_{ST} and derivatives) can be related to genome-wide estimates of fitness variation. Since most adaptive characters are typically under polygenic control, and the expression of such genes is subject to genotype–environment interactions, the dynamics of quantitative traits is highly complex, and often independent of single-locus molecular measures (Butlin & Tregenza 1998; Reed & Frankham 2001). Moreover, the genetic architecture of complex traits, including dominance and epistasis, although dependent upon population-level processes (Barton & Turelli 1989), may respond to selection and evolutionary history in fundamentally different ways to molecular variation (Lynch 1996; Hedrick 2001).

Notwithstanding, the integration of molecular and quantitative genetics within an adaptive context is increasingly possible (Karhu *et al.* 1996; Kuittinen *et al.* 1997; Merilä 1997; Lynch *et al.* 1999; Kruuk *et al.* 2000; Pfrender *et al.* 2000), affording valuable insights into the role of different microevolutionary forces in shaping single-locus and polygenic variation. For example, a study on molecular variation using a combination of protein and DNA markers in Scots pine (*Pinus sylvestris*) (Karhu *et al.* 1996) and adaptive variation in the date of bud-set, allowed comparative assessments of population subdivision in molecular and quantitative traits. Here, molecular variation was a poor indicator of the quantitative trait measured. Such studies not only assay adaptive variation in organisms ranging from the waterflea, *Daphnia*, pine trees to birds and mammals, but also show how molecular information can be used to infer the nature of the adaptive process itself, including the effects of colonization history and demographic events. Before considering some key factors in the design of laboratory- and field-based studies to assaying adaptive variation, the relationship between molecular and quantitative genetic estimates of diversity will be highlighted.

Relationship between molecular and quantitative genetic variation

Evolutionary dynamics

A major motivation in the molecular estimation of adaptive variation arises from the frequently quoted Fisher's (1958) fundamental theorem of natural selection, predicting that the rate of change of mean fitness in a population should be equal to the additive genetic variance in fitness. In other words, adaptive evolutionary change is positively correlated with genetic diversity. If so, in the absence of information on trait-specific effects on fitness, high levels of molecular diversity might be used to indicate correspondingly high levels of adaptive trait variation, and vice versa. Since adaptive traits are responsive primarily to selection, and molecular markers mainly to random genetic drift and other stochastic processes, it is clear, however, that such a relationship is likely to be complex.

There are additional reasons for expecting only a weak covariance between the two character states in most circumstances, including the effects of non-additive genetic variation, different mutation rates, environmental effects on quantitative characters and impact of regulatory variation. Since non-additive variation may be

Plate 6.1 A maternal half-sib family of guppy males (*Poecilia reticulata*). The brood was produced by crossing a single female (mother) to two unrelated males (fathers). The male Y chromosome containing many colour genes is transmitted in its entirety from the father to his sons. Full sibs and fathers are therefore phenotypically very similar, which explains the close resemblance between full siblings (from the top of the photograph) 1 and 4, and 2 and 3.

converted to additive variation following population bottlenecks (Bryant *et al.* 1986), a reduction in molecular diversity may be concurrent with an increase in adaptive variation. The amount of within-population genetic diversity may also differ between molecular markers and quantitative traits because of the higher mutation rates of polygenic characters (Lynch 1985). It follows therefore that small populations should retain more diversity in quantitative characters, and that variation released during recombination recovers variability more rapidly following a population bottleneck than at molecular loci. Heritability and measures of phenotypic variance are environment dependent, whereas most protein and DNA markers are not. Moreover, since the amount of additive genetic variance for a trait, and hence its heritability, should decrease with the trait's association with fitness (Falconer 1989), those traits under strong selection are predicted to have lower heritabilities. Such an effect is likely to reduce further any correlation between heritability measures and molecular diversity. Finally, most molecular markers do not detect mutations in regulatory genes and gene rearrangements, though these are known to impact significantly on variation in the anatomy, physiology, behaviour and ecology of species (Arnone & Davidson 1997).

Nevertheless, there are also convincing arguments why some correlation between molecular and quantitative genetic variation would be expected. Both systems intimately depend on genetic drift and inbreeding, and thus the effective population size is a major determinant of both molecular heterozygosity and heritability at adaptive traits. Such effects are clearly shown in many inbreeding experiments (e.g. van Oosterhout & van Brakefield 1999; van Oosterhout *et al.* 2000a), and are also evident from field studies on guppies where small headwater populations clearly show lower heritability as well as lower heterozygosity than large lowland populations (C. van Oosterhout *et al.*, unpublished data). Because of the complicating factors reducing such correlations, however, molecular markers should not be used as a predictor for quantitative variation in isolation from population size and history; rather they can be applied to reconstruct demographic histories, which can then help to assess the likely extent of adaptive genetic variation.

Despite the convincing theoretical basis for expecting only weak correlations in the diversity of molecular and quantitative trait loci, the topic remains controversial (Crandall *et al.* 2000). For example, many assumptions concerning response to selection are formulated for equilibrium situations, whereas many wild populations exhibit regular or catastrophic fluctuations in population size (Elton 1924). Moreover, the levels of diversity at both phenotypic levels depend markedly on the stage in the demographic cycle at which measures are taken; levels of quantitative and molecular diversity are likely to be significantly different immediately following a population crash (Bryant *et al.* 1986) than those following a prolonged phase of relative stability (Hedrick 2001). Localized effects such as isolation and gene flow may also confound equilibrium predictions (Latta 1998). It also remains unclear whether a weak relationship may arise from the low power of individual studies undertaken. Lewontin (1984) stated that 'differences between species and populations in quantitative characters cannot be compared with differences in gene frequencies at individual loci, because the power of statistical tests to discriminate populations are vastly different for the two kinds of characters'. Recently, a plethora of studies have tackled this ensuing uncertainty (Bonnin *et al.* 1996; Karhu *et al.* 1996; Kuittinen *et al.* 1997; Butlin & Tregenza 1998; Lynch *et al.* 1999; Pfrender *et al.* 2000; Flint & Mott 2001; Hedrick 2001; Reed & Frankham 2001).

Molecular estimates of fitness variation: patterns from nature

Molecular estimates of adaptive variation can be made at two distinct levels: first, the levels detected within populations and second, the extent of differentiation among populations (Hard 1995). Both aspects are relevant to population ecology and conservation: the first usually considers relationships between gene diversity (e.g. heterozygosity) and heritabilities, and is related directly to perceived evolutionary potential under environmental change, whereas estimates of population divergence, usually quantified as subdivision for nuclear gene markers (F_{ST}) or subdivision for traits (Q_{ST}), are commonly used to identify units of conservation (Milligan *et al.* 1994).

A recent meta-analysis of 71 data sets encompassing simultaneous measures of within-population molecular and quantitative variability (Reed & Frankham 2001) indicated only a weak correlation between the two estimates. Interestingly, correlations were higher for morphological traits of uncertain adaptive significance than of life-history traits, suggesting that in common with molecular markers, the dynamics of genetic variation at morphological traits may be influenced predominantly by drift. It appears therefore, based on these data, that molecular estimates have only a limited ability to predict quantitative genetic variability (max. 4% of variation in quantitative traits), and that estimates of short-term evolutionary potential should be measured directly by assaying trait variation and heritabilities. A recent survey of within-heritabilities and allozyme and microsatellite diversity in *Daphnia* (Pfrender *et al.* 2000) detected no significant relationship, thus reinforcing the generality of the pattern.

Estimates of genetic differentiation among populations, however, appear to be highly correlated between molecular markers and quantitative traits. A meta-analysis of 18 independent studies on plant, invertebrate and vertebrate species indicated that molecular marker differentiation (F_{ST}) could predict adaptive differentiation (Q_{ST}) to a high degree (Merilä & Crnokrak 2001, Figure 6.1). However, Q_{ST} values were usually higher than F_{ST} values, probably due to divergent selection in different populations.

It appears therefore that molecular estimates of adaptive variation may be most informative at the among-population level, though additional studies that employ more molecular loci and larger sample sizes are required to further explore within-population correlations (Reed & Frankham 2001). The extent of divergence among populations in neutral and selective traits will, however, depend critically upon such factors as population size, selection pressures and levels of gene flow (Karhu *et al.* 1996). The apparently weak information content of molecular markers as indicators of adaptive variation does not negate their use in population and conservation biol-

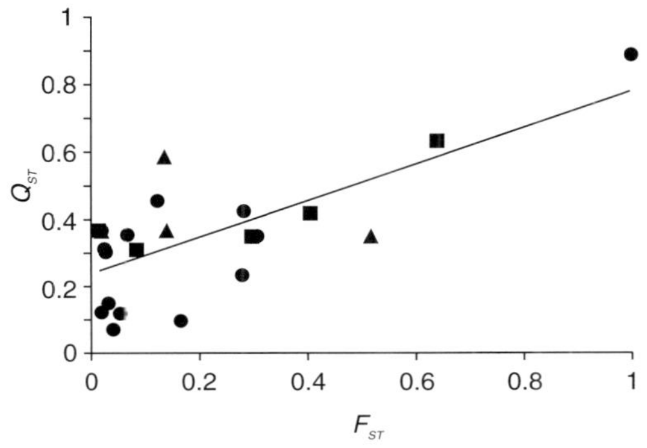

Figure 6.1 Relationship between genetic population differentiation estimated from molecular markers (F_{ST}) and quantitative traits (Q_{ST}). Molecular markers are indicated with symbols (filled circles, allozymes; filled squares, microsatellites; filled triangles, other DNA markers (RAPD, rDNA, mtDNA)), the line shows the linear regression for all marker systems ($r^2 = 0.49$). (Data from Merilä & Crnokrak 2001.)

ogy. Such neutral markers may provide significant insights into the determinants of processes impacting adaptive variation. These include effective population size (N_e), past bottlenecks, founder contributions, sex-specific gene flow, and historical and geographical relationships between groups (Milligan *et al.* 1994; Hedrick 2001). Moreover, molecular data on pedigrees, routinely available for endangered and captive populations, may be employed to yield information on relatedness that can be combined with phenotypic measures to estimate variance in quantitative traits and heritabilities. Such studies are essential to determine, for example, whether species that appear depauperate in molecular variation are also devoid of adaptive variation (Wielebnowski 1996).

Assessing adaptive variation with molecular tools

Which molecular tool?

Despite the difficulties of using molecular markers to assess adaptive diversity, they may be the only option where quantitative genetic variation cannot be measured, for example in species that are rare, long-lived or cannot be reared. It is thus worth considering which molecular markers, if any, are best suited to estimate adaptive variation at which traits, or indeed whether such a question is even realistic.

Until recently, the most commonly used molecular markers were allozymes, which typically involve relatively few loci with few alleles. The typically observed low variability at few loci is, however, unlikely to represent genomic heterozygosity accurately (Leberg 1992). The weak correlation between allozyme and adaptive trait variation (Reed & Frankham 2001) may partly be due to the low precision of heterozygosity estimates from allozymes. Furthermore, most allozymes are involved in cell physiology, and may thus represent a biased sample of the genome. Allozymes may also be under direct selection, or more commonly, may be linked to selected loci, thus affecting estimates of heterozygosity and weakening any correlations with quantitative genetic variation.

Recently, most population studies have employed microsatellites (Goldstein & Schlötterer 1999), which due to their higher variability allow more powerful tests of diversity and differentiation. The ubiquity of microsatellites, their high variability and wide and relatively even distribution in eukaryotic genomes may render them ideal for the assessment of total genomic variability, and potentially also of adaptive variability. Indeed, it has been proposed that variation at micro- and minisatellite DNA is a major source of quantitative genetic variation through changes in gene expression and function (Kashi *et al.* 1997; Kashi & Soller 1999), although such an assertion requires empirical testing. However, since the deterioration of linkage does not only depend on recombination, but also on the number of alleles (Reich & Goldstein 1999), the high mutation rates of microsatellites render them less likely to be affected by selection at linked loci. The main disadvantage of microsatellites is the cost and effort needed for the isolation of species-specific primers for amplification, limiting the number of loci in most species. Many microsatellite loci are available for some species (e.g. zebrafish, Pacific salmon, *Drosophila*), but to our knowledge no correlation between microsatellite and quantitative variability has been investigated.

Multilocus markers, such as amplified fragment length polymorphism (AFLP) or random amplified polymorphic DNA (RAPD) (Loxdale & Lushai 1998), may represent genomic variability more accurately, though they are limited by dominance that allows only indirect estimators of population heterozygosity. Possibly for this reason, the few direct comparisons available to date failed to detect a difference between allozymes and RAPD in their ability to predict adaptive variation (Karhu *et al.* 1996; Pfrender *et al.* 2000; Reed & Frankham 2001). However, more empirical data are required to test patterns further.

The genetic divergence detected between populations often varies considerably depending on the molecular marker used (e.g. Hauser & Ward 1998; Shaw *et al.* 1999), and a general relationship between molecular and quantitative variation seems unlikely in these cases. Depending on mutation rate, levels and patterns of variability and the nature of data, molecular markers reveal different aspects of genetic variation. For example, while diversity patterns of the relatively slowly mutating allozymes arise over long periods of evolutionary history, the more rapidly mutating microsatellites may reflect more recent demographic history (Nauta & Weissing 1996). Similarly, while mtDNA data are useful in reconstructing deep phylogenies, recently diverged species are often indistinguishable due to shared polymorphisms (Moran & Kornfield 1993). Multilocus markers such as AFLP may be preferable (Albertson *et al.* 1999). Some of the recently developed techniques, such as hypervariable microsatellites and single nucleotide polymorphisms (SNPs), are so sensitive and rapidly mutating, that insufficient time may have elapsed to allow the generation of adaptive differentiation (Hedrick 2001). Such differentiation does, however, usually represent limited gene flow and therefore the potential for adaptive differentiation.

It follows that it may be too ambitious to look for a general correlation between molecular and quantitative variation, which is confounded by different dynamics and properties of markers and traits. Instead, it may be more advisable to investigate specific marker systems that may be able to predict specific traits or groups of traits. For example, molecular markers may more adequately represent genetic diversity and differentiation at selectively neutral traits than that of traits under strong selection (Reed & Frankham 2001). The interpretation of such correlations requires considerable knowledge of the genetic mechanisms involved, the functions of gene products and the ecology of the species, and therefore is unlikely to provide generally applicable predictors for adaptive variation. For individual species or populations, however, they may not only provide useful marker systems, but also allow insights into the genetic determination of the traits under investigation.

A marker for each trait?

Although most variation at allozyme loci is thought to be neutral (Avise 1994), selection at specific allozyme loci has been described for a variety of species, and thus could theoretically be used as markers of physiological traits. Clines in allele frequency are an indication for selection acting at a particular locus (e.g. Schopf 1974; Corbin 1977; Hoffman 1981), though the adaptive significance of the respective polymorphisms was not determined unambiguously. In a few species, the kinetic superiority of specific alleles has been shown, for example, *Salmo trutta* (Henry & Ferguson 1985), *Fundulus heteroclitus* (Crawford & Powers 1989; Powers *et al.* 1991), *Poeciliopsis monacha* (Vrijenhoek *et al.* 1992), *Drosophila* (reviewed in Avise 1994), *Tigriopus californicus* (Burton & Feldman 1983) and *Mytilus edulis* (Hilbish & Koehn 1985), but functional differences between allelic products, upon which selection acts, are usually much smaller than the experimental errors of *in vitro* measurements of enzyme activity (Ferguson 1980). Correspondingly, most selection differentials may be undetectable by biochemical methods. If, however, such selective differences between alleles are found, the respective locus may be used as an indicator for the physiological trait in question.

Another group of genes that has recently received increased attention as marker loci is the major histocompatibility complex (MHC). Gene products at these loci are involved in the response of the immune system to diseases and parasites, and may be selected for high variability (Klein 1991). There are indications that low MHC variability increases susceptibility to disease and parasitism (Hedrick *et al.* 2001), which in some species may be linked to secondary sexual traits (Ditchkoff *et al.* 2001). MHC variability may therefore be useful not only to reconstruct demographic histories, but also as an indicator of adaptive traits. However, the adaptive significance of MHC loci renders their use in the assessment of genetic differentiation and reconstruction of phylogenetic relationships as limited because of convergent evolution and balancing selection (Kriener *et al.* 2000; Hedrick *et al.* 2001).

Since most quantitative traits are controlled by many loci (Lynch & Walsh 1998), it may seem impractical to isolate molecular markers for specific traits. However, in some traits, only a few loci have large effects and cause most variation (Mitchell-Olds 1995; Mackay 1996). Such bias in the distribution of gene effects on traits has been used in selective breeding of animals and plants to identify markers for genes controlling desired traits (quantitative trait loci, QTL). QTL markers are usually found by examining the association between molecular markers (usually microsatellites or AFLP) co-segregating with the desired trait in crosses between inbred lines. Since inbred lines represent only a fraction of the genetic variation of a species, the method probably underestimates the number of QTL, and may not be predictive of those found in other crosses of the same species (Flint & Mott 2001). In some instances, however, QTL markers could be applied to different genetic backgrounds (Fishback *et al.* 2000), and used to predict heterosis (Joshi *et al.* 2001). The usual specificity of QTL, however, constrains their application to wild populations.

With the development of new molecular markers and high-throughput molecular technology, it may soon become feasible to investigate the molecular basis of specific traits using high-density mapping and sequencing of the respective QTL regions (Flint & Mott 2001). The identity of candidate genes can then be verified by comparison of sequences with fully sequenced model organisms, mutagenesis or gene expression studies. The aim would be to identify QTNs (quantitative trait nucleotides), nucleotide variation contributing to the quantitative variation of a phenotypic trait (Lyman *et al.* 1999). Although such studies are new, and so far largely restricted to a few model organisms, they may provide powerful insights into the molecular basis of quantitative variation.

Approaches for investigating adaptive variation

Assessment in the laboratory and field

Several approaches are commonly employed to explore adaptive variation, often in combination, including studies on adaptation at molecular level (Mitton 1997), genotypic selection experiments (Mackay *et al.* 1994), transplantation experiments (Madsen *et al.* 1999), estimation of population differentiation (Magurran *et al.* 1995), quantitative genetic estimates of heritability and inbreeding (van Oosterhout & Brakefield, 1999; van Oosterhout *et al.* 2000b) and detecting inheritance with inferred relatedness in nature (Ritland 2000a,b). Among the alternatives, some present tractable logistics for adoption by ecologists both in the field and laboratory.

Field studies

Advances have been made by estimation of the heritability of lifetime reproductive success in the wild (Kruuk *et al.* 2000; Merilä & Sheldon 2000). The value of such an approach is that by studying organisms in the wild for which realistic estimates of fitness can be made, direct estimates of the relationship between different traits and fitness are possible, rather than assuming that certain trait classes are under putative forms of selection. Analysis of long-term data sets on the collared flycatcher, *Ficedula albicollis*, confirmed that traits closely associated with fitness tend to have lower heritability (Kruuk *et al.* 2000). Interestingly, however, and in contrast to the predictions of Fisher (1958), estimates of additive genetic variation (V_A) revealed that traits closely associated with fitness exhibited higher levels of additive genetic variance than traits with less impact on fitness. The negative relationship between trait fitness and heritability arose not from lower levels of V_A in fitness traits, but from their higher residual variance due either to environmental variance or non-additive effects. Such observations highlight the complex genetic architecture of fitness-related traits, as well as the need to determine the relative contribution of non-additive variance to residual variance. By contrast, and in accordance with theory, a negative relationship between a trait's heritability and its association with fitness was observed in red deer (*Cervus elaphus*) (Kruuk *et al.* 2000), though here, the decline in heritability likely arose from environmentally mediated effects on longevity. Other life-history traits not associated with longevity, such as average annual breeding success, had higher heritabilities. Estimates of additive genetic variance differed markedly among traits, with measures such as male breeding success exhibiting higher variance than morphometric traits.

Both of these long-term studies were dependent upon extensive field data on fitness variation, together with parallel behavioural or genetic information on pedigrees. The approach is effective not only in exploring the determinants of relationships among traits, but also the effects on heritability directly. Long-term data acquisition from the wild is therefore to be encouraged, and may (Wigensberg & Roff 1996), or may not (Hoffmann 2000), reveal similar patterns obtained from laboratory-based estimates on heritability (see also Lynch & Walsh 1998).

New advances have also been made through the employment of highly variable genetic markers to estimate the heritability of phenotypic traits in the field by using inferred relatedness measures (Ritland 2000a,b). At its simplest level, a positive association between allele sharing and phenotypic resemblance among pairs of individuals would indicate a genetic basis to a quantitative trait. Such an approach removes the need for breeding and rearing of progeny, as well as allowing estimates in nature, thereby removing laboratory artefacts such as lowered competition and lack of stress. Such an approach employing VNTR genetic fingerprinting has been used, for example, to estimate heritabilities in weight, length and precocious male maturation in Chinook salmon (*Oncorhynchus tshawytscha*) (Mousseau *et al.* 1998). Estimated heritabilities were in accord with values obtained using classical quantitative genetic approaches, demonstrating the efficacy of the field-based approach.

Inbreeding studies provide a third approach that has yielded especially informative data recently on both the relative contributions of non-additive effects and the links between specific traits and fitness consequences. Although the negative effects of inbreeding on population size are well documented in captive populations, there are few empirical data demonstrating that genetic factors cause a decline in wild populations. Saccheri *et al.* (1998) studied the effect of inbreeding on local extinction in a large metapopulation of the Glanville fritillary butterfly (*Melitaea cinxia*) and found that extinction risk increased significantly with decreasing heterozygosity. Another demonstration was provided by a manipulative field study on adders

(*Vipera berus*, Madsen *et al.* 1999), whereby outbred males were released into a highly isolated and inbred population. Reproductive success of the introduced males was high, leading to a marked elevation in molecular diversity and a rapid increase in recruitment due to increased survival of juvenile adders. The introduction of new genes from a different population led to a marked population expansion, so halting the precipitous decline towards extinction. A similar study examined the impact of immigrants on the survival of outbred offspring in a formerly inbred, wild *Daphnia* metapopulation that had undergone frequent genetic bottlenecks (Ebert *et al.* 2002). Outbred offspring increased in frequency as the result of hybrid vigour, suggesting that gene flow provided a 'genetic rescue' of populations from extinction. Such empirical demonstration of elevated fitness after inbreeding provides a mechanism for exploring impacts of inbreeding and genetic load on adaptive variation and fitness in populations.

Two important questions remain insufficiently investigated in the study of polygenic variation in the survival and extinction risk of natural populations: First, what is the role of heritable genetic variation for long-term survival of populations and species? Second, does a high mutation load increase the extinction risk of natural populations? Future effort is required on natural populations where long-term data are available and field manipulations are feasible.

Laboratory studies

Quantitative genetic experiments compare the phenotypic resemblance between one or more levels of relatives with that of the total phenotypic variation. For example, a powerful quantitative genetic approach is a North Carolina Design 1 (Lynch & Walsh 1998), in which the variance between half sibs, full sibs and unrelated individuals are compared to estimate additive and dominance variation. We employed such a breeding design for Trinidadian guppies (unpublished data), which involved the crossing of one male to several females, so producing full-sib families nested within paternal half-sib families (Figure 6.2). Full sibs share (on average) half their alleles in common, and additionally, have an identical genotype for one-quarter of all their loci. Resemblance between full sibs thus not only reflects the additive affects of alleles, but also some genetic variation from dominance, whereas resemblance between relatives may sometimes arise due to a shared common environment. More complex patterns, however, may arise when traits are sex linked. For example, half-sib male guppies sharing the same mother exhibit little similarity, whereas full sibs exhibit a highly similar colour phenotype (Figure 6.3; Plate 6.1, facing p. 94), demonstrating that genes on the Y chromosome are largely responsible for phenotypic resemblance in male sexual traits.

One intrinsic difficulty with quantitative genetic experiments is that the phenotypic variation observed arises from both genetic and environmental factors. Heritable genetic variation is measured in one environment (e.g. in the laboratory), and may thus be distinctly different from heritability measured elsewhere (e.g. in the field; Hoffmann 2000). Furthermore, phenotypic resemblance of individuals does not necessarily reflect similarity in their genes or genotypes. When full-sib offspring

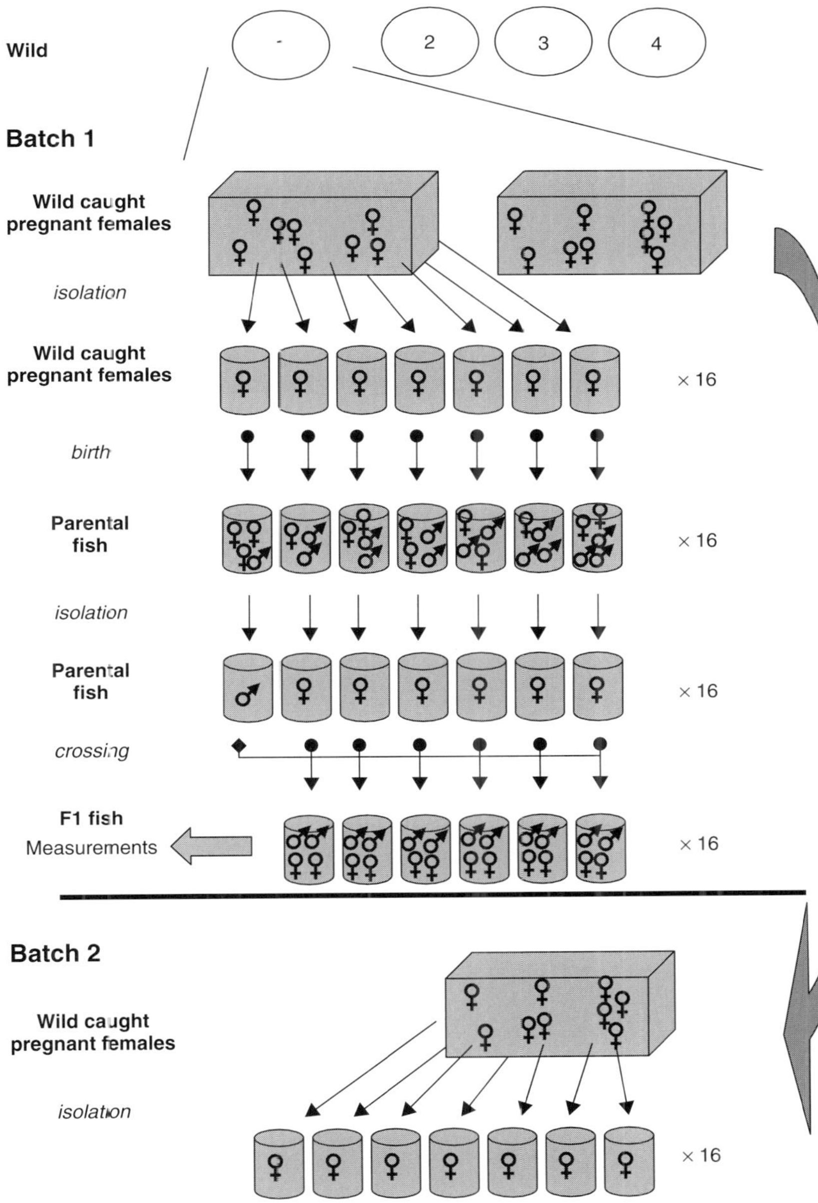

Figure 6.2 Schematic plan of the North Carolina I design used for estimating heritabilities in Trinidadian guppies. Pregnant females were collected from Trinidad, and kept in tanks for quarantine and acclimatization. Females were then kept individually in bottles within a recirculation system until the birth of offspring. These offspring were then used as the parental generation in the breeding experiments. Each male was mated to six females, producing six full-sib F_1 families which were reared to maturity. Measurements from these families were used to calculate heritability estimates.

Figure 6.3 A maternal half-sib family of guppy males (*P. reticulata*). Note the close similarity between full siblings (1 and 4 male vs. 2 and 3 male from top). See text and Plate 6.1 (facing p. 94).

are reared in a common environment (e.g. cage, aquarium) at one time or place, the spatial or temporal correspondence may lead to a phenotypic resemblance. Such a component of environmental variation (or common environment effects) is particularly troublesome in quantitative genetic studies with live-bearing vertebrates where long gestation periods and maternal care may result in phenotypic resemblance between full sibs due to sharing the same mother (i.e. maternal effects) (Lynch & Walsh 1998).

Small pilot experiments are often conducted in order to quantify the effects of the environment on the phenotypic resemblance between individuals. Such experiments may reveal whether the laboratory conditions are indeed adequate to breed the experimental organism and generate reproducible data. Pilot experiments are not only useful to optimize feeding regimes and disentangle temporal and spatial variation, but also provide hands-on experience with the model organism under experimental conditions. Such preliminary observations allow investigators to make more accurate estimates on the sample size needed to detect statistical differences, and may further help to identify potential logistical difficulties. Such constraints may arise, for instance, due to the fact that experimental animals need to be fed at regular intervals. Failing to provide food during just a few days of development may seriously compromise quantitative genetic estimates. For instance, in a pilot study on the guppy *Poecilia reticulata* we found that over 70% of the variation in standard length was explained by the number of days fish were fed on *Artemia* nauplii (Figure 6.4). Clearly, such data indicate that quantitative genetic experiments on guppy growth rates require a constant quantity and quality of food. The pilot experiment also provided valuable information on the maximum density with which sibling groups of guppies could be reared in a common environment, thereby influencing the decision on optimal design. Several other logistical, ethical and statistical difficulties may constrain investigators utilizing quantitative genetic techniques. Large samples sizes are required to estimate genetic variance components and heritabilities reliably, requiring complex experimental set-ups, and resulting in an excess of animals (or plants) that often need to be sacrificed and discarded.

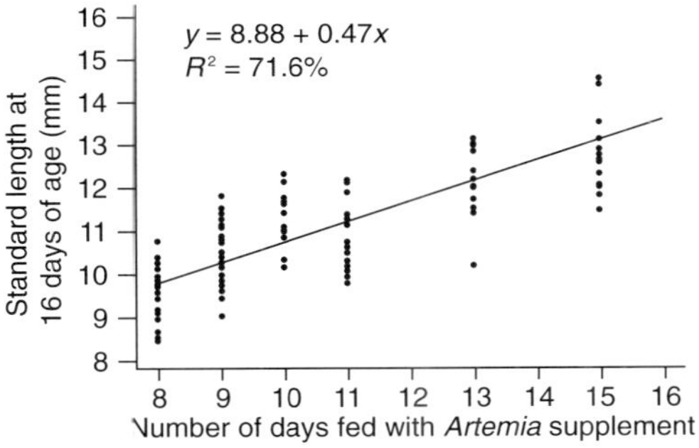

Figure 6.4 The effect of feeding *Artemia* nauplii on the growth of 16-day-old guppies. All fish were photographed on day 16 and received *Artemia* for at least 8 days prior to photographing.

Besides studies on adaptive variation, laboratory experiments can be used to estimate the number of mutations accumulated in the genome of individuals, and analyse their effects on population fitness (Charlesworth & Charlesworth 1987; Lynch & Walsh 1998). By inbreeding individuals deliberately, the recessive mutations present on some loci will be exposed in their homozygous state, revealing deleterious consequences on fitness. By monitoring inbreeding depression, the load of mutations can be estimated. For instance, a regression technique developed by Morton *et al.* (1956) uses the loss in fitness observed during inbreeding to estimate the mutation load for characteristics that can be classified by incidence (e.g. mortality). Morton *et al.* (1956) expressed the deleterious consequences of inbreeding by estimating the number of mutations present in a gamete that would cause on average one death when expressed as homozygotes. This measure was labelled a 'lethal equivalent', which may consist of a number of slightly deleterious mutations or a single lethal mutation. van Oosterhout *et al.* (2000a) used this regression technique to estimate the number of mutations responsible for mortality in fragmented populations of the butterfly *Bicyclus anynana*. Presence and absence data of traits other than survival can also be used in these analyses.

Whereas for many quantitative genetic analyses, large sample sizes are required to estimate genetic variation reliably, or detect differences between populations (Figure 6.5), the mutation load can be estimated with relatively small sample sizes of one to a few hundred individuals. Estimating the mutation load becomes more accurate with larger sample sizes and more variation in inbreeding coefficients (e.g. two or more generations of full-sib inbreeding). Lynch and Walsh (1998) suggested such a method using an outbred sample and a single generation of inbred individuals with known *F*-coefficients. In a study on the mutation load of guppies, C. van Oosterhout *et al.* (unpublished data) used 63 inbred lines and monitored fitness, courtship behaviour and colour pattern over three generations of inbreeding. Whereas the relatively small sample size of just over 200 individuals was generally insufficient to detect significant differences in heritable genetic variation (see Figure 6.5), the inbreeding experiment was statistically sufficiently robust to detect a 5–20% reduction in phenotypic values over three generations.

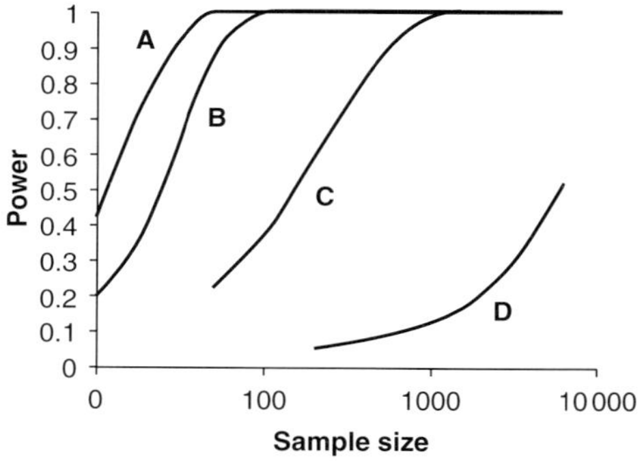

Figure 6.5 Power analysis for molecular and quantitative analysis of genetic diversity, based on Monte Carlo simulation of molecular and quantitative variation estimates from an upland and a lowland population of Trinidadian guppies. Heritability estimates used are typical for life-history traits in guppies from the Upper ($h^2 \approx 0.2$) and Lower Aripo ($h^2 \approx 0.4$). Note the logarithmic scale of the x-axis (sample size). A: Power to detect differences in microsatellite heterozygosity ($H_1 = 0.23, H_2 = 0.45$, 7 loci) with an ANOVA on individual multilocus heterozygosities. B: Power to detect differences in allozyme heterozygosity ($H_1 = 0.07, H_2 = 0.17$, 7 loci) with the same statistical method as in A. C: Power to detect a significant heritability in a nested half-sib design with a true heritability of $h^2 = 0.25$ (Lynch & Walsh 1998). D: Power of tests for differences in heritability in a nested half-sib design, if true heritabilities are $h_1^2 = 0.2$ and $h_2^2 = 0.4$, respectively (Klein 1974).

Logistic considerations in breeding experiments

One crucial consideration in quantitative genetic experiments is the low power of tests, and thus the requirement for large sample sizes. When investigating adaptive variation within a conservation biology context, it is important not only to obtain significant estimates of heritability (i.e. significantly different from zero), but also to determine whether populations differ in their heritability. Although the former requires considerable sample sizes (Klein 1973, Lynch & Walsh 1998, see Figure 6.5), the sample size requirements of the latter are often prohibitive. Even if true differences in heritability between populations are considerable, as with some life-history characters of guppies (approx. 0.2 vs. 0.4), very large samples indeed are required to obtain a reasonable power of tests (Klein 1974, see Figure 6.5). Such experiments may thus be impossible except in easily reared model organisms with short generation time (e.g. *Drosophila*, guppies), although modern statistical approaches may reduce the sample size requirements somewhat (Shaw 1991). By contrast, molecular markers can detect differences in heterozygosity with relatively small samples (see Figure 6.5): typical differences in microsatellite diversity among Trinidadian guppy populations could be detected using sample sizes of about 20 individuals per population (L. Hauser *et al.*, unpublished data). Furthermore, the time required to obtain data from one individual is usually considerably smaller for molecular markers than for quantitative genetic traits, even considering the effort required to develop some markers. Despite the problems and difficulties in the correlation between genetic variation at the molecular and adaptive levels, molecular markers may remain the only tractable way of assessing genetic diversity in most species.

Concluding remarks

The need to assess and predict the dynamics of adaptive variation in the wild becomes more urgent as environmental changes accelerate and the numbers of threatened habitats and species increase. Foremost among the priority questions posed were and remain the impact of environmental heterogeneity in space and time on selection pressures, heritability, phenotypic plasticity and the genetic correlations among traits (Endler 2000). In particular, the uncertainty about relationships between short-term vulnerability of populations and heritability and genetic diversity, and the impact of genetic load on population persistence remain key challenges for empirical demonstration. In contrast to such requirements, the majority of available information to date on adaptive diversity is indirect, and largely in the form of molecular estimates of genetic diversity. Evidence suggests that although measures of adaptive and molecular divergence among populations are closely related, within-population correlates are usually weak owing to their contrasting evolutionary dynamics. It is thereby increasingly clear that direct measures of adaptive trait variation and their associated heritabilities comprise the most effective way of predicting adaptive response.

Surveys of molecular variation can, however, be useful in several ways. First, they afford data on contemporary and historical aspects of population structure, so providing a context for the interpretation of quantitative genetic variation; second, they enable the estimation of relatedness among individuals to infer heritability in the wild; and third, they facilitate comparisons of population differentiation at molecular (subdivision for nuclear gene markers, F_{ST}) and quantitative genetic (subdivision for traits, Q_{ST}) levels, so allowing inferences on the relative roles of selective and stochastic evolutionary forces. The recently disclosed generally positive correlation between among-population estimates of differentiation in marker loci and quantitative traits (Merilä & Crnokrak 2001) (see Figure 6.1) suggests that the degree of differentiation in neutral marker loci is closely predictive of differentiation in loci coding quantitative variation.

Advances in statistical, molecular and quantitative genetic theory and approaches do now provide feasible opportunities to undertake direct studies on adaptive variation, even in some formerly intractable species, including vertebrates. The most significant insights are likely to accrue from those species where long-term data sets on ecology and trait variation exist, and where field- or laboratory-determined pedigrees can be readily obtained. Although such integrated studies will continue to be rare in relation to molecular genetic studies, if chosen carefully in relation to their attributes and representation among taxa, they may identify the constraints and opportunities afforded by diversity in development, physiology, life history, behaviour and morphology in specific environments. Molecular markers will continue to provide a valuable means for examining individual, population and species-level processes that have an impact on adaptive variation, as well as providing a direct means of pedigree determination. Moreover, developments in QTL and genomics (Sankoff & Nadeau 2000) are likely to advance significantly our understanding of the genomic architecture underlying specific traits.

Acknowledgements

This work was funded by the Natural Environment Research Council's thematic programme EDGE (Ecological Dynamics and Genes).

References

Albertson, R.C., Markert J.A., Danley P.D. & Kocher T.D. (1999) Phylogeny of a rapidly evolving clade: the cichlid fishes of Lake Malawi, East Africa. *Proceedings of the National Academy of Sciences of the United States of America* **96**, 5107–5110.

Arnone, M.I. & Davidson, E.H. (1997) The hardwiring of development: organisation and function of genomic regulatory systems. *Development* **124**, 1851–1864.

Avise, J.C. (1994) *Molecular Markers, Natural History and Evolution.* Chapman & Hall, London.

Barton, N.H. & Turelli, M. (1989) Evolutionary quantitative genetics: how much do we know? *Annual Reviews in Genetics* **23**, 337–370.

Bonnin, I., Prosperi, J.-M. & Olivieri, I. (1996) Genetic markers and quantitative genetic variation in *Medicago truncatula* (Leguminosa): a comparative analysis of population structure. *Genetics* **143**, 1795–1805.

Bryant, E.H.S., McCommas, A. & Combs, L.M. (1986) The effect of an experimental bottleneck upon quantitative genetic variation in the housefly. *Genetics* **114**, 1191–1211.

Burton R.S. & Feldman M.W. (1983) Physiological effects of an allozyme polymorphism: glutamate-pyruvate transaminase and response to hyperosmotic stress in the copepod *Tigriopus californicus. Biochemical Genetics* **21**, 238–251.

Butlin, R.K. & Tregenza, T. (1998) Levels of genetic polymorphism: marker loci versus quantitative traits. *Philosophical Transactions of the Royal Society Series B* **353**, 187–198.

Charlesworth B. & Charlesworth D. (1998) Some evolutionary consequences of deleterious mutations. *Genetica* **103**, 3–19.

Charlesworth D. & Charlesworth B. (1987) Inbreeding depression and its evolutionary consequences. *Annual Review of Ecology and Systematics* **18**, 237–268.

Corbin K.W. (1977) Phosphoglucose isomerase polymorphism and natural selection in the sand crab *Emerita talpoida. Evolution* **31**, 331–340.

Crandall, K.A., Binind-Emonds, O.R.P., Mace, G.M. & Wayne, R.K. (2000) Considering evolutionary processes in conservation biology. *Trends in Ecology and Evolution* **15**, 290–295.

Crawford, D.L. & Powers, D.A. (1989) Molecular basis of evolutionary adaptation in two latitudinally extreme populations of *Fundulus heteroclitus. Proceedings of the National Academy of Sciences of the United States of America* **86**, 9365–9369.

Ditchkoff, S.S., Lochmiller, R.L., Masters, R.E., Hoofer, S.R. & Van Den Bussche, R.A.(2001) Major-histocompatibility-complex-associated variation in secondary sexual traits of white-tailed deer (*Odocoileus virginianus*): evidence for good-genes advertisement. *Evolution* **55**, 616–625.

Ebert, D., Haag, C., Kirkpatrick, M. *et al.* (2002) A selective advantage to immigrant genes in a *Daphnia* metapopulation. *Science* **295**, 485–488.

Elton, C. (1924) Periodic fluctuations in the numbers of animals: their causes and effects. *British Journal of Experimental Biology* **2**, 119–163.

Endler, J.A. (1986) *Natural Selection in the Wild.* Princeton University Press, Princeton, NJ.

Endler, J.A. (2000) Adaptive genetic variation in the wild. In: *Adaptive Genetic Variation* (eds T. A. Mousseau, B. Sinervo & J. Endler), pp. 251–260. Oxford University Press, Oxford.

Falconer, D.S. (1989) *Introduction to Quantitative Genetics*. Longman, New York.

Ferguson, A. (1980) *Biochemical Systematics and Evolution*. Blackie, Glasgow.

Fishback, A.G., Danzmann, R.G. & Ferguson, M.M. (2000) Microsatellite allelic heterogeneity among hatchery rainbow trout maturing in different seasons. *Journal of Fish Biology* **57**, 1367–1380.

Fisher, R.A. (1958) *The Genetical Theory of Natural Selection*. Dover, New York.

Flint, J. & Mott, R. (2001) Finding the molecular basis of quantitative traits: successes and pitfalls. *Nature Reviews Genetics* **2**, 437–445.

Goldstein, D.B. & Schlötterer, C. (1999) *Microsatellites: Evolution and Applications*. Oxford University Press, Oxford.

Grant, P.R. & Grant, R. (2000) Quantifying genetic variation in populations of Darwin's finches. In: *Adaptive Genetic Variation* (eds T. A. Mousseau, B. Sinervo & J. Endler), pp. 3–40. Oxford University Press, Oxford.

Haldane J. B. S. (1937) The effect of variation on fitness. *American Naturalist* **71**, 337–349.

Hard, J.J. (1995) A quantitative genetic perspective on the conservation of intraspecific diversity. *American Fisheries Society Symposium* **17**, 304–326.

Hauser, L. & Ward, R.D. (1998) Population identification in pelagic fish: the limits of molecular markers. In: *Advances in Molecular Ecology* (ed. G. R. Carvalho), pp. 191–224. IOS Press, Amsterdam.

Hedrick, P.W. (2001) Conservation genetics: where are we now? *Trends in Ecology and Evolution* **16**, 629–636.

Hedrick, P.W., Kim, T.J. & Parker, K.M. (2001) Parasite resistance and genetic variation in the endangered Gila topminnow. *Animal Conservation* **4**, 103–109.

Henry T. & Ferguson A. (1985) Kinetic studies on the lactate dehydrogenase (Ldh–5) isozymes of brown trout, *Salmo trutta* L. *Comparative Biochemistry and Physiology* 82B, 95–98.

Hilbish, T.J. & Koehn, R.K. (1985) The physiological basis of natural selection at the LAP locus. *Evolution* **39**, 1302–1317.

Hoffman, R.J. (1981) Evolutionary genetics of *Metridium senile*. I. Kinetic differences in phosphoglucose isomerase allozymes. *Biochemical Genetics* **19**, 129–144.

Hoffmann, A.A. (2000) Laboratory and field heritabilities: some lessons from *Drosophila*. In: *Adaptive Genetic Variation* (eds T. A. Mousseau, B. Sinervo & J. Endler), pp. 200–218. Oxford University Press, Oxford.

Joshi S.P., Bhave S.G., Chowdari K.V. *et al.* (2001) Use of DNA markers in prediction of hybrid performance and heterosis for a three-line hybrid system in rice. *Biochemical Genetics* **39**, 179–200

Karhu, A., Hurme, P., Karjalainen, M. *et al.* (1996) Do molecular markers reflect patterns of differentiation in adaptive traits of conifers? *Theoretical and Applied Genetics* **93**, 215–221.

Kashi, Y. & Soller, M. (1999) Funtional roles of microsatellites and minisatellites. In: *Microsatellites: Evolution and Applications* (eds D. B. Goldstein & C. Schlötterer), pp. 10–23. Oxford University Press, Oxford.

Kashi, Y., King, D. & Soller, M. (1997) Simple repeats as a source of quantitative genetic variation. *Trends in Genetics* **13**, 74–78.

Klein, T.W. (1973) Heritability and genetic correlation: standard error of estimates and sample size. *Behavior Genetics* **3**, 355–364.

Klein, T.W. (1974) Heritability and genetic correlation: statistical power, population comparisons and sample size. *Behavior Genetics* **4**, 171–189.

Klein, J. (1991) Of HLA, Tryps and selection—an essay of coevolution of MHC and parasites. *Human Immunology* **30**, 247–258.

Kriener, K., O'hUigin C., T.H. & Klein J. (2000) Convergent evolution of major histocompatibility complex molecules in humans and New World monkeys. *ImmunoGenetics* **51**, 169–178.

Kruuk, L.E.B., Clutton-Brock, T.H., Slate, J., Pemberton, J.M., Brotherstone, S. & Guinness, F.E. (2000) Heritability of fitness in a wild mammal population. *Proceedings of the National Academy of Science of the United States of America* **97**, 698–703.

Kuittinen, H., Mattila, A. & Savolaienen, O. (1997) Genetic variation at marker loci and in

quantitative traits in natural populations of *Arabidopsis thaliana*. *Heredity* **79**, 144–152.
Lande, R. (1994) Risk of population extinction from fixation of new deleterious mutations. *Evolution* **48**, 1460–1469.
Latta, R.G. (1998) Differentiation of allelic frequencies at quantitative trait loci affecting locally adaptive traits. *American Naturalist* **151**, 283–292.
Leberg, P.L. (1992) Effects of population bottlenecks on genetic diversity as measured by allozyme electrophoresis. *Evolution* **46**, 477–494.
Lewontin, R.C. (1984) Detecting population differences in quantitative characters as opposed to gene frequencies. *American Naturalist* **123**, 115–124.
Lindström, J., Ranta, E., Kokko, H, Lunberg, P. & Kaitala, V. (2001) From arctic lemmings to adaptive dynamics: Charles Elton's legacy in population ecology. *Biological Reviews* **76**, 129–158.
Loxdale, H.D. & Lushai, G. (1998) Molecular markers in entomology. *Bulletin of Entomological Research* **88**, 577–600.
Lyman, R.F., Lai, C. & Mackay, T.F. (1999) Linkage disequilibrium mapping of molecular polymorphisms at the scabrous locus associated with naturally occurring variation in bristle number in *Drosophila melanogaster*. *Genetical Research* **74**, 303–311.
Lynch, M. (1985) Spontaneous mutation rates for life history characters in an obligate parthenogen. *Evolution* **39**, 804–818.
Lynch, M. (1996) A quantitative-genetic perspective on conservation issues. In: *Conservation Genetics: Case Studies from Nature* (eds J. Avise & J. Hamrick), pp. 471–501. Chapman & Hall, New York.
Lynch, M. & Gabriel, W. (1983) Phenotypic evolution and parthenogenesis. *American Naturalist* **122**, 745–764.
Lynch, M. & Walsh, B. (1998) *Genetic Analysis of Quantitative Traits*. Sinauer Associates, MA.
Lynch, M., Conery, J. & Bürger, R. (1995) Mutation accumulation and the extinction of small populations. *American Naturalist* **146**, 489–518.
Lynch, M., Pfrender, M., Spitze, K. *et al.* (1999) The quantitative and molecular genetic architecture of a subdivided species. *Evolution* **53**, 100–110.
Mackay, T.F.C. (1996) The nature of quantitative genetic variation revisited: lessons from *Drosophila* bristles. *BioEssays* **18**, 113–121.
Mackay, T.F.C., Fry, J.D., Lyman R.F. & Nuzhdin, S.V. (1994) Polygenic mutation in *Drosophila melanogaster*: estimates from response to selection of inbred strains. *Genetics* **136**, 937–951.
Madsen, T., Shine, R., Olsson, M. & Wittzell, H. (1999) Restoration of an inbred adder population. *Nature* **402**, 34–35.
Magurran, A. E., Seghers, B. H., Shaw, P. W. & Carvalho, G. R. (1995) The behavioral diversity and evolution of guppy, *Poecilia reticulata*, populations in Trinidad. *Advances in the Study of Behavior* **24**, 155–202.
Merilä, J. (1997) Quantitative trait and allozyme divergence in the Greenfinch (*Carduelis chloris*, Aves: Fringilliadae). *Biological Journal of the Linnean Society* **61**, 243–266.
Merilä, J. & Crnokrak, P. (2001) Comparison of genetic differentiation at marker loci and quantitative traits. *Journal of Evolutionary Biology* **14**, 892–903.
Merilä, J. & Sheldon, B.C. (2000) Lifetime reproductive success and heritability in nature. *American Naturalist* **155**, 301–310.
Milligan, B.G., Leebens-Mack, J. & Strand, A.E. (1994) Conservation genetics: beyond the maintenance of marker diversity. *Molecular Ecology* **3**, 423–435.
Mitchell-Olds, T. (1995) The molecular basis of quantitative genetic variation in natural populations. *Trends in Ecology and Evolution* **10**, 324–328.
Mitton, J.B. (1997) *Selection in Natural Populations*. Oxford University Press, New York.
Moran, P. & Kornfield, I. (1993) Retention of an ancestral polymorphism in the mbuna species flock (Teleostei, Cichlidae) of Lake Malawi. *Molecular Biology and Evolution* **10**, 1015–1029.
Morton, N.E., Crow, J.F. & Muller, H.J. (1956) An estimate of the mutational damage in man from data on consanguineous mating. *Proceedings of the National Academy of Sciences of the United States of America* 42: 855–863.

Mousseau, T.A., Ritland, K. & Heath, D.D. (1998) A novel method for estimating heritability using molecular markers. *Heredity* **80**, 218–224.

Mousseau, T.A., Sinervo, B. & Endler, J. (2000) *Adaptive Genetic Variation in the Wild.* Oxford University Press, Oxford.

Nauta, M.J. & Weissing, F.J. (1996) Constraints on allele size at microsatellite loci: implications for genetic differentiation. *Genetics* **143**, 1021–1032.

van Oosterhout, C. & Brakefield, P.M. (1999) Quantitative genetic variation in *Bicyclus anynana* metapopulations. *Netherlands Journal of Zoology* **49**, 67–80.

van Oosterhout, C., Smit, G., Heuven, M.K. & Brakefield, P.M. (2000a) Pedigree analysis on small laboratory populations of the butterfly *Bicyclus anynana*: the effects of selection on inbreeding and fitness. *Conservation Genetics* **1**, 321–328.

van Oosterhout, C., Zijlstra, W.G., Heuven, M.K. & Brakefield, P.M. (2000b) Inbreeding depression and genetic load in laboratory metapopulations of the butterfly *Bicyclus anynana. Evolution* **54**, 56–64.

Pfrender, M.E., Spitze, K., Hicks, J. *et al.* (2000) Lack of concordance between genetic diversity estimates at the molecular and quantitative-trait levels. *Conservation Genetics* **1**, 263–269.

Powers, D.A., Lauerman, T., Crawford, D. *et al.* (1991) The evolutionary significance of genetic variation at enzyme synthesizing loci in the teleost *Fundulus heteroclitus. Journal of Fish Biology* 39 (Suppl. A), 169–184.

Reed, D.H. & Frankham, R. (2001) How closely correlated are molecular and quantitative measures of genetic variation? A meta-analysis. *Evolution* **55**, 1095–1103.

Reich, D.E. & Goldstein, D.B. (1999). Estimating the age of mutations using variation at linked markers. In: *Microsatellites: Evolution and Applications* (eds D. B. Goldstein & C. Schlötterer), pp. 129–138. Oxford University Press, Oxford.

Ritland, K. (2000a) Marker-inferred relatedness as a tool for detecting heritability in nature. *Molecular Ecology* **9**, 1195–1204.

Ritland, K. (2000b) Detecting inheritance with inferred relatedness in nature. In: *Adaptive Genetic Variation* (eds T. A. Mousseau, B. Sinervo & J. Endler), pp. 187–199. Oxford University Press, Oxford.

Saccheri, I.J., Kuussaari, M, Kankare, M. *et al.* (1998) Inbreeding and extinction in a butterfly metapopulation. *Nature* **392**, 491–494.

Sankoff, D. & Nadeau, J.H. (2000) *Comparative Genomics: Empirical and Analytical Approaches to Gene Order Dynamics, Map Alignment and the Evolution of Gene Families.* Kluwer Academic Publishers, London.

Schopf, T.J.M. (1974) Survey of genetic differentiation in a coastal zone invertebrate: the ectoproct *Schizoporella errata. Biological Bulletin* **146**, 78–87.

Shaw, P.W., Turan, C., Wrigth, J.M., O'Connell, M. & Carvalho, G.R. (1999) Microsatellite DNA analysis of population structure in Atlantic herring (*Clupea harengus*), with direct comparison to allozyme and mtDNA RFLP analyses. *Heredity* **83**, 490–499.

Shaw, R.G. (1991) The comparison of quantitative genetic parameters between populations. *Evolution* **45**, 143–151.

Vrijenhoek, R.C., Pfeiler, E. & Wetherington, J. (1992) Balancing selection in a desert stream dwelling fish, *Poeciliopsis monacha. Evolution* **46**, 1642–1657.

Wielebnowski, N. (1996) Reassessing the relationship between juvenile mortality and genetic monomorphism in captive cheetahs. *Zoo Biology* **15**, 353–369.

Wigensberg, I. & Roff, D.A. (1996) Natural heritabilities: can they be reliably estimated in the laboratory? *Evolution* **50**, 2149–2157.

Chapter 7

Have recent developments in molecular techniques led to greater insight into evolutionary and ecological processes in plant populations?

A. Smithson and M. R. Macnair*

Introduction

New molecular genetics techniques have offered exciting opportunities for researchers in many diverse fields including evolution and ecology to ask novel and interesting questions that were previously impossible. These techniques can, for example, be used to identify different species where visual identification is very difficult, as for many soil microorganisms, and can identify genetic relationships between individuals where exact pedigrees or ancestry are unknown, as in studies of paternity. The aim of EDGE was to encourage and develop the use of molecular genetic and mathematical tools to reveal a deeper understanding of the dynamics of genes in populations and communities. A previous BES symposium held 10 ago (Berry *et al.* 1992) hoped that new concepts and techniques would lead to major advances in the understanding of 'genes in ecology'. At this BES-sponsored symposium marking the culmination of EDGE, we have decided to look at the literature and quantify what impact novel molecular genetics techniques have really had on our own field of plant population biology over the last 10 years. A large number of recent reviews have summarized in detail the wide range of different molecular techniques that are currently available for use in plant evolution and ecology, how they can potentially be used, and the advantages and disadvantages of each (e.g. Cruzan 1998). In this paper, we do not aim to repeat these reviews. Instead, we aimed to quantify how neutral molecular markers have been used to answer questions in plant evolution and ecology. Has the availability of new molecular genetics techniques affected the questions addressed by researchers over the past 10 years, and are these techniques being fully utilized in their analyses of plant populations and communities?

* Correspondence author: A.Smithson@exeter.ac.uk

Have the uses of neutral markers in plant evolution and ecology changed between 1990 and 2000?

Methods

In order to compare how the questions addressed and methodologies used in studies using neutral molecular markers in plant evolution and ecology have changed in the past 10 years, we reviewed in detail the literature in 1990 and 2000. We looked at the relevant issues of six core journals addressing a broad range of evolutionary and ecological topics: *Evolution*, *Heredity*, *Journal of Evolutionary Biology*, *American Journal of Botany*, *Biological Journal of the Linnean Society* and *Molecular Ecology*. We extracted those papers whose main topic was a plant (including angiosperms, ferns, mosses and algae but excluding fungi) and also included directly the use of a neutral molecular marker. We excluded papers that referred indirectly to the results of molecular techniques studied in detail in other papers.

We classified each relevant plant study into the following question categories:

1 Studies quantifying interspecific genetic differentiation including phylogeny construction where many species were included in the study, the use of molecular techniques in investigating species identity, and also those studies describing the extent of hybridization between different species.

2 Studies quantifying intraspecific genetic differentiation within and between populations, including studies that compared between-population genetic structure across two subspecies or sibling species, and also intraspecific historical phylogeography.

3 Papers reporting predominantly new techniques or methodologies.

4 'Others', incorporating everything else.

We also classified the papers by molecular method used, defining 13 different categories of molecular methods (see Table 7.1 for the categories we considered, and also

Table 7.1 Categories of molecular techniques used in the literature analysis, including definitions of abbreviations.

Category	Definition
Allozyme	Different allelic forms of enzymes
Microsatellite	Simple sequence repeats (single locus)
AFLP	Amplified fragment length polymorphism
RAPD	Random amplified polymorphic DNA
RFLP	Restriction fragment length polymorphism
SSCP	Single-stranded conformation polymorphism
ISSR	Intersimple sequence repeats
SCAR	Sequence-characterized amplified region
Minisatellite	Variable number tandem repeats (multilocus)—classical DNA fingerprinting
mtDNA	Markers based on mitochondrial DNA
cpDNA	Markers based on chloroplast DNA (excluding RFLP based on chloroplasts)
ITS	Sequencing of internal transcribed spacer regions
Other	Any other method used

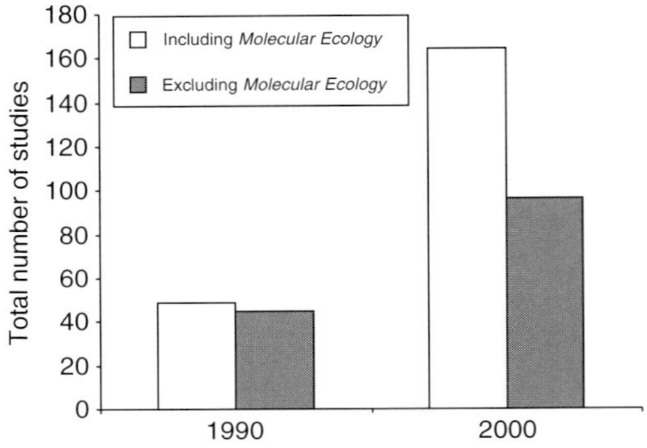

Figure 7.1 The total number of studies found in a literature review that used molecular methods to answer questions in plant ecology and evolution during 1990 and 2000 in six key journals. Results are presented both including and excluding *Molecular Ecology*.

the definitions of abbreviations used in this paper), and also by the numbers of different categories of methods used in each paper. The years were then compared, using standard χ^2 tests to look for significant differences where appropriate.

Note that although *Molecular Ecology* (ME) is probably now a journal of central importance in this field, it was challenging to include in such a study for two reasons. First, ME was not published until 1992, and the journal was not indexed on ISI databases until 1993. Second, volumes up to 2000 included methodological papers, while the journal was split in 2001 to give a separate ME Notes volume for such studies. Thus while we included the 1993 and 2000 issues of ME in our 1990 and 2000 data sets, we analysed results that both include and exclude ME to ensure that there were no inherent biases in including this journal in our study. More general ecological journals were not incorporated into the study, because we found few/no papers that incorporated molecular genetics techniques in them.

Results

Have the overall numbers of studies changed?

The total number of plant studies recorded that used neutral molecular markers is shown in Figure 7.1. There has been a rapid increase in the number of studies—in excess of a threefold increase (twofold, excluding ME) across the last 10 years. The overall number of studies published in these journals has also increased over the last 10 years, but this has only been by a factor of 1.7 (1.4 excluding ME). Indeed there has been a significantly faster increase in the number of published plant studies utilizing molecular markers than there has been in the number of published studies overall (including ME: $\chi^2_1 = 14.71$, $P < 0.001$; excluding ME: $\chi^2_1 = 5.10$, $P < 0.05$).

Have the questions addressed by the studies changed?

Figure 7.2 compares the proportions of the four types of questions addressed by the studies between 1990 and 2000, with and without ME. The results were rather

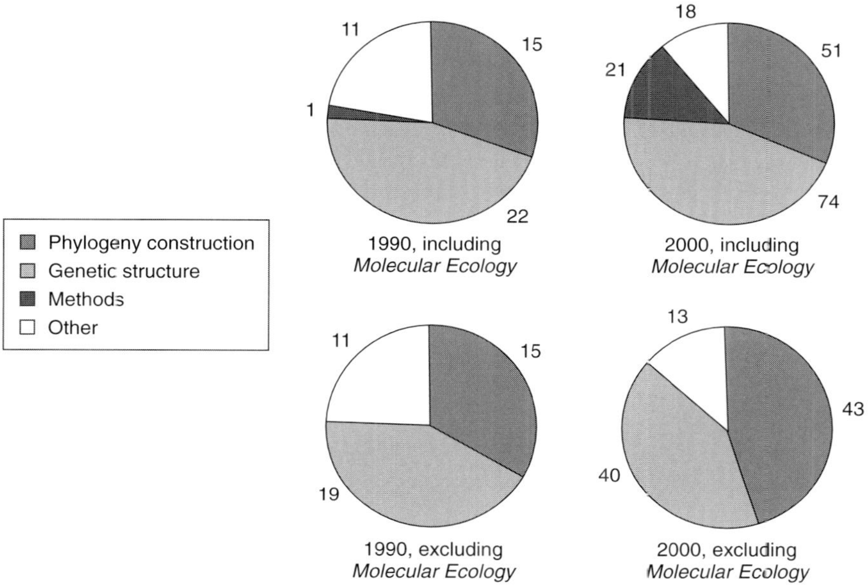

Figure 7.2 The proportion of classes of questions answered by papers in plant ecology and evolution using molecular methods during 1990 and 2000. Questions were categorized into four classes (see text for further details). Results are presented both including and excluding *Molecular Ecology*. Actual numbers of studies are given in figures next to the relevant pie chart segment.

surprising. We found that, including ME, there was a decrease in the proportion of studies addressing 'Other' topics, and a concomitant increase in the proportion of methodological papers. This change is significant ($\chi^2_2 = 7.88$, $P < 0.05$). When ME is excluded we found an identical decrease in the proportion of 'Other' topics, with an increase in studies covering the alternative topics of phylogeny construction and genetic structure. There were no methodological papers outside ME. In terms of absolute numbers of studies, Figure 7.2 shows that the number of 'Other' studies published has remained static or increased slightly, whereas all other study types have increased by at least two- to four-fold. 'Other' studies included a wide range of subjects—paternity analyses, analyses of selfing rates and breeding systems, direct quantification of rates of gene flow, analysis of clonality and comparison of adaptive characters with neutral genetic markers, with no one question predominating. By contrast, the majority (69% including ME) of phylogenetics and species identity questions reported phylogeny resolution, and similarly the majority (78% including ME) of genetic structure and historical biogeography questions reported analyses of genetic differentiation between populations of a single species.

We also, in addition to classifying by type of question, quantified the numbers of studies that used neutral molecular markers as part of an experiment or a comparative analysis, as opposed to studies that were purely descriptive. The number of experimental papers we found was two in 2000 and one in 1990. All experimental

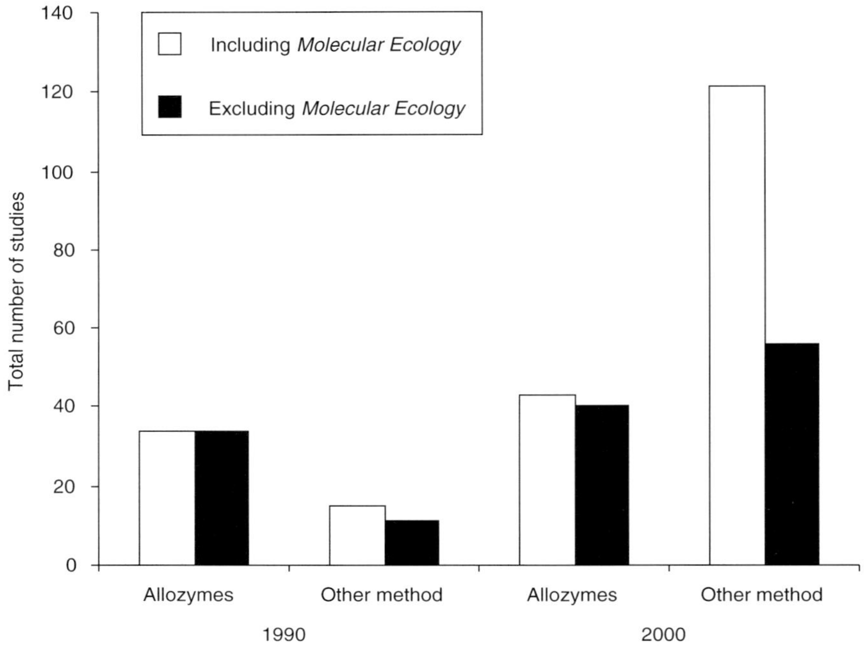

Figure 7.3 The total number of studies in plant ecology and evolution carried out using either allozymes, or other molecular methods combined, in both 1990 and 2000. Results are presented both including and excluding *Molecular Ecology*.

papers that we recorded were classified into the 'Other' question category. We recorded one study in 1990 that included use of the comparative method.

Have the techniques used in molecular studies changed?

Figure 7.3 shows the total number of studies in 1990 and 2000 that used either allozymes or other molecular techniques, both including and excluding ME. Interestingly, the numbers of studies carried out both for allozymes and other methods increased. In 1990, the majority (69% including or 76% excluding ME) of studies analysed used allozymes, with only 31% (24% excluding ME) using other molecular techniques. By contrast, by 2000 the majority of studies used molecular methods other than allozymes (74% including ME, 58% excluding ME). These differences are significant (including ME: $\chi^2_1 = 28.62$, $P < 0.001$; excluding ME: $\chi^2_1 = 22.00$, $P < 0.001$). In 1990 these other techniques consisted almost entirely of cpDNA (5% of all studies), RAPDs (3%) and RFLPs (2%). Indeed, we recorded only six types of molecular methods used in total in 1990 (five excluding ME). In 2000, all 13 different types of molecular methods were recorded (11 excluding ME). By far the predominant method used other than allozymes (26% of all studies) was microsatellites (21% of all studies), followed by RAPDs (12%) and AFLPs (10%). In 1990, the aver-

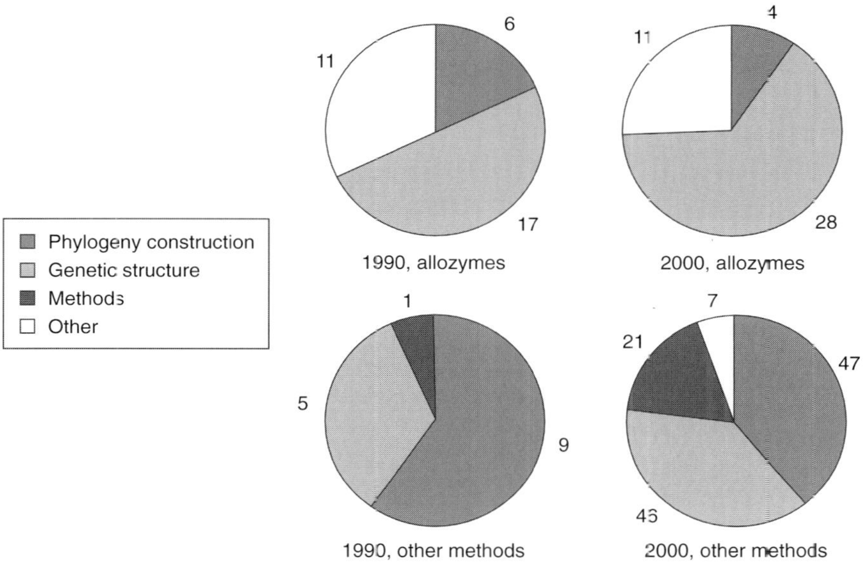

Figure 7.4 The proportion of classes of questions answered by papers in plant ecology and evolution, subdivided according to the molecular method used, either allozymes or other molecular methods combined, in 1990 and 2000. Actual numbers of studies are given in figures next to the relevant pie chart segment.

age study used 1.01 different molecular methods per paper (1.01 excluding ME), but by 2000 this had increased to 1.30 methods per paper (1.22 excluding ME).

We also classified the type of question studied according to the method used (Figure 7.4). To us, the most interesting pattern was that 'Other' questions were analysed entirely by allozymes in 1990 and, despite the observed increase in the diversity and use of molecular methods other than allozymes, almost all 'Other' questions continued to use allozymes in 2000 (Figure 7.4). All methods papers addressed techniques other than allozymes both in 1990 and 2000, probably not surprising given that allozymes have been used extensively in this field since 1966. Phylogenetic studies have increasingly switched to the use of molecular methods other than allozymes. Both allozymes and other molecular methods are still being used to answer genetic structure questions, however.

What did the literature study reveal?

Our analysis of the literature revealed a considerable increase in the number of evolutionary and ecological studies employing neutral molecular markers. We also found a substantial increase in the sophistication of the molecular techniques used, both in the overall numbers of techniques used across journals and also in the numbers used within individual papers. Despite this, the diversity of questions addressed using these techniques has apparently decreased over the last 10 years. Indeed, the great majority of studies, 99% in 2000, were descriptive, in particular quantifying genetic structure between populations and resolving phylogenetic relationships. We also found that most studies predominantly focus on one species (54% and 61% of studies in 2000 and 1990 respectively, including ME). The almost complete lack of studies employing molecular techniques as part of an experiment was surprising.

We do not suggest, from these results, that descriptive studies of genetic diversity are intrinsically less interesting than experimental ones. Resolution of phylogenetic relationships among extant plants is vital for understanding their evolutionary history, and also to provide the potential to test hypotheses concerning trait evolution. Phylogenetic studies may be the precursors to future studies that require a robust and unbiased phylogeny for hypothesis testing or alternatively simply require confirmation of the status of a taxon classified on morphological grounds. We found in our literature study that only one phylogenetic study used the comparative approach in addition to the use of molecular markers. Of course, we cannot exclude the possibility that further such studies would be undertaken subsequently utilizing the published phylogeny. Quantifying relationships between potentially hybridizing species has also allowed deeper consideration of species concepts and investigation of speciation processes.

Studies investigating genetic structuring of populations are of importance in assessing population differentiation as a measure of the potential for adaptive evolution, as indirect estimates of gene flow, to identify species history and range changes, and as an aid to identify genetically and adaptively differentiated subspecies or populations for conservation purposes (Avise 1994). The incorporation of a geographical element into studies of intraspecific genetic differentiation and phylogeography has allowed identification of physical barriers to gene flow and also tentative localization of the origin of a species or its refugium during inhospitable climatic conditions such as interglacial periods. It is possible to test a phylogeographical hypothesis by exploring changes in the past distribution of the species concerned by correlating genetic data with those from fossils, for example by pollen analysis. However, there are a number of problems in utilizing an entirely descriptive approach to these issues. Extant patterns of genetic differentiation are open to a number of interpretations, in that they may reflect current barriers to gene flow, changes in the past regarding species distribution or gene flow or adaptive evolution, and it has been argued that descriptive studies often fail to address these issues of multiple interpretation with interpretation potentially reflecting researcher bias (Bossart & Prowell 1998). As shown elsewhere in this volume (see Chapter 5), however, varying spatial geometries can substantially affect predicted patterns of pollen dispersal and thus gene flow in plants. Indirect estimates of gene flow are often invalidated by assumptions made during analysis that populations are infinite, and thus genetic differentiation does not necessarily reflect dispersal behaviour (see Chapter 1). Despite this, we revealed only a tiny fraction of studies in 2000 (2%) that attempted to quantify gene flow using direct methods, such as by paternity analysis, as recommended by recent reviewers (Bossart & Prowell 1998). Measures of quantitative genetic variation are not necessarily correlated with measures of genetic variation between populations (Reed & Frankham 2001; see Chapter 6), and thus molecular markers may be poor indicators of the potential for adaptation (see also Linhart & Grant 1996). As argued by Carvalho *et al.* (see Chapter 6), there is a complex relationship between molecular and adaptive variation as a result of polygenic control of adaptive traits, dominance, epistasis and finite population size, but despite this only a tiny minority of studies in 2000 (1%) attempted to compare molecular and quantitative variation.

Thus we find that the great range of molecular tools available now are increasingly being focused on a smaller subset of questions, and we found that these questions predominantly involved descriptive analyses of phylogenetics and genetic structure. But even if the diversity of questions is less, is the power with which they are being addressed much greater? For phylogenetic questions, it is clear that direct sequencing of DNA will generally have higher power than a study of allozymes, since the number of informative loci will be greater, and often the range of phylogenetic distances between species included within studies can be much greater. For genetic structure questions, however, it is not always true that the more modern techniques will give 'better' answers than allozymes. As noted below, there are theoretical problems associated with the analysis and interpretation of the data generated using some techniques. The greater cost of many of the techniques means that sample sizes may be smaller than typical with allozyme studies. However, studies using both nuclear markers (allozymes or microsatellites) and cytoplasmic markers (particularly cpDNA) can shed light on the relative contribution of seed movement and pollen transport to gene flow (e.g. Ennos 1994; Latta & Mitton 1999; Comes & Abbott 1999). In addition, anonymous techniques such as RAPDs, AFLPs or ISSRs may be used on any species without the development time required for either allozymes or microsatellites, and may reveal variation in species (such as highly selfing species) in which there is little variation in allozymes (e.g. Bonnin *et al.* 1996). Nevertheless, these techniques introduce other problems such as dominance and unknown homology between fragments of equal size (Vekemans *et al.* 2002).

Why has there been a decline in the diversity of questions addressed by neutral molecular markers in plant ecology and evolution?

It is unlikely that there is a simple or single explanation to this question. Our results could be merely an artefact of the journals that we surveyed, but since we feel that these represent a broad spectrum of journals that report studies utilizing neutral molecular markers in plant evolution and ecology, we think it is unlikely that this result is strongly biased. In particular, results were very similar whether ME was included or excluded from analyses. Further, our study supports similar suggestions by other authors—for example Parker *et al.* (1998) suggest that 'the rate at which new molecular techniques are being developed far outstrips their efficient and reasoned incorporation into studies of population ecology'. An alternative explanation is that all the interesting questions in more diverse areas of plant evolution and ecology have already been answered, or are not worth asking. We feel that this is unlikely.

A number of recent studies have indeed used molecular markers in extremely novel and interesting ways. For example, an experimental analysis showed that gene flow through pollen is increased into plant populations consisting of highly related individuals compared to controls using unrelated individuals (Richards 2000). Manipulation of flower number followed by paternity analysis of progeny arrays using allozymes has revealed that large inflorescences have higher selfing rates and lower outcrossed siring success than models had predicted (Harder & Barrett 1995). Intelligent use of nuclear and cytoplasmic markers has also allowed quantification of the relative impact of seeds and pollen on gene flow, since levels of population genetic differentiation differ for maternally or paternally inherited markers (Ennos 1994).

In animal evolutionary ecology, one spur to the increased use of molecular markers was data showing that extra-pair paternity was pervasive in natural populations (e.g. Burke *et al.* 1989), together with a novel molecular test for sex in birds (Ellgren & Sheldon 1997). There has been a proliferation of studies testing theories on animal sexual selection and lifetime reproductive success that incorporated molecular techniques as a tool for assessing paternity and reproductive success (e.g. Möller 1997; Coulson *et al.* 1998; Petrie *et al.* 1998; Coltman *et al.* 1999; Oddie 2000; Whittingham & Dunn 2000). There was no such spur in plant evolutionary ecology, and while some plant paternity analyses have been attempted (reviewed in Snow & Lewis 1993), these have been an order of magnitude fewer than for animal populations. This is despite the increased realization that paternal and maternal reproductive success are equally important in determining overall fitness in plant populations (Willson 1994). Further, it was noticeable from our literature review that most studies in our 'other' category tend to be those looking at fine-scale genetic diversity questions such as paternity, selfing rates and breeding system analyses. Are there insurmountable difficulties that researchers in plant reproductive ecology experience in applying molecular techniques to their field over and above those experienced by researchers studying animals?

While our data suggest that there are many questions in plant evolution and ecology to which molecular markers are not being applied, looking at the wider literature suggests that there has been an explosion in the use of neutral molecular markers for QTL mapping in plants. This is certainly an area that has only been made possible by the development of modern techniques and plentiful markers. QTL mapping is easier in plants than in animals, because of the larger family sizes, and possibility of developing recombinant inbred lines readily in selfing species. Powerful experimental designs are also easier to achieve. While much QTL mapping has concentrated on agronomically important characters in crops, there has also been a number of studies of ecologically relevant characters in the model plant species *Arabidopsis thaliana* (e.g. Mitchell Olds 1996; Van der Schaar *et al.* 1997) and even in 'wild' species such as *Microseris* (Bachmann & Hombergen 1997) and *Mimulus* (Bradshaw *et al.* 1998; Lin 2000).

Colleagues have suggested to us that the pattern of decline in diversity of questions asked with molecular markers in plant evolution and ecology, despite the increase in sophistication of techniques, may just be due to the way that this field has developed over time, and that it is incorrect to 'pathologize' this observation. It is certainly true that, in another 10 years, the patterns we have observed may have changed, and over time and with the encouragement of programmes such as EDGE researchers may have begun to incorporate increasingly these new technologies into their research. However, unfortunately we do not have the data for 2010, and also the task we were set when writing this paper was to address the state of our field now! Further, new molecular technologies are continually being developed. For example there has been a great increase in interest in the application of the relatively new field of genomics to evolutionary and ecological studies, as shown by NERC's new thematic research topic of 'Environmental Genomics'. It is clearly important that these new techniques should be used to answer questions that could not be answered by other methods: research should be driven by questions, not the availability of techniques. Thus we feel that these patterns do provide a starting point to consider what problems people may be having in applying neutral genetic markers to a wider diversity of questions in plant evolution and ecology, and perhaps also what could be done to address these problems.

Deficiencies in molecular techniques

Recent reviews of techniques available to plant ecologists and evolutionary biologists detail the large range of potential techniques currently available to researchers in plant evolution and ecology (Avise 1994; Jarne & Lagoda 1996; Cruzan 1998; Parker *et al.* 1998; Mueller & Wolfenbarger 1999; Sunnucks 2000; Provan *et al.* 2001). The most important factor in determining the applicability of a molecular technique to a specific question in plant ecology and evolution is the amount of heterozygosity that is expressed, and the level at which that heterozygosity is expressed—at scales between individuals, between populations or between species (Parker *et al.* 1998; Sunnucks 2000). Reviews indicate that the range of molecular techniques available today are not limited, but truly cover all regions of this spectrum (Avise 1994; Jarne & Lagoda 1996; Cruzan 1998; Parker *et al.* 1998; Mueller & Wolfenbarger 1999; Sunnucks 2000; Provan *et al.* 2001). For looking at broad-scale relationships between species, such as through phylogeny resolution and species identification, relatively slowly evolving loci are appropriate. Our literature review showed that in 1990 most such studies were carried out using allozymes, while in 2000 we found that 11 different techniques had been used. By far the most popular technique used at this broad scale in 2000 was direct sequencing, particularly of cpDNA or ITS. This probably reflects the availability of universal primers for such regions, which were developed widely during the 1990s (e.g. Demesure *et al.* 1995). For questions looking at more fine-scale resolution such as genetic structuring of populations and historical phylogeography, more rapidly evolving loci are more appropriate. In 1990, most studies at this scale again used allozymes, while we recorded 11 different molecular techniques used in 2000. While allozymes are still widely used at this scale in 2000, microsatellites, RAPDs, AFLPs and RFLPs, particularly for the latter focusing on the chloroplast genome, were also commonly used. At the scale of identifying relationships between individuals, rapidly evolving loci are suitable, as the resultant high variability can result in efficient resolution of relationships. Techniques such as nuclear and chloroplast microsatellites, AFLPs, SSCPs and ISSRs have all been shown to be potentially suitable at such a scale of resolution (Jarne & Lagoda 1996; Cruzan 1998; Ge & Sun 1999; Mueller & Wolfenbarger 1999; Sunnucks *et al.* 2000; Provan *et al.* 2001), although markers that are highly variable and co-dominant like microsatellites prove most efficient (Gerber *et al.* 2000). Despite this, we found that nearly all of these studies in both 1990 and 2000 used allozymes, although allozymes often lack sufficient variability for individual identification (Hamrick & Godt 1990). In 2000 only three molecular techniques were used in total for these types of analyses.

Our analyses thus suggest that a very wide range of techniques are currently available to be used at each scale of genetic variation. For broader scale resolution, such as phylogeny resolution and analysis of genetic structure, the wide range of techniques available appear to be being fully used to address these questions. At fine-scale levels of genetic variation, however, a lower diversity of molecular techniques is being used, and it appears that possibly less appropriate techniques like allozymes are still preferred. It could certainly be argued that this reflects particular difficulties in obtaining the most appropriate fine-scale genetic markers in plants. For example microsatellites have been shown to be five times less abundant in plants than in mammals (Lagercrantz *et al.* 1993). Further, allozymes have been shown to be more polymorphic in plants than in animals on average (Parker *et al.* 1998). However, we found that markers recommended for fine-scale resolution like microsatellites and AFLPs are frequently being used to study questions at broader-scale resolution such as genetic structuring, despite studies indicating that highly hypervariable loci are not the markers best suited to quantifying genetic distances between populations (e.g. Freville *et al.* 2001). Thus we suggest that difficulty in obtaining appropriate molecular markers is not a crucial limiting factor in studies of plant evolution and ecology, and that there must be reasons other than availability of techniques that have limited their wider application to questions in plant evolution and ecology.

Deficiencies in mathematical analysis techniques

In order to answer questions in plant evolutionary and ecological studies using neutral molecular markers, we not only need the correct molecular technique for gathering the appropriate data, but also the appropriate mathematical methods for analysing statistically the data that are generated. All methods of statistical analysis are based on an assumption that the underlying genetic mechanisms of inheritance and mutation are understood, and that the markers used are truly neutral. Some allozyme loci show clines in gene frequency along environmental gradients suggesting non-neutrality (Hamrick & Godt 1990; Parker *et al.* 1998). DNA-based markers are normally assumed to be non-coding and therefore neutral, although there is accumulating evidence that microsatellites, for example, may serve as regulatory elements during transcription and variation in the numbers of repeats may not be neutral (Kashi & Soller 1999). For most markers, variation is presumed to be generated by random mutation in the DNA sequence, which either directly modifies the size of the marker or in the case of allozymes modifies the translated products of the sequence. The most notable exception is that of microsatellites, where the mechanism of mutation is clearly not random, but where agreement has yet to be reached on a single appropriate mutational model. Tandemly repeated units are believed to increase predominantly by single repeat units through slip-strand mispairing during meiosis (Sia *et al.* 1997). Mutation often results in the gain or loss of a single repeat unit with some degree of bias favouring increasing total length, but with an apparent constraint on maximum size and also with dependence on the type of repeat unit (reviewed by Estoup & Cornuet 1999). The rapid rate of slip-strand mutation, combined with the possibility either to gain or lose repeat units, indicate that homoplasy is highly likely, and analyses such as those involved in assessing the degree of population structure depend crucially on the degree of homoplasy and assigning the correct mutational model (Jarne & Lagoda 1996). However, there appears to be no agreement on a single model (Jarne & Lagoda 1996; Estoup & Cornuet 1999), and most analyses of genetic structure using microsatellites ignore both the information provided by a precise knowledge of the length of the repeat unit and the possibility for homoplasy. Resolution of this issue is important, given that microsatellite markers were, according to our literature review, the most popular DNA-based marker used in plant studies in 2000.

Each scale of resolution that we considered in the previous section not only requires appropriate molecular markers but also appropriate analysis methods. There have been a number of proposed statistical models for analysing gene flow indirectly through population genetic structuring (see Bossart & Prowell 1998). However, most of these models assume infinite population size, stable demographic and genetic equilibria and constant dispersal rates over space and time, assumptions that are clearly unrealistic but are ignored by most researchers (Lewontin 1985; Bossart & Prowell 1998). Novel analysis techniques, for example those derived from the coalescent process as reviewed by Taneyhill *et al.* (see Chapter 1) combined with more experimental validation may, we hope, address these issues.

Other potential problems of analysis of marker data in plant populations apply, we feel, particularly to paternity analyses. For animals with internal fertilization, there is typically a limited pool of potential fathers for a given set of offspring, and often potential fathers can be assessed by direct observation. For most plants, paternity pools are likely to be large and multiple paternity pervasive, due to the extent of long-distance pollen flow, both for plants with wind-dispersed pollen and where pollinators can carry pollen from many individuals on their bodies (Snow & Lewis 1993; Dow & Ashley 1996, 1998). Thus the likelihood of misassigning of paternity and the required sample sizes would be expected to be considerably higher in plants than for comparable studies in animals with internal fertilization. Further, studies suggest that parentage tests can underestimate or overestimate the often extensive amounts of gene flow from other populations, depending on marker type, which can lead to biased estimates of parentage (Gerber *et al.* 2000). Improved statistical techniques including Bayesian inference (Emery *et al.* 2001), which allow for a proportion of unknown fathers or fathers and mothers, may allow paternity analysis in populations with extensive pollen and gene flow.

Lack of resources

Molecular biology is very expensive, both in researcher time, consumables and equipment. In such a field, it is inevitable that funding becomes an issue of importance. Allozymes are the cheapest of all available molecular techniques in both consumables and time, and it is potentially possible to utilize allozymes as part of a studentship or other research with limited consumable funding. Once more sophisticated DNA-based molecular genetics techniques are employed, research grant support becomes more important. Success rates for NERC non-thematic grants continue to decrease, particularly at the highest graded levels (Anon 2001). This suggests that one explanation for a decrease in diversity of questions addressed by molecular markers could be that research grants asking innovative questions are more likely to be rejected by funding bodies. Alternatively, researchers might be playing a safe strategy by submitting applications that may be more ambitious in terms of applying novel technologies developed in other fields, but more conservative in terms of the biological questions asked. Carrying out manipulative experiments, particularly those based in the field, are usually very time and resource consuming. Performing manipulative field experiments with subsequent analysis using molecular markers clearly becomes difficult to accomplish within the normal 3-year span of a research grant, and may also be perceived as too risky a strategy, both by researchers and by reviewers or grant-giving bodies. Longer timespans for research grants might assist in this area.

When developing a technique for a new organism, researchers need to consider not only running costs but also development costs. For many marker types, techniques developed for one species can be directly applied to others with a small amount of further development time (e.g. allozymes, Hamrick & Godt 1990) or with a small investment in optimization (e.g. AFLPs, Mueller & Wolfenbarger 1999). Some techniques, in particular microsatellites, require a great investment in both resources and time for initial development (Parker *et al.* 1998), with no guarantee of obtaining the required markers (A. Smithson, personal observation). Such a high investment in initial development costs and time may force researchers to publish more methodological papers (most 2000 methods papers reported new microsatellite primers), encourage 'hypersplitting' of results amongst papers or to focus the following work on descriptive but safe studies that are straightforward and quick to publish.

As discussed above, analysis of paternity in plant populations is more difficult than for animals, as the potential paternity pool is so large. Thus analyses of plant populations, particularly at a fine scale, are likely to require high sample sizes. Given that the costs and time involved in screening large numbers of samples are high, this may directly constrain some studies. Hopefully, new high-throughput screening technology will soon be developed that are sufficiently cost effective for plant ecologists and evolutionary biologists to allow such studies to be pursued efficiently.

Difficulties in obtaining training

The techniques involved in applying molecular methods to plant ecological and evolutionary questions are complex, and clearly require researchers to be trained. In the UK, obtaining training should, however, be relatively simple, due to the NERC Molecular Genetics Facility in Sheffield, which aims to provide facilities and training for students and researchers falling within the NERC remit. We have noticed that the applications to use this facility are principally from researchers working on animals, particularly birds, with the numbers of applications from plant scientists being very limited. Indeed, only four out of 40 applications to the Sheffield facility have been for projects involving plant species. It seems, therefore, that the availability of training, within the UK at least, is unlikely to be a factor limiting the application of neutral molecular markers in plant evolution and ecology. Indeed, we feel that, from our experience of student training, it is possibly more essential to ensure that learning tricky molecular genetics techniques does not become so all-consuming that training in experimental techniques and testing ecological and evolutionary hypotheses do not become forgotten!

Lack of researcher interest

Are researchers more interested a priori in descriptive studies in plants than they are in animals? One possibility is that plants have an 'image problem'—they maintain a fixed position in a population, and thus it might be perceived that plant ecology, in particular plant sex, is determined by non-dynamic processes dependent on simple spatial factors such as distance from pollen donor. Animals, particularly of the cute and fluffy variety often featured in TV wildlife programmes, are highly mobile, are perceived to interact in complex ways with other organisms, and in particular they have interesting sex lives! Despite these prejudices, plant sex is, to us, fascinating, in that irrespective of their stationary position plants manipulate and sometimes deceive the organisms they interact with (e.g. Smithson & Gigord 2001). We have no doubt that it is easier to interest students in animals than plants, and the number of animal population biologists in the UK certainly outnumber their plant counterparts. Perhaps we need a 'plants are interesting' campaign?

It is clear that the pressure on researchers to publish their results has increased, particularly in the UK since the advent of the Research Assessment Exercise. The observation that increasingly sophisticated molecular techniques are being used to answer a narrower range of questions leads us to ask whether this field is currently being driven by hypothesis-led science, technique-led science or easy-paper-based science? As discussed above, new techniques require a large investment in resources, both in terms of equipment and consumables and also in terms of researcher time. Is the drive for publication combined with the unexpectedly long and expensive development times leading to researchers answering straightforward descriptive questions that will guarantee a paper? Particular problems in this respect may relate to the 3-year periods of funding normally awarded by grant-giving bodies, as discussed above. If the increase in descriptive studies at the present time is due to easy-paper-based science, will more hypothesis-driven experimental questions emerge in the future once new techniques are perfected? Or, in the pursuit of further funding and easy papers, will we merely continue to move in the direction of the latest new techniques?

Where now?

The use of neutral molecular markers in plant evolution and ecology has grown rapidly over the last 10 years, and increasingly sophisticated molecular techniques are being used by researchers. However, the diversity of questions addressed by researchers has decreased, and there has been increasing focus on more descriptive questions with a very small proportion of studies incorporating experimental analyses. We have attempted to suggest reasons why researchers may not be applying neutral molecular markers more widely in evolution and ecology. We argue that there is no evidence that lack of appropriate molecular markers is restricting their wider application, and indeed we are currently suffering more from an 'embarrassment of riches' in terms of the number of available techniques. There is also no strong evidence for deficiencies in available training, at least within the UK. We suggest some areas where researchers may be experiencing difficulties with analysis of molecular data, and much could be done by theoreticians to assist the mathematically challenged in this area. Funding is clearly a limiting factor in an expensive field, although in our experience researcher time may be equally limiting, particularly when large sample sizes are essential to ensure robust results. Further development of high-throughput techniques that are affordable for plant evolutionary biologists and ecologists are likely to prove highly beneficial. For some researchers, the cost and length of time required for the development of new molecular markers to answer interesting questions may be prohibitive. The time required to accomplish both experiments and molecular analysis may also be viewed as too time consuming or risky. Thought should perhaps be given by grant-giving bodies to fund promising but challenging studies for 5 years rather than 3, to allow greater chances of successful completion of the work. It is also clear that there is a danger that the development of techniques may be driving the allocation of resources, rather than the quality of the questions being asked. As techniques become more sophisticated, it may become increasingly difficult for any one individual or institution to possess both the technical ability and ecological insight to use these techniques to their best advantage. It is to be hoped that more consortia of researchers with complementary skills will be funded, and the new NERC policy of providing funds specifically for consortium projects is to be welcomed. However this should not be at the expense of other research funding, nor limit the originality of individual researchers. As an alternative, the central provision of resource-intensive equipment to be shared amongst departments could be more widely considered. Finally, it is clear that these new techniques, including the 'omics technologies, QTL mapping, etc., will vastly increase our ability to investigate the deep processes of gene function underlying ecological and evolutionary processes. However, we must not lose sight of the fact that there are many fundamental processes that cannot be answered by these approaches, and researchers and funders must recognize that these are equally worthy of support. Our particular interest lies in unravelling the 'black-box' that lies between the production of pollen in anthers and the germination of a viable seed, and which determines the fitness of a plant through male function. The study of the genes underlying pollen production, floral form or even bumble-bee brains will tell us very little about the behavioural and ecological factors which influence this process in insect pollinated plants. This is but one example of many higher order interactions in ecology that need to be understood, and which can benefit from the intelligent use of modern techniques, but where the question must be more important than the technique adopted.

Acknowledgements

Ann Smithson was funded by NERC Research Fellowship GT5/98/12/TS. Our funding on the use of molecular markers in plant evolutionary ecology has been supported by NERC standard grants GR3/12106, GR3/11052, GR9/2554 and GR9/3507.

References

Anon (2001) *Natural Environment Research Council Annual Report 2000–2001.*

Avise, J.C. (1994) *Molecular Markers, Natural History and Evolution.* Chapman & Hall, New York.

Bachmann, H.D. & Hombergen, E.J. (1997) From phenotype via QTL to virtual phenotype in *Microseris* (Asteraceae): predictions from multilocus marker genotypes. *New Phytologist* **137**, 9–18.

Berry, R.J., Crawford, T.J. & Hewitt, G.M. (1992) *Genes in Ecology.* Blackwell Scientific Publications, Oxford.

Bonnin, I., Prosperi, J.M. & Olivieri, I. (1996) Genetic markers and quantitative genetic variation in *Medicago truncatula* (Leguminosae): a comparative analysis of population structure. *Genetics* **143**, 1795–1805.

Bossart, J.L. & Prowell, D.P. (1998) Genetic estimates of population structure and gene flow: limitations, lessons and new directions. *Trends in Ecology and Evolution* **13**, 202–206.

Bradshaw, H.D., Otto, K.G., Frewen, B.E., McKay, J.K. & Schemske, D.W. (1998) Quantitative trait loci affecting differences in floral morphology between two species of monkeyflower (*Mimulus*). *Genetics* **149**, 367–382.

Burke, T., Davies, N.B., Bruford, M.W. & Hatchwell, B.J. (1989) Parental care and mating behaviour of polyandrous dunnocks *Prunella modularis* related to paternity by DNA fingerprinting. *Nature* **338**, 249–251.

Coltman, D.W., Bancroft, D.R., Robertson, A. *et al.* (1999) Male reproductive success in a promiscuous mammal: behavioural estimates compared with genetic paternity. *Molecular Ecology* **8**, 1199–1209.

Comes, H.P. & Abbott, R.J. (1999) Population genetic structure and gene flow across arid versus mesic environments; a comparative study of two parapatric *Senecio* species from the Near East. *Evolution* **53**, 36–54.

Coulson, T.N., Pemberton, J.M., Albon, S.D. *et al.* (1998) Microsatellites reveal heterosis in red deer. *Proceedings of the Royal Society of London Series B.* **265**, 489–495.

Cruzan, M. (1998) Genetic markers in plant evolutionary ecology. *Ecology* **79**, 400–412.

Demesure, B., Sodzi, N. & Petit, R.J. (1995) A set of universal primers for amplification of polymorphic non-coding regions of mitochondrial and chloroplast DNA in plants. *Molecular Ecology* **4**, 129–131.

Dow, B.D & Ashley, M.V. (1996) Microsatellite analysis of seed dispersal and parentage of saplings in bur oak, *Quercus macrocarpa. Molecular Ecology* **5**, 615–627.

Dow, B.D. & Ashley, M.V. (1998) High levels of gene

flow in bur oak revealed by paternity analysis using microsatellites. *Journal of Heredity* **89**, 62–70.

Ellegren, H. & Sheldon, B.C. (1997) New tools for sex identification and the study of sex allocation in birds. *Trends in Ecology and Evolution* **12**, 255–259.

Emery, A.M., Wilson, I.J., Craig, P.R., Boyle, P.R. & Noble, L.R. (2001) Assignment of paternity groups without access to parental genotypes: multiple mating and developmental plasticity in squid. *Molecular Ecology* **10**, 1265–1278.

Ennos, R.A. (1994) Estimating the relative rates of pollen and seed migration among plant populations. *Heredity* **72**, 250–259.

Estoup, A. & Cornuet, J. (1999) Microsatellite evolution: inferences from population data. In: *Microsatellites: Evolution and Applications* (eds D.B. Goldstein & C. Schlötterer), pp. 49–65. Oxford University Press, Oxford.

Freville, H., Justy, F. & Olivieri, I. (2001) Comparative allozyme and microsatellite population structure in a narrow endemic plant species, *Centaurea corymbosa* Pourret (Asteraceae). *Molecular Ecology* **10**, 879–889.

Ge, X.J. & Sun, M. (1999) Reproductive biology and genetic diversity of a cryptoviviparous mangrove *Aegiceras corniculatum* (Myrsinaceae) using allozyme and intersimple sequence repeat (ISSR) analysis. *Molecular Ecology* **8**, 2061–2069.

Gerber, S., Mariette, S., Streiff, R., Bodénès C. & Kremer A. (2000) Comparison of microsatellites and amplified fragment length polymorphism markers for parentage analysis. *Molecular Ecology* **9**, 1037–1048.

Hamrick, J.L. & Godt, M.J.W. (1990) Allozyme diversity in plant species. In: *Plant Population Genetics, Breeding and Genetic Resources* (eds A. H. D. Brown, M. T. Clegg, A. L. Kahler & B. S. Weir), pp 43–63. Sinauer, Sunderland, MA.

Harder, L.D. & Barrett, S.C.H. (1995) Mating cost of large floral displays in hermaphrodite plants. *Nature* **373**, 512–515.

Jarne, P. & Lagoda, P.J.L. (1996) Microsatellites, from molecules to populations and back. *Trends in Ecology and Evolution* **11**, 424–429.

Kashi, Y. & Soller, M. (1999) Functional roles of microsatellites and minisatellites. In: *Microsatellites: Evolution and Applications* (eds D.B. Goldstein & C. Schlötterer), pp. 10–23. Oxford University Press, Oxford.

Lagercrantz, U., Ellegren, J. & Andersson, L. (1993) The abundance of various polymorphic microsatellite motifs differs between plants and vertebrates *Nucleic Acids Research* **21**, 1111–1115.

Latta, R.G. & Mitton, J.B. (1999) Historical separation and present gene flow through a zone of secondary contact in ponderosa pine. *Evolution* **53**, 769–776.

Lewontin, R.C. (1985) Population genetics. In: *Essays in Honour of John Maynard Smith* (eds P. J. Smith Greenwood, P. H. Harvey & M. Slatkin), pp. 3–18. Cambridge University Press, Cambridge.

Lin, J.Z. (2000) The relationship between loci for mating system and fitness-related traits in *Mimulus* (Scrophulariaceae): a test for deleterious pleiotropy of QTLs with large effects. *Genome* **43**, 628–633.

Linhart, Y.B. & Grant, M.C. (1996) Evolutionary significance of local genetic differentiation in plants. *Annual Review of Ecology and Systematics* **27**, 237–277.

Mitchell Olds, T. (1996) Genetic constrains on life-history evolution: quantitative trait loci influencing growth and flowering in *Arabidopsis thaliana. Evolution* **50**, 140–145.

Möller, A.P. (1997) Immune defence, extra-pair paternity and sexual selection in birds. *Proceedings of the Royal Society of London Series B.* **264**, 561–566.

Mueller, U.G. & Wolfenbarger L.L. (1999) AFLP genotyping and fingerprinting. *Trends in Ecology and Evolution* **14**, 389–394.

Oddie, K.R. (2000) Size matters: competition between male and female great tit offspring. *Journal of Animal Ecology* **69**, 903–912.

Parker, P. G., Snow, A. A., Schug, M. D., Booton, G. C. & Fuerst, P. A. (1998) What can molecules tell us about populations: choosing and using a molecular marker. *Ecology*, **79**, 361–382.

Petrie, M., Doums, C. & Möller, A.P. (1998) The degree of extra-pair paternity increases with genetic variability. *Proceedings of the National Academy of the United States of America* **95**, 9390–9395.

Provan, J., Powell, W. & Hollingsworth, P.M. (2001)

Chloroplast microsatellites: new tools for studies in plant ecology and evolution. *Trends in Ecology and Evolution* **16**, 142–147.

Reed, D.H. & Frankham, R. (2001) How closely correlated are molecular and quantitative measures of genetic variation? A meta-analysis. *Evolution* **55**, 1095–1103.

Richards, C.M. (2000) Inbreeding depression and genetic rescue in a plant metapopulation. *American Naturalist* **155**, 383–394.

Sia, E.A., Jinks-Robertson, S. & Petes, T.D. (1997) Genetic control of microsatellite stability. *Mutation Research* **383**, 61–70.

Smithson, A. & Gigord, L.D.B. (2001) Are there fitness advantages in being a rewardless orchid? Reward supplementation experiments with *Barlia robertiana*. *Proceedings of the Royal Society of London Series B* **268**, 1435–1441.

Snow, A.A. & Lewis, P.O. (1993) Reproductive traits and male fertility in plants: empirical approaches. *Annual Review of Ecology and Systematics* **24**, 331–352.

Sunnucks, P. (2000) Efficient genetic markers for population biology. *Trends in Ecology and Evolution* **15**, 199–203.

Sunnucks, P., Wilson, A.C.C., Beheregaray, L.B. *et al.* (2000) SSCP is not so difficult: the application and utility of single-stranded conformation polymorphism in evolutionary biology and molecular ecology. *Molecular Ecology* **9**, 1699–1710.

Van der Schaar, W., Alonso Blanco, C., Leon Kloosterzeil, K.M., Jansen, R.C. & Koornneef, M. (1997) QTL analysis of seed dormancy in *Arabidopsis* using recombinant inbred lines and MQM mapping. *Heredity* **79**, 190–200.

Vekemans, X., Beauwens, T., Lemaire, M. & Roldán-Ruiz, I. (2002) Data from amplified fragment length polymorphism (AFLP) markers show indication of size homoplasy and of a relationship between degree of homoplasy and fragment size. *Molecular Ecology* **11**, 139–151.

Whittingham, L.A. & Dunn, P.O. (2000) Offspring sex ratios in barn swallows: females in better condition produce more sons. *Molecular Ecology* **9**, 1123–1129.

Willson, M.F. (1994) Sexual selection in plants: perspective and overview. *American Naturalist* **144**, S13-S39.

Part 2

Microbial community dynamics in soils and water

Chapter 8

The role of genomics in methylotrophic bacteria for understanding biogeochemical cycling

*L. Chistoserdova and M. E. Lidstrom**

Introduction

The availability of genome sequences for both prokaryotes and eukaryotes is laying the foundation for a revolution that will ultimately transform biology from a largely descriptive and reductionist discipline into a fundamentally predictive science. The genomic-based insights that are already beginning to be gained into the cell as a complex dynamic system will translate into a greater understanding of the environmental role of specific functional types of bacteria, when coupled to studies that target evolutionary relationships and metabolic diversity. We are interested in one such functional type, the methylotrophic bacteria. These bacteria are important in global and local biogeochemical cycling of carbon, nitrogen, sulfur and halogenated compounds. The increasing availability of methylotrophic genome sequences is providing a major opportunity to address evolutionary and metabolic diversity of methylotrophic bacteria in natural habitats. Comparative genomics approaches are expected to lay the groundwork for environmental genomic studies of methylotrophic communities. Such studies will identify a core set of genes whose expression will be diagnostic for methylotrophic function in natural environments, and a set of accessory genes whose expression will be diagnostic for specific subgroups of methylotrophs. Conversely, approaches that link our understanding of diagnostic metabolic genes to partial genomic sequencing of environmental DNA will further our knowledge of the metabolic diversity of methylotrophic bacteria in natural habitats.

Methylotrophic bacteria

Methylotrophic bacteria are capable of utilizing one-carbon compounds as their sole carbon and energy source (Anthony 1982; Lidstrom 2001). The ability to grow aerobically on reduced one-carbon compounds is fairly widespread throughout the

* Correspondence author: lidstrom@u.washington.edu

bacteria. Both gram-positive and gram-negative methylotrophs have been isolated and studied (Anthony 1982; Lidstrom 2001), and the gram-negative methylotrophs span the alpha, beta and gamma subgroups of the *Proteobacteria* (Hanson *et al.* 1993; Lidstrom 2001).

Importance and distribution of methylotrophic bacteria

A number of C_1 compounds commonly found in the environment are consumed by aerobic methylotrophs, including methane, methanol, methylated amines such as methylamine and trimethylamine, methylated glycines such as betaine and sarcosine, halomethanes, and methylated sulfur compounds such as dimethylsulfide and methanesulfonate (King 1992; Kelly & Murrell 1999; Lidstrom 2001; Miller *et al.* 2001). Many of these compounds are important greenhouse gases, and in some cases methylotrophs are the main biological sink for these gases, controlling the net emission to the atmosphere (King 1992, Oremland & Culbertson 1992; Miller *et al.* 2001). Methylotrophs are widespread in nature, occurring in a variety of freshwater, marine, and terrestrial habitats, including acidophilic, alkaline, halophilic, psychrophilic and thermophilic environments (King 1992; Kalyuzhnaya *et al.* 1999; Dedysh *et al.* 2000; Doronina *et al.* 2000). In some environments, methylotrophs can be a significant portion of the total aerobic bacteria, contributing significantly to the total oxygen consumption at specific sites and effecting major biogeochemical cycling under certain conditions (King 1992). For instance, during times of turnover in stratified lakes, massive methane oxidation and O_2 consumption occur (Rudd & Taylor 1980), which contribute substantially to short-term trophic dynamics in these environments and can result in suboxic conditions in the water column. In addition, many methylotrophs contain nitrogenase and are predicted to play a significant role in some habitats in nitrogen cycling (King 1992).

Very little information is available on environmental distribution of methylotrophic groups other than methanotrophs. However, a number of recent studies have targeted methanotrophs (Murrell & Radajewski 2000; Eller *et al.* 2001). In general, methanotrophs and methane oxidation are present wherever methane and oxygen meet, often at a narrow interface with steep concentration gradients. Most terrestrial and aquatic environments appear to be dominated by strains containing the ribulose monophosphate pathway for assimilation (γ-*Proteobacteria*), but organic-rich and/or acidophilic environments such as rice paddies and peat bogs are dominated by strains containing the serine cycle for assimilation (α-*Proteobacteria*). The reason for this difference is not yet understood, but presumably it reflects the different physiological capabilities of these two phylogenetically separate groups.

Recently uncovered diversity of methylotrophic bacteria

Recent interest in methylotrophs as important participants in the global carbon cycle (DeLong 2000), as well as natural sinks for halogenated methanes (Schaefer & Oremland 1999; Vanelli *et al.* 1999), has stimulated progress in isolation and characterization of novel methylotrophic strains from a variety of natural environments.

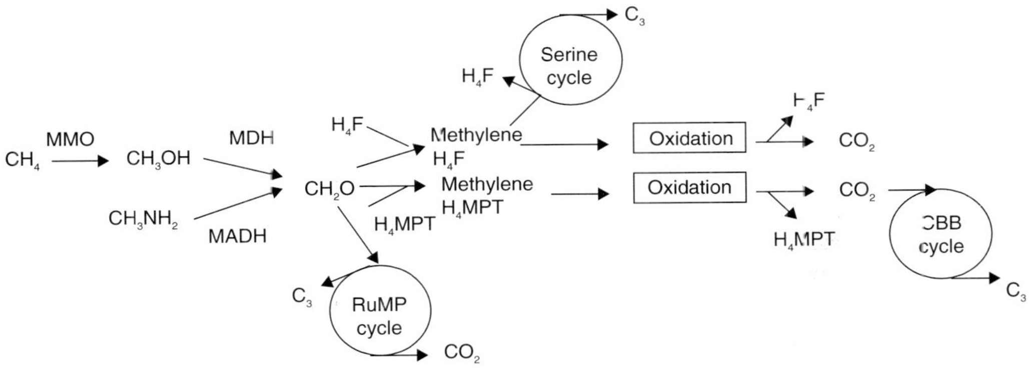

Figure 8.1 Major C_1 pathways in methylotrophs used in this study. The tetrahydrofolate (H_4F)-linked pathway for formaldehdye oxidation functions in the serine cycle facultative methylotrophs, the tetrahydromethanopterin (H_4MPT)-linked pathway functions in most gram-negative methylotrophs. MMO, methane monooxygenase; MDH, methanol dehydrogenase; MADH, methylamine dehydrogenase.

These new isolates have considerably broadened our knowledge about the variety of natural habitats of methylotrophs, as well as the wealth of their specialization and diversity. Interesting recent methylotrophic isolates include rhizobial strains, both symbiotic and free living (Connell Hancock *et al.* 1998; Sy *et al.* 2001), methylotrophic halophiles (Doronina *et al.* 2000), acidophilic (Dedysh *et al.* 2000) and psychrophilic methylotrophs (Kalyuzhnaya *et al.* 1999). Substrate specialization of these novel isolates, besides methanol and methane, includes chloromethane, bromomethane, dichloromethane and methanesulfonic acid. Some of the novel isolates do not cluster with previously characterized methylotrophs, based on 16S rRNA sequences, but they rather cluster with other specialists adapted to certain natural environments. For example, an acidophilic methanotroph species clusters with the non-methylotrophic acidophiles *Beijerinckia* and *Acidobacterium* (Dedysh *et al.* 2000). Representatives of these diverse groups of methylotrophs provide a significant opportunity to study methylotrophic functions in divergent bacterial groups that occupy a variety of ecological niches.

Methylotrophic metabolism

General metabolism

Methylotrophs obtain energy and reducing power by oxidizing methylated compounds to carbon dioxide via a series of two-electron steps (Anthony 1982; Lidstrom 2001; Figure 8.1). The primary C_1 substrates methane, methanol and methylamine are oxidized by, respectively, methane monooxygenase (MMO), methanol dehydrogenase (MDH) and methylamine dehydrogenase (MADH) (Anthony 1982; Lidstrom 2001). Carbon is assimilated at the level of formaldehyde through one of two unique pathways: the ribulose monophosphate pathway or the serine cycle, or, in the case of autotrophic methylotrophs, carbon is assimilated at the level of carbon dioxide via the ribulose bisphosphate (Calvin–Benson–Bassham) cycle (CBB cycle) (Figure 8.1; Anthony 1982; Lidstrom 2001). Methylotrophs are an unusual group of prokaryotes, in that all of their cell carbon for both assimilatory and dissimilatory processes flows through a toxic intermediate, formaldehyde. Although these bacteria face a number of metabolic challenges, including building all of their cell carbon from one-carbon units and extracting energy from simple one-carbon compounds, the most difficult of these is how to handle the flow of carbon through a toxic intermediate like formaldehyde.

Different methylotrophs have specialized metabolic capabilities that place them in physiological subtypes. For instance, MMO is found in methane-oxidizing bacteria, MDH is universally found in all gram-negative methylotrophs, an NAD-linked methanol dehydrogenase is found in the gram-positive methylotrophs, and MADH is present in methylotrophs capable of growth on methylamine (Anthony 1982; Lidstrom 2001). The serine cycle is used by the α-proteobacterial methylotrophs, which are either obligate methane utilizers or facultative methanol-utilizers (Lidstrom 2001). The ribulose monophosphate cycle (RuMP) cycle operates in obligate methanol-utilizers belonging to the β-*Proteobacteria*, in obligate methane-utilizers of the γ-*Proteobacteria*, and in gram-positive methylotrophs (Lidstrom 2001).

Formaldehyde oxidation

Formaldehyde oxidation is more complex. In collaboration with R. Thauer, J. Vorholt and colleagues, we recently discovered a novel formaldehyde oxidation pathway in methylotrophs, involving a cluster of genes encoding enzymes that are similar in sequence and function to the first three steps of the CO_2 reduction pathway found in archaeal methanogens (Figure 8.2; Chistoserdova *et al.* 1998; Vorholt *et al.* 1999). This pathway utilizes a co-factor previously thought to be specific to methanogens and related archaea, tetrahydromethanopterin (H_4MPT) and it has been suggested that this is an example of functional pathway acquisition by lateral gene transfer (Chistoserdova *et al.* 1998). H_4MPT is analogous to H_4F and the reactions carried out are similar to H_4F-dependent reactions (Figure 8.2). This archaeal-like oxidation pathway is found in almost all methylotrophs, being absent only in some autotrophic methylotrophs (e.g. *Paracoccus* and *Rhodobacter*), which employ a glutathione-linked formaldehyde oxidation pathway (Harms *et al.* 1996; Barber & Donohue 1998), and gram-positive methylotrophs which possess a mycothiol-dependent formaldehyde dehydrogenase (Norin *et al.* 1997). An additional formaldehyde oxidation pathway exists in the β-proteobacterial methylotrophs, which involves a cyclic version of the RuMP pathway (Anthony 1982). The serine cycle facultative methylotrophs use the H_4F- and the H_4MPT-linked pathways simultaneously, while the β-proteobacterial methylotrophs use the H_4MPT-linked pathway in addition to the dissimilatory RuMP cycle (Vorholt *et al.* 1999; Chistoserdova *et al.* 2000). In most obligate methanotrophs and in the autotrophic

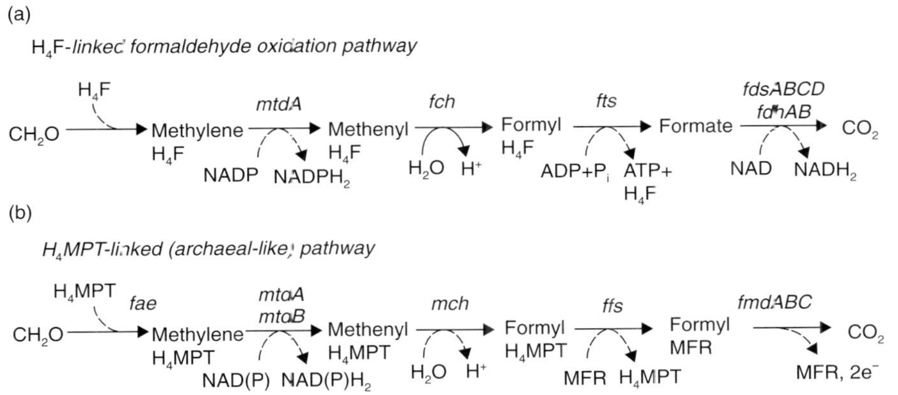

Figure 8.2 (a,b) Reactions of the two alternate formaldehyde oxidation pathways operating in methylotrophs and the respective genes, as found in *Methylobacterium extorquens* AM1. *fae*, formaldehyde activating enzyme; *mtdA*, methylene H_4MPT/methylene H_4F dehydrogenase (NADP); *mtdB*, methylene H_4MPT dehydrogenase (NAD(P)); *fch*, methenyl H_4F cyclohydrolase; *mch*, methenyl H_4MPT cyclohydrolase; *fts*, formyl H_4F synthase; *ffs*, formyl methanofuran (MFR)-H_4MPT formyltransferase; *fdsABCD*, *fdhAB*, subunits of the NAD-linked formate dehydrogenases; *fmdABC*, subunits of formyl MFR dehydrogenase.

Xanthobacter strains, the H_4MPT pathway is the only formaldehyde-oxidizing pathway (Vorholt *et al.* 1999).

Mutants in these pathways are unable to grow on C_1 compounds due to the inability to generate energy from formaldehyde oxidation, but in addition they are extremely sensitive to formaldehyde indicating a major role in formaldehyde detoxification (Chistoserdova *et al.* 1998, 2000). Our results so far suggest that formaldehyde handling by these oxidative pathways is the crucial function in methylotrophy for allowing the high flux of formaldehyde necessary to grow aerobically on C_1 compounds. It seems likely that acquisition of an effective formaldehyde oxidation pathway was the enabling step for the development of aerobic methylotrophy as a metabolic mode. This hypothesis is supported by evidence suggesting the H_4MPT-linked pathway may be an ancient acquisition (see below).

Methylotrophic genetics

Genetic tools have been available for methylotrophs for decades and their use has contributed substantially to our understanding of methylotrophic metabolism. Broad-host-range vectors, conjugation and electroporation have been applied successfully to carry out studies of genes and gene expression, and directed insertional mutagenesis using allelic exchange techniques is routine (Chistoserdov *et al.* 1994; Harms *et al.* 1996; Lidstrom *et al.* 1998; Serebrijski *et al.* 1998; Vanelli *et al.* 1999). However, the only vectors that were successful in most methylotrophs were large, cumbersome and unsequenced. Recently, a suite of small, completely sequenced

IncPI vectors has been developed for cloning, promoter identification and analysis, and expression (Marx & Lidstrom 2001). Application of these genetic techniques to the study of methylotrophy has resulted in the identification of over 100 genes in *Methylobacterium extorquens* AM1, approximately 75 of which have been shown to be involved in at least one of the branches of methylotrophic metabolism (Chistoserdova *et al.* 1998; Lidstrom *et al.* 1998). Although no complete physical map is yet available for the chromosome, approximately 200 kb of DNA have been mapped with regard to the 18 known methylotrophy clusters (L. Chistoserdova, Korotkova and M.E. Lidstrom, unpublished results). Pulsed field gel electrophoresis (PFGE) analysis and also genome sequence analysis have shown that *M. extorquens* AM1 contains a chromosome of 7.5 Mbp, and previous work had shown the presence of two plasmids (65 and 33 kb) of unknown function (Kim & Lidstrom 1989).

Clustering of methylotrophy genes

Genes involved in different segments of methylotrophy are often found clustered on the chromosomes of methylotrophs. For instance, the genes involved in the production of methanol dehydrogenase and methylamine dehydrogenase are clustered and the order is generally conserved in a variety of methylotrophs (Chistoserdov *et al.* 1994; Gak *et al.* 1997; Lidstrom *et al.* 1998; Lidstrom 2001). The bacterium that has been most well studied in this regard is *M. extorquens* AM1, an α-proteobacterial serine cycle facultative methylotroph. Eighteen methylotrophy islands have been found so far. Eight of them contain one methylotrophy gene each, while 10 others contain groups of tightly linked methylotrophy genes. The largest methylotrophy island covers about 61 kb of the chromosome and contains 61 genes, 31 of which, when mutated, produce a C_1-minus phenotype (Chistoserdova *et al.* 1998; Lidstrom *et al.* 1998; L. Chistoserdova & M.E. Lidstrom, unpublished data). In *Methylobacillus flagellatus* KT, at least four genes for the RuMP cycle are found tightly linked (Serebrijski *et al.* 1998). Some genes of the H_4MPT-linked pathway for formaldehyde oxidation are clustered in methylotrophs studied so far, and some conservation of gene order is found (Chistoserdova *et al.* 1998; Vorholt *et al.* 1999). An analysis of the *M. capsulatus* 3X sequence reveals that these genes are clustered and ordered in a similar way in that methanotroph (see below).

Hypothesis of lateral transfer (LGT) of archaeal-like genes

Genes encoding enzymes of the H_4MPT-mediated C_1 transfer reactions (see Figure 8.2) have only been found in archaeal methanogens, *Archaeoglobus* (an archaeal sulfate reducer) and methylotrophic bacteria so far. They are not recognized in other available genome sequences including those of other *Proteobacteria* and other archaea. This result makes the hypothesis that this major formaldehyde oxidative pathway arose in methylotrophs as a result of LGT from an archaeon an attractive one. A scenario in which these genes were present in the last common ancestor before separation of archaea and bacteria, and then were lost without a trace from all lineages of bacteria except for methylotrophs, seems less likely. None of the archaea for which genome sequences are available could make a prospective donor of the

H_4MPT pathway gene cluster, as these genes are scattered in those genomes. However, when more organisms are included in the comparisons, missing links in a chain of events that happened long ago may emerge. A recent discovery of a novel archaeaon (a *Methanosarcina* species) that does not produce methane but instead oxidizes it (Boetius *et al.* 2000), most probably in a manner similar to methylotrophic bacteria, may turn out to show different gene clustering.

Genomics in methylotrophs

Genomics in methylotrophs has recently become a reality, with two (so far incomplete) methylotrophic genomes now publicly available, of the α-proteobacterial methylotroph *M. extorquens* AM1 (University of Washington) and of the γ-proteobacterial methanotroph *Methylococcus capsulatus* Bath (The Institute for Genomics Research), long known model methylotrophs (Anthony 1982). Although limited to just two strains so far, methylotrophic genomics have the strong advantage of being combined with extensive functional analysis in these bacteria, therefore providing an efficient means for identification of genes that play key roles in methylotrophic pathways. In addition, comparative genomic analysis will allow us to distinguish between the core methylotrophy functions shared by different methylotrophs and functions specialized in different groups of methylotrophs. One output of such analyses is identification of potential targets for novel environmental probes, to become test genes for either methylotrophs in general or for specific groups of methylotrophs.

Comparative genomic analyses

So far, initial comparative analysis of the two methylotroph genomes confirms the pre-existing biochemical, physiological and genetic evidence suggesting that the two phylogenetically separate methylotrophs are distinct and only share a few core methylotrophy functions (Anthony 1982). Such functions fall into two different categories, based on protein sequence conservation. Functions such as the MDH complex including MDH subunits, a specific cytochrome and a number of accessory proteins represent the first group, with high protein conservation (>70% at amino acid level) indicative of a relatively recent evolution. The second group of shared functions consists of enzymes with low to very low conservation (20–70%), represented primarily by the 'archaeal' functions in the H_4MPT-linked formaldehyde oxidative pathway, functions suspected to be involved in archaeal co-factor biosynthesis, other archaeal-like functions that may be involved in formaldehyde handling, and H_4F-linked C_1 transfer functions (*mtdA*, *fch*, see below).

Despite the low protein conservation, clustering of the core archaeal genes in the two organisms is remarkably conserved (Figure 8.3), albeit bearing signs of separate evolution. An example is *mtdB*, a non-archaeal gene encoding a novel methylene-H_4MPT dehydrogenase involved in the formaldehyde oxidation pathway (see Figure 8.2). This gene is located in the cluster of core archaeal genes in *M. extorquens* AM1 and apparently is co-transcribed with *mch*, which encodes the step in the pathway

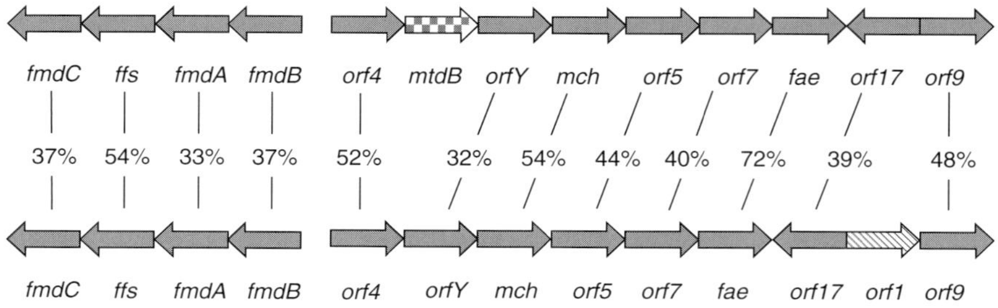

Figure 8.3 Alignment of archaeal-like gene clusters in *Methylobacterium extorquens* AM1 (top) and *Methylcoccus capsulatus* (bottom). % refer to amino acid identity for translated polypeptides. Checkered arrow represents a deletion, striped arrow represents an insertion in the *M. capsulatus* cluster with respect to the *M. extorquens* AM1 cluster.

immediately following the MtdB reaction (Chistoserdova *et al.* 1998). By contrast, two copies of *mtdB* are found elsewhere in the *M. capsulatus* genome and not linked to *mch* (L. Chistoserdova and M. E. Lidstrom, unpublished observations). The clustering in *M. extorquens* AM1 may reflect more recent evolution. It is possible that placing *mtdB* and *mch* under the same promoter might have been advantageous in formaldehyde handling, in which *M. extorquens* AM1 is more proficient than *M. capsulatus*.

While in *M. extorquens* AM1 the archaeal genes form a part of a larger methylotrophy cluster including genes for the assimilatory serine cycle, genes involved in pyrroloquinoline quinone (PQQ) biosynthesis and in regulation of the MDH complex (Lidstrom *et al.* 1998), the archaeal genes in *M. capsulatus* are linked to one of the two copies of pMMO genes (L. Chistoserdova and M. E. Lidstrom, unpublished observations). Clustering of this type may also be reflective of specific selective pressures during the evolution of methylotrophy functions in the two organisms. Such selection may have led to gathering of C_1 genes together on the chromosomes, possibly for more efficient regulation or as a result of acquisition by multiple lateral transfers. As more information becomes available regarding the relationship between gene clustering and metabolic mode, specific sets of genes may become diagnostic for specific physiological types.

Methanopterin-linked C_1 oxidation pathway genes

Comparative analysis of genes involved in archaeal-like functions in M. *extorquens* AM1 and *M. capsulatus* indicate that the genes involved in methanopterin-linked C_1 oxidative reactions, while clustered in a similar way, show different degrees of identity at the amino acid level (from as low as 32% identity for OrfY to 72% identity for Fae). Mutant analysis of the archaeal gene cluster in *M. extorquens* AM1 indicates that some genes in the cluster do not appear to have a function in C_1 metabolism (Chistoserdova *et al.* 1998). These genes (*orfY*, *orf5*) might be the remnants of the suggested lateral transfer from an archaeon *en masse*, and their divergence may be a reflection of mutation not restricted by function. Such genes would be unlikely targets for environmental probing as these may be prone not only to mutations, but deletions as well. Indeed *orfY* was not found upstream of *mch* in *Xanthobacter autotrophicus* (Vorholt *et al.* 1999), and it lacks a start codon in *M. flagellatus* (L. Chistoserdova and M. E. Lidstrom, unpublished observations). Although *orf5* shows more conservation (this gene has also been detected in *M. flagellatus* and *X. autotrophicus*, Vorholt *et al.* 1999), its function in methylotrophy is doubtful, as in *M. flagellatus* the gene is truncated and showing signs of genetic decay. The likely targets for environmental probe development are archaeal genes with known functions in methylotrophy (formaldehyde oxidation for energy or detoxification), such as the genes for the three subunits of formyl-MFR dehydrogenase (*fmdABC*), formyl-MFR-H_4MPT formyltransferase (*ffs*), methylene-H_4MPT dehydrogenase (*mtdB*), methenyl-H_4MPT cyclohydrolase (*mch*) and formaldehyde activating enzyme (*fae*), and these should be the primary targets for probe development. Although the functions of other conserved genes are not known at present (*orf4, orf7, orf17, orf9*; Chistoserdova *et al.* 1998), mutants in these genes generated in *M. extorquens* AM1 have specific phenotypes indicative of their importance for H_4MPT-linked functions. These may be involved in methanopterin biosynthesis or the regulation of the pathway. Therefore these genes are also prospective as potential environmental probe targets.

Folate-linked C_1 oxidation pathway genes

While methylotrophs share the H_4MPT-linked functions with archaeal methanogens, they have developed novel, non-orthologous functions for the H_4F-linked C_1 metabolism. So far, three species of *Methylobacterium* that have been investigated in this regard, as well as *M. capsulatus* (judged from its genome sequence) all contain *mtdA* and *fch* genes encoding, respectively, methylene-H_4F dehydrogenase and methenyl-H_4F cyclohydrolase (see Figure 8.2, Vorholt *et al.* 1998; Pomper *et al.* 1999; L. Chistoserdova, unpublished results; S. Vuilleumier, personal communication). The classical *folD* gene encoding the bifunctional methylene-H_4F dehydrogenase/methenyl-H_4F cyclohydrolase normally found in all bacteria and some archaea is not present in the genomes of *M. extorquens* AM1 or *M. capsulatus*. *folD* is present, however, in the chloromethane-utilizing *Methylobacterium* strain, *M. chloromethanicum* (Vanelli *et al.* 1999), in addition to *mtdA*. In this strain, *folD* is specifically involved in chloromethane utilization, and mutation in this gene does not affect growth on other C_1 substrates (Vanelli *et al.* 1999). Remarkably, MtdA has an additional co-factor specificity besides the one towards H_4F, it also uses H_4MPT, like MtdB (Vorholt *et al.* 1998). Therefore, MtdA appears to have a second function, as an NADP-linked methylene H_4MPT dehydrogenase in the archaeal-like formaldehyde oxidation pathway (see Figure 8.2). It is not clear at this time why both *mtdA* and *mtdB* are involved in this pathway. MtdA and MtdB are paralogous proteins with about 30% identical amino acid residues (Chistoserdova *et al.* 1998). At this time, for the reason of limited data it is not possible to conclude which one arose earlier and which one arose later in their evolution. So far, *mtdA* has only been found in methylotrophs but not in other bacteria or in archaea, therefore *mtdA* might be a good target for environmental probe design. The gene seems to be ancient, as the amino acid conservation between *Methylobacterium* strains and *M. capsulatus* is only about 57%. However, for probe development, information is required on the presence and sequence conservation of this gene from more methylotrophs.

While the evolutionary history of *mtdA* is not known at this point, the other novel gene in the H_4F-linked C_1 pathway, *fch*, may have been present in the last common ancestor, as orthologs of this gene are found in diverse deeply branching bacteria, some archaea and eucaryotes, including mammals (Stanhope *et al.* 2001). The ortholog of *fch* has a different function in higher organisms (a formiminotransferase/cyclodeaminase), while functions of the bacterial and archaeal orthologs in non-methylotrophs are not known. In most of these organisms, *folD* is present, and it is unlikely that *fch* orthologs specialize in the cyclohydrolase function found in methylotrophs. Only two *fch* sequences are available now for methylotrophs, with about 58% identity, and *fch* and *mtdA* are clustered together in both *M. extorquens* AM1 and *M. capsulatus*. When more methylotroph sequences are available, *fch* is a good candidate for development of environmental probes to target diverse methylotrophs.

Environmental probes

Two types of environmental probes have been in use for methylotroph detection, phylogenetic (16S rRNA-based oligonucleotides; McDonald *et al.* 1999; Bourne *et al.* 2000; Murrell & Radajewski, 2000; Eller *et al.* 2001) and functional probes. The latter have been developed so far based on highly conserved enzymes for primary methylotrophic substrate oxidation, such as methane monooxygenase (*mmoX*, *pmoA*; McDonald *et al.* 1999; Holmes *et al.* 1999; Jensen *et al.* 2000; Murrell & Radajewski 2000), corrinoid-linked methyltransferase (*cmuA*; McAnulla *et al.* 2001), methanesulfonic acid monooxygenase (*msmA*; Kelly & Murrell 1999) and methanol dehydrogenase (*mxaF*; Murrell & Radajewski 2000). Both types of probes, although very effective in detecting groups of closely related specialists in specific environments, would not account for the variety and diversity of methylotrophs as a whole in a given environment. In addition, they would not be useful for identifying new types of methylotrophs using novel substrates.

Comprehensive knowledge of microbial populations is important for evaluating the metabolic capacities of specific environmental niches and their dynamics. The results emerging from methylotrophic genomics are starting to provide the opportunities for developing probes with lesser specificity, directed to the broader variety of existent methylotrophs. Such potential probes will be based on ancient methylotrophic enzymes, the enzymes involved in archaeal-like H_4MPT-linked formaldehyde oxidation and H_4F-linked C_1 metabolism mediated by *mtdA* and *fch*. The advantage of such environmental probes, provided they are robust, will be in detecting a specific metabolic capability (formaldehyde handling) that is central to most methylotrophs and is likely to be an enabling function for broad types of methylotrophs, rather than a specific, predicted type of methylotroph. Probes with broad specificity will detect microbes capable of formaldehyde metabolism, be it methylotrophic bacteria belonging to known phylogenetic groups, novel methylotrophs, methylotrophic or methanogenic archaea or microbes not grouping with known methylotrophs based on their 16S rRNA, but possessing such functions.

Methenyl H_4MPT cyclohydrolase (Mch)

The gene encoding methenyl H_4MPT cyclohydrolase (*mch*) has been tested as one of the first targets for detecting a broad range of methylotrophs. Pairs of degenerate oligonucleotide primers have been developed towards regions of high conservation in bacterial and archaeal Mch enzymes (Vorholt *et al.* 1999). These primers were used for the detection of *mch* in a variety of laboratory strains of methylotrophs belonging to α-, β- and γ-*Proteobacteria*, via PCR amplification. Use of these primers produced positive results with most tested methylotrophs. These included facultative methylotrophs with serine cycle or CBB cycle for C_1 assimilation (α-*Proteobacteria*), obligate non-methane-utilizing methylotrophs (β-*Proteobacteria*) and methanotrophs belonging to the γ-*Proteobacteria* (Type I) or α-*Proteobacteria* (Type II). The only exceptions negative for *mch* were some autotrophic methylotrophs, such as *Paracoccus* and *Rhodobacter* in which the GSH-linked formaldehyde oxidation pathway was shown to fulfil the functions in energy generation and formaldehyde detoxification (Harms *et al.* 1996; Barber & Donohue 1998) and gram-positive methylotrophs in which a similar mycothiol-dependent formaldehyde dehydrogenase activity is present (Norin *et al.* 1997). Another exception was a chloromethane-utilizing strain, *Aminobacter* IMB1, which does not possess MDH or MADH and therefore does not produce formaldehyde as a major metabolic intermediate (L. Chistoserdova and M.E. Lidstrom, unpublished observations). PCR amplification results were confirmed by Mch activity measurements (Vorholt *et al.* 1999).

Comparisons of partial Mch sequences indicated significant divergence. A phylogenetic tree built for Mch from representatives of bacterial methylotrophs and methanogenic archaea indicated that: (i) bacterial and archaeal enzymes are clearly separated, implying a long-term evolution of *mch* within bacteria; and (ii) positions of some methylotrophic Mchs on the tree are ambiguous (*Xanthobacter*, *Hyphomicrobium*), suggesting a complex evolution of *mch* within bacteria (Vorholt *et al.* 1999). Recently we tested the degenerate primers designed previously (Vorholt *et al.* 1999) with a number of novel methylotrophic isolates belonging to *Methylomonas*, *Methylocystis*, *Methylobacter*, *Methylosinus*, *Methylocella*, *Methylomicrobium* and *Methylobacterium*, to search for strains possessing more divergent Mch enzymes, as an indicator of potential genomic diversity. A total of 16 new sequences (complete or partial) were obtained, in addition to the 11 previously determined sequences (Vorholt *et al.* 1999), and these were subjected to phylogenetic analysis. This analysis revealed that Mch clustering patterns do not always follow 16S rRNA clustering patterns, further supporting the idea that evolution of the *mch* gene within different groups of methylotrophs has not followed a strictly linear path. Besides *X. autotrophicus* and *H. methylovorum*, two other strains showed significant divergence for Mch sequences, a nodulating methylotroph *Methylobacterium nodulans* and a

```
Methylophilus methylotrophus    CLGSQYAGWSLSHEK----FFSLGSGPGRALAG---------REELYKELGYKD-SADAAVLVLESD-KVPPQEVVE
Methylobacillus flagellatus     CLGSQYAGWSLSHEK----FFSLGSGPGRALAG---------REELYKELGYKD-SADAAVLVLESD-KVPPQEVIE
Strain S1                       CLGSQYAGWSLSHKSEDSKFFALGSGPGRALAA---------REEIFKEFGYKD-EAESTVIVLEVD-SFPPVEVAE
Xanthobacter autotrophicus      CLGSQYAGWSLAAGD----FFALGSGPGRAIAA---------VETLYGELGYRD-RGEKVVLVLETA-VVPPAEVVD
Methylocystis LW2               CLGSQYAGWSLQHEK----FFALGSGPGRAQAR---------VEKLFEELGYKD-HADRVTIVLESG-SPPPKEVVE
Methylocystis LW5               CLGSQYAGWSLSHEK----FFALGSGPGRAQAR---------VEKLFEELSYKD-QAGRVTIVLESG-SPPPKEVVE
Methylosinus LW4                CLGSQYAGWSLSHEK----FFALGSGPGRSLAR---------KEALFQELPYKD-SADRSTIVLEAG-APPPEAVVA
Methylosinus LW8                CLGSQYAGWSLSHEK----FFALGSGPGRSLAR---------KEALFQELPYKD-SADRSTIVLEAG-APPPEAVVA
Methylosinus LW3                CLGSQYAGWSLSHEK----FFALGSGPGRSLAR---------KEALFQELPYKD-SADRSTIVLEAG-APPPEAVVA
Methylosinus PW1                CLGSQYAGWSLSHEK----FFALGSGPGRSLAR---------KEALFQELPYKD-AADRSTIVLEAG-APPPEAVVA
Methylosinus trichosporium      CLASQYAGWSLSHEK----FFALGSGPGRSLAR---------KEPLFQELPYED-SASRATIVLEAG-APPPEPVVA
Methylocella palustris          CLGSQYAGWALSHGEGKGSFFALGSGPARALAQ---------KEPLFHDLGYHD-RAHRATLVLEGDRAPPPE-ITD
Methylobacterium extorquens     CLGSQYAGWSLADEEGDSGFFALGSGPGRAVAV---------VEELYKELGYRD-NATTTALVLESG-SAPPASVVN
Methylobacterium organophilum   CLGSQYAGWSLADETGDSGFFALGSGPGRAVAA---------VEDLYQELGFRD-SATKTALVLEAA-GGPPESVVA
Methylomicrobium 5G             CLGSQYAGWSLSHEK----YYALGSGPARAMAVKIKDGQQEPVEELYKELAYRD-DAENTVLVIEND-AIPPLEIVE
Methylomicrobium 20Z            CLGSQYAGWSLSHEK----YYALGSGPARAMAVKIKDGQQEPVEELYKELAYRD-DAENTVLVIEND-AIPPLEIVE
Methylomicrobium N1             CLGSQYAGWSLSHEK----YYALGSGPARAMAVKIKDGQQEPVEELYKELAYRD-NAENTVLVIEND-AIPPLEIVE
Methylomonas LW13               CLGSQYAGWSLSHGK----YYALGSGPARALATKLKDGAVEPVEELYKELEYRD-SAEQTALVIEND-ALPPIEIVE
Methylomonas LW21               CLGSQYAGWSLSHGK----YYALGSGPARAMATKTKEGVVEPVEELYKELEYRD-SCEKTVLVIEND-ALPPVEIVE
Methylomonas rubra              CLGSQYAGWSLSHGK----YYALGSGPARAMATKVKDGAVEPVEELYKELEYRD-SHDKTVLVIEND-AVPPVEIVE
Methylobacter LW14              CLGSQYAGWSLSHEK----YYALGSGPARALATKLKEGKEESVEELYKELDYRD-RFDSTVLVIEND-AIPPLEIVE
Methylomicrobium albus          CLGSQYAGWSLSHEK----YYALGSGPARAMAAKIKDGNEEPVEELYKELGYRD-SFDSTVLVIEND-AVPPPAIVE
Methylobacter A45               CLGSQYAGWSLSHEK----YYALGSGPARAMATKVKDGKEEPVEELYKELEYRD-EAESTALVIEND-AVPPVELVE
Methylococcus thermophilus      CLGSQYAGWSLSHGEGKNAFYALGSGPARALATKVKDGNEEPVEELYKELGYRD-HSNETAIVMEVD-KVPPVEVIE
Methylococcus capsulatus        CLGSQYAGWSLSHGEGKDGFYALGSGPARAMATKIKDGNEEPVEELYKELNYRD-KHTETAIVMEVD-KVPPVEVTD
Hyphomicrobium methylovotum     ----------TITEETTGFFALGSGPARALSR---------VEDLFKELNYAD-QGPNTALVIEGD-KAPPAAVIE
Methylobacterium nodulans       CLGSQYAGWNLSREK----YFAMGSGPARTLAR---------VEPLFDEIAYREPEAKSSVIVLETT-TPPPVAIVE

Methylophilus methylotrophus    KVARDTGVKAENLTFILTPTRSLAGTVQIVARVLEVALHKIHTLH-FPLEHVVDGMASAPLPPPAPDFLIGMGRTND
Methylobacillus flagellatus     KVARDTGVNPENLTFILTPTRSLAGTVQIVARVLEVALHKIHTLH-FPLEHVVDGAASAPLPPPAPDFLIGMGRTND
Strain S1                       KVANNCGIKPENLTFILTPTSSLAGVMQIAIRVLEVAMHKAHELH-FPLEKIVDGYGATPVAPPGGDFMTAMGRTND
Xanthobacter autotrophicus      EIAARCAVAPSDITLILTPTSSLAGTVQIVARVLEVALHKAHALH-FPLEHIADGVGSAPICPPSPDFLTAMGRTND
Methylocystis LW2               KVASKSGLTPDKVAFIYAPTQSLAGGVQVVGRVLEVALHKAHELH-FPLENIVDGIVTAPLSPPHPDFVQAMGRTND
Methylocystis LW5               KVASKSGVSPDKVAFVYAPTQSLAGGVQVVGRVLEVALHKTHELH-FPLENVVDGIATAPLSPPHPDFVTAMGRTND
Methylosinus LW4                KVANDCGVSPDKLAFVYAPTQSLTGSVQVVGRVLEVALHKAHELK-FPLEHIVDGIATAPLSPPHPDFVTAMGRTND
Methylosinus LW8                KVANDCGVSPDKLAFVYAPTQSLAGSVQVVGRVLEVALHKAHELK-FPLEHIVDGIATAPLSPPHPDFVTAMGRTND
Methylosinus LW3                KVANDCGVSPDKLAFVYAPTQSLAGSVQVVGRVLEVALHKAHELK-FPLEHIVDGIAAAPLSPPHPDFVTAMGRTND
Methylosinus PW1                KVANDCGVSPDKLAFVYAPTQSLAGSVQVVGRVLEVALHKAHELK-FPLEHIVDGIATAPLSPPHPDFVTAMGRTND
Methylosinus trichosporium      KVAESCGVSPDKLAFIYAPTQSLAGSVQVVGRVLEVALHKAHELK-FPLEHIVDGIATAPLSPPHPDFVTAMGRTND
Methylocella palustris          KVAAACGVKPENVAFIFAPTQSLAGSVQVVARVLEVALHKAHELK-FPLERIVDGLGAAPLSPPHPNFVEAMGRTND
Methylobacterium extorquens     KVAAATGLAPENVTFIYAPTQSLAGSTQVVARVLEVALHKAHTVG-FDLHKILDGIGSAPLSPPHPDFIQAMGRTND
Methylobacterium organophilum   KVAEASGLKPEDLTFIYAPTQSLAGSTQVVARVLEVALHKAHTVG-FDLHAIVDGIGSAPLSPPHPDFIKAMGRTND
Methylomicrobium 5G             KVANACQVDPCNLTIIVTPTSSLAGGVQVVARVLEVAMHKAHALH-FPLENIVDGSGSAPICPPHPDFIKAMGRTND
Methylomicrobium 20Z            KVANACQVDPCNLTIIVTPTSSLAGGVQVVARVLEVAMHKAHALH-FPLENIVDGSGSAPICPPHPDFIKAMGRTND
Methylomicrobium N1             KVANACQVDPSNLTIIVTPTSSLAGGVQVVARVLEVAMHKAHALH-FPLKNIVDGSGSAPICPPHPDFIKAMGRTND
Methylomonas LW13               KVAAACGVSPANLTIIVTPTSSLAGGVQVVGRVLEVAMHKAHALH-FPLENIVDGSGSAPICPPHPNFVKTMGRTND
Methylomonas LW21               KVATACGLSPANLTIIVTPTSSLAGGVQVVGRVLEVAMHKAHALH-FPLENIIDGSGSAPICPPHPNFVKAMGRTND
Methylomonas rubra              KVAAACGVSPADLTIIVTPTSSLAGGVQVVGRVLEVAMHKAHALH-FPLENIVDGTGSAPVCPPHPNFVKAMGRTND
Methylobacter LW14              RVASACKVAPSKLTIIVTPTSSLAGGLQVVARVLEVAMHKAHALH-FPLENIIDGSGSAPVCPPHPNFVKAMGRTND
Methylomicrobium albus          KVAAACKVDPGKLTVIVTPTSSLAGGLQVVARVLEVAMHKAHALH-FPLENIVDGFGSAPLCPPHPNFVKAMGRTND
Methylobacter A45               KVAAACKVDPSRLTIIVTPTSSLAGGVQVVARVLEVAMHKAHAQH-FPLENIIDGSGSAPVCPPHPNFVKAMGRTND
Methylococcus thermophilus      KIAKACKVDASGVHVILTPTSSLAGGMQVVGRVLEVALHKAHSLH-FPLGNIIDGTGTAPVPPPHPNFVKAMGRTND
Methylococcus capsulatus        KIARACKLDPSHLTVILTPTSSLAGGMQVVARVLEVALHKAHALH-FPLENIIDGTGCAPVPPPHPNFIKAMGRTND
Hyphomicrobium methylovorum     NIAKHCGINPKGLSILYATTWSLAGTVQIAARVLEVAMHKAHALH-FALENIIDGTATTPIAPPFPDFVKAMGRTND
Methylobacterium nodulans       KVSKATGFAPENLTFIYAPTQSLTGAVQIVGRVLEVAMHKAHDLK-FPLGDIVDGAGRAPIPAPHPDFLKAMGRTND
```

Figure 8.4 Multiple alignment of partial amino acid sequences for Mch (amino acids 103–245 in *Methylobacterium extorquens* AM1). Conserved residues are shown in bold.

novel methanotroph *Methylocella palustris*. Divergence of Mch enzymes in these strains might be a reflection of their respective unique environments: *M. nodulans* is a specific symbiont of *Crotolaria* legumes in Africa (Sy *et al.* 2001), while *M. palustris* is a dominant methanotroph species in the acidic soils of tundra (Dedysh *et al.* 2000). Although we have not tested our primers on any environmental samples yet, this preliminary analysis indicates that *mch* is a potential target gene for environmental probing, and the alignment of the sequences available at this point (Figure 8.4) provides clues for developing *mch* probes for specific groups. For example, an insertion of 10 amino acids is only found in γ-proteobacterial

```
M.e.   MARS--ILHMLTPLKHMSPFDVNMAIDAGFETLIPYTGVDLTDVVSLTQDSIFSRAPQDGVRTGIFIGGKNAELALD
X.a.   MAESAPILHIITPLNHVSPFDVNMAVDAGYTTIAPYSRVALAEVRDLTQDMMFSRAPQSAPRTGLFIGGKDAALALD
M.c.1  MEKR-AVLHMFDPMPHVSPFDINMAVDAGFEVIVPYSNVKAEEVHGLVQDAIFSRGPAGVKRTGIFIGGRDLAVALT
M.c.2  VEKP-YILHMFTPEKNLSPFDANMALDAGWLSLVPYTNLEAGEITALVQDAIFSRSPAGVKRTAVFIGGRDVGTAMA
                         ******

M.e.   MVDRAKKAFVPPFVNHVFADPAGSFTTGAAMVAEVNRALKARFSTDLKGKRIVIFGGAGVVAYVAAVIGALEGAQTV
X.a.   MAAEAKASIFPPFEISIFPDPAGSFTTAAAMIAKVEQALLKTRPKGLEGVAVQVYGATGVVGSIAAVIAAQAGALVT
M.c.1  MLETAKGAMVPPFEVSVLADPSGGFTTAAALVALVEKQLKLKHGAELAGQRAVVFGGTGPVGIATGVIASLAGADVT
M.c.2  MLRTARKSMVPPFEISAFADPSGAFTTAAAMVAMAEGELRKHYQTGLEGKSVLALGGTGPVGQIAAVLAARAGAQVK

M.e.   LVGHDGEERVSKIAFTMKWRFGIDVGAVDGTLPEARRAAITDADVILSAGPAGVSILTAEDLESAPKLLVASDVNAV
X.a.   LVSHREVAAVEEKAKDLKARFGVELACTAAVTDSEKAKLVPEAEVVLAAGKAGVEILTAAHLQAATRLLVAADVNAV
M.c.1  LVDPFNVETALAKADEYNRRCGARLHGTFAGSEADKARLLSNADVVFCTAKAGVEVLNASVLADAKRLKVAGDVNAV
M.c.2  IIGR-QLEKAQKVATLCNAEFGEGQTAIEGDADANKGELLKTADVVFATAAAGIQVLSAELLAASPRLKVAADVNAV
                                                                                ****

M.e.   PPAGIAGIDVNAVDVPLPT----GKGVGIGALAVGNVKYQTQCRMFRKMLEAQEPLCLDFRDAYKLAVEIAG--
X.a.   PPAGIAGIGAHDAAAVLP------HGIGIGALAIGNLKYAVQHTLLKQMCVTDTPLFFDLPAIFKLARELDA--
M.c.1  PPLGIEGIKLKHNGEPLVHAVNSPGAVGVGALAVGNVKYQLQNHLLALLLKSESPVFFDFRAAFERAREIV---
M.c.2  PPSGIAGLDPHHKGTPIE--GSPSGAIGIGALAIGNVKYQAQHRLLLRMRSTDKPVYLAFNEAFEIAREVVRGA
       **
```

Figure 8.5 Alignment of MtdB amino acid sequences from *Methylobacterium extorquens* (M.e.), *Xanthobacter autotrophicus* (X.a.) and *Methylcoccus capsulatus* (M.c.1, copy 1; M.c.2, copy 2). Asterisks indicate conserved amino acid residues used for designing degenerate oligonucleotide primers for PCR amplification.

methanotrophs, and amino acid conservation within this insert seems to be species-specific.

Methylene H_4MPT dehydrogenase (MtdB)

MtdB is another key enzyme in the H_4MPT-linked formaldehyde oxidation pathway. Like MtdA, MtdB is only found in methylotrophs. In *M. extorquens* AM1 and *X. autotrophicus*, *mtdB* is found in the archaeal-like gene clusters, while in *M. capsulatus*, two copies of *mtdB* are found not linked to the archaeal-like genes, but closely linked together. Whether both of them are functional remains to be determined. The polypeptides translated from these two genes are only about 50% identical. These two sequences, aligned with the sequences from *M. extorquens* and *X. autotrophicus* MtdBs (Figure 8.5), showed only two regions of strong conservation, which were used to design degenerate primers for PCR analysis. The resulting primers were tested on a variety of methylotrophic isolates, generating new sequences that were subjected to phylogenetic analysis. The MtdB phylogenetic trees showed some difference in patterns compared to the Mch trees. These results are consistent with the hypothesis that MtdB has a different evolutionary history compared to Mch. Remarkably, MtdBs translated from the two copies of *mtdB* in *M. capsulatus* did not cluster together, implying that the two copies have evolved differently. This might be because they carry out different functions, or because one of the copies is a result of a more recent lateral transfer. Further work will be necessary to determine the roles of these two genes.

Future directions

Ancient methylotrophy enzymes, as described above, present excellent targets for environmental probing and prediction, being widespread within diverse methylotroph populations and suitable for recognition of divergent functional gene/enzyme sequences. These targets are not found in all methylotrophs, but they appear to occur in most of the strains with high formaldehyde flux. Therefore, these probes should allow the detection of novel methylotrophs, methylotrophs that are difficult to propagate in laboratory conditions, or methylotrophs representing a minority in a given population. However, the current data for developing such probes are limited to just a few sequences, and more sequences are needed for key formaldehyde handling genes/enzyme. Although complete genomes of more methylotrophs may be sequenced in the future, it is unlikely to happen soon. An alternative approach to complete genome sequencing is comparative analysis of entire clusters of functional genes (Xiong *et al.* 2000). This is especially attractive for genes of the archaeal-like pathway, since so far they are found in clusters in methylotrophs. Identifying and sequencing such clusters in cosmid libraries would quickly provide a rich database of information on the archaeal genes in diverse organisms, which will lead to the development of robust probes.

These probes can be combined with current functional and phylogenetic probes and used on total DNAs from specific environments, to evaluate the diversity and relative abundance of methylotrophs as a whole as well as the methylotrophs belonging to specific groups. This analysis has the potential to detect novel species of methylotrophic bacteria, bacteria not capable of growth on C_1 substrates but possessing formaldehyde detoxification systems shared with methylotrophs, or archaea specialized in either methane production or methane utilization. For example, surprisingly little is known about mesophilic archaea capable of C_1 metabolism.

When novel species are detected by the divergence of their core methylotrophy enzymes, these can then be further addressed by the direct approach of metagenome analysis (Rondon *et al.* 2000). In this approach, large fragments of environmental DNA are cloned into BAC vectors, clones of interest are detected with target PCR products generated from the same environmental site, and the positive BAC inserts (up to 100 kb) are analysed by sequencing. Gene clusters can be examined and compared to known sets of genes, not only adding to the database but also providing clues as to the physiological type of the original host. In addition, these sequences will provide valuable information on the surrounding genes, indicative of the nature and phylogenetic make up of the uncultivated organism. Such information can then be applied to isolation of these strains, targeting capabilities revealed through the cluster analysis.

The approaches described here direct the power of comparative genomics and evolutionary analysis to environmental problems. By focusing on clusters of co-evolving genes that carry out a specific, enabling function within a metabolic mode, a broader understanding of community structure and function can be gained. Sampling of 100 kb 'bytes' of genomes that contain functional gene clusters can provide novel insights into metabolic diversity and community structure directed at environmental function.

References

Anthony, C. (1982) *The Biochemistry of Methylotrophs.* Academic Press, London.

Barber, R.D. & Donohue, T.J. (1998) Function of a glutathione-dependent formaldehyde dehydrogenase in *Rhodobacter sphaeroides* formaldehyde oxidation and assimilation. *Biochemistry* **37**, 330–337.

Boetius, A., Ravenschlag, K., Schubert, C. J. C. J. *et al.* (2000) A marine microbial consortium apparently mediating anaerobic oxidation of methane. *Nature* **407**, 623–626.

Bourne, D.C., Holmes, A.J., Iversen, N. & Murrell, J.C. (2000) Fluorescent oligonucleotide rDNA probes for specific detection of methane oxidising bacteria. *FEMS Microbiology Ecology* **31**, 29–38.

Chistoserdov, A.Y., Chistoserdova, L.V., McIntire, W.S. & Lidstrom, M.E. (1994) Genetic organization of the *mau* cluster in *Methylobacterium extorquens* AM1: complete nucleotide sequence and generation and characteristics of *mau* mutants. *Journal of Bacteriology* **176**, 4052–4065.

Chistoserdova, L., Vorholt, J.A., Thauer, R.K. & Lidstrom, M.E. (1998) C_1 transfer enzymes and coenzymes linking methylotrophic bacteria and methanogenic archaea. *Science* **281**, 99–102.

Chistoserdova, L., Gomelsky, L., Vorholt, J. A. *et al.* (2000) Analysis of two formaldehyde oxidation pathways in *Methylobacillus flagellatum* KT, a ribulose monophosphate cycle methylotroph. *Microbiology* **146**, 3762–3769.

Connell Hancock, T.L., Costello, A.M., Lidstrom, M.E. & Oremland, R.S. (1998) Strain IMB–1, a novel bacterium for the removal of methyl bromide in fumigated agricultural soils. *Applied and Environmental Microbiology* **64**, 2899–2905.

Dedysh, S.N., Leisack, W., Khmelenina, V. *et al.* (2000) *Methylocella palustris* gen. nov., sp. nov., a new methane-oxidizing acidophilic bacterium from peat bogs, representing a novel subtype of serine-pathway methylotrophs. *International Journal of Systematic and Evolutionary Microbiology* **50**, 955–969.

DeLong, E.F. (2000) Resolving a methane mystery. *Nature* **407**, 577–579.

Doronina, N.V., Trotsenko, Y.A. & Tourova, T.P. (2000) *Methylarcula marina* gen. nov., sp. nov, and *Methylarcula terricola* sp. nov.: novel aerobic, moderately halophilic, facultatively methylotrophic bacteria from coastal saline environments. *International Journal of Systematic and Evolutionary Microbiology* **50**, 1849–1859.

Eller, G., Stubner, S. & Frenzel, P. (2001) Group-specific 16S rRNA targeted probes for the detection of type I and type II methanotrophs by fluorescence in situ hybridisation. *FEMS Microbiology Letters* **198**, 91–97.

Gak, E.R., Tsygankov, Y.D. & Chistoserdov, A.Y. (1997) Organization of methylamine utilization genes (*mau*) in '*Methylobacillus flagellatus*' KT and analysis of *mau* mutants. *Microbiology* **143**, 1827–1835.

Hanson, R.S., Bratina, B.J. & Brusseau, G.A. (1993) Phylogeny and ecology of methylotrophic bacteria. In: *Microbial Growth on C_1 Compounds* (eds J. C. Murrell & D. P. Kelly), pp. 285–302. Intercept, Andover.

Harms, N., Ras, J., Koning, S. *et al.* (1996) Genetics of C_1 metabolism regulation in *Paracoccus denitrificans.* In: *Microbial Growth on C_1 Compounds* (eds M. E. Lidstrom & F. R. Tabita), pp. 126–132. Kluwer Academic Publishers, Dordrecht.

Holmes, A.J., Rosley, P., McDonald, I.R. *et al.* (1999) Characterization of methylotrophic bacterial populations in soils showing atmospheric methane uptake. *Applied and Environmental Microbiology* **65**, 3312–3318.

Jensen, S., Holmes, A.J., Olsen, R.A. & Murrell, C.J. (2000) Detection of methane oxidizing bacteria in forest soil by monooxygenase PCR amplification. *Microbial Ecology* **39**, 282–289.

Kalyuzhnaya, M.G., Khmelenina, V.N., Kotelnikova, S. *et al.* (1999) *Methylomonas scandinavica* sp. nov., a new methanotrophic psychrotrophic bacterium isolated from deep igneous rock ground water of Sweden. *Systems in Applied Microbiology* **22**, 565–572.

Kelly, D.P. & Murrell, J.C. (1999) Microbial metabolism of methanesulfonic acid. *Archives of Microbiology* **172**, 341–348.

Kim, Y.M. & Lidstrom, M.E. (1989) Plasmid analysis in pink facultative methylotrophic bacteria using a modified acetone-alkaline hydrolysis method. *FEMS Microbiology Letters* **51**, 125–130.

King, G.M. (1992) Ecological aspects of methane oxidation, a key determinant of global methane dynamics. *Advances in Microbial Ecology* **12**, 431–474.

Lidstrom, M.E. (2001) The methylotrophic bacteria. In: *The Prokaryotes* (eds M. Dworkin *et al.*), Springer-Verlag, New York. Available online.

Lidstrom, M.E., Chistoserdova, L., Stoliar, S. & Springer, A.L. (1998) Genetics and regulation of C_1 metabolism in methylotrophs. In: *NATO Symposium on Electron Transfer Chains* (eds E. Vijgenboom & G. Canters), pp. 89–97.

McAnulla, C., Woodall, C.A., McDonald, I.R. *et al.* (2001) Chloromethane utilization gene cluster from *Hyphomicrobium chloromethanicum* strain CM2 and development of functional gene probes to detect halomethane-degrading bacteria. *Applied and Environmental Microbiology* **67**, 307–316.

McDonald, I.R., Upton, M., Pickup, R.W. *et al.* (1999) Molecular ecological analysis of methanogens and methanotrophs in blanket bog peat. *Microbial Ecology* **38**, 225–233.

Marx, C.J. & Lidstrom, M.E. (2001) Development of improved versatile broad-host-range vectors for use in methylotrophs and other Gram-negative bacteria. *Microbiology* **147**, 2065–2075.

Miller, L.G., Kalin, R.M., McCauley, S.E. *et al.* (2001) Large carbon isotope fractionation associated with oxidation of methyl halides by methylotrophic bacteria. *Proceedings of the National Academy of Sciences of the United States of America* **98**, 5833–5837.

Murrell, J.C. & Radajewski, S. (2000) Cultivation-independent techniques for studying methanotroph ecology. *Research in Microbiology* **151**, 807–814.

Norin, A., van Ophem, P.W., Piersma, S.R. *et al.* (1997) Mycothiol-dependent formaldehyde dehydrogenase, a prokaryotic medium-chain dehydrogenase/reductase, phylogenetically links different eucaryotic alcohol dehydrogenases. Primary structure, conformational modeling and functional correlations. *European Journal of Biochemistry* **248**, 282–289.

Oremland, R.S. & Culbertson, C.W. (1992) Importance of methane-oxidizing bacteria in the methane budget as revealed by the use of a specific inhibitor. *Nature* **356**, 421–423.

Pomper, B.K., Vorholt, J.A., Chistoserdova, L., Lidstrom, M.E. & Thauer, R.K. (1999) A methenyl tetrahydromethanopterin cyclohydrolase and a methenyl tetrahydrofolate cyclohydrolase in *Methylobacterium extorquens* AM1. *European Journal of Biochemistry* **261**, 475–480.

Rondon, M.R., August, P.R., Bettermann, A.D. *et al.* (2000) Cloning the soil metagenome: a strategy for accessing the genetic and functional diversity of uncultured microorganisms. *Applied and Environmental Microbiology* **66**, 2541–2547.

Rudd, J.W.M. & Taylor, C.D. (1980) Methane cycling in aquatic environments. *Advances in Aquatic Microbiology* **1**, 77–150.

Schaefer, J.K. & Oremland, R.S. (1999) Oxidation of methyl halides by the facultative methylotroph strain IMB–1. *Applied and Environmental Microbiology* **65**, 5035–5041.

Serebrijski, I., Gomelsky, M., Bashkirova, E., Chistoserdov, A. & Tsygankov, Y.D. (1998) Refined genetic map of the obligate methylotroph *Methylobacillus flagellatus*. *Molecular and General Genetics* **258**, 133–138.

Stanhope, M.J., Lupas, A., Italia, M.J. *et al.* (2001) Phylogenetic analyses do not support horizontal gene transfers from bacteria to vertebrates. *Nature* **411**, 940–944.

Sy, A., Giraud, E., Jourand, P. *et al.* (2001) Methylotrophic *Methylobacterium* bacteria nodulate and fix nitrogen in symbiosis with legumes. *Journal of Bacteriology* **183**, 214–220.

Vanelli, T., Messmer, M., Studer, A., Vuilleumier, S. & Lesinger, T. (1999) A corrinoid-dependent catabolic pathway for growth of a *Methylobacterium* strain with chloromethane. *Proceedings of the National Academy of Sciences of the United States of America* **96**, 4615–4620.

Vorholt, J.A., Chistoserdova, L., Lidstrom, M.E. & Thauer, R.K. (1998) The NADP-dependent methylene tetrahydromethanopterin dehydrogenase in *Methylobacterium extorquens* AM1. *Journal of Bacteriology* **180**, 5351–5356.

Vorholt, J.A., Chistoserdova, L., Stolyar, S.M., Lidstrom, M.E. & Thauer, R.K. (1999) Distribution of tetrahydromethanopterin-dependent enzymes in methylotrophic bacteria and phylogeny of methenyl tetrahydromethanopterin cyclohydrolases. *Journal of Bacteriology* **181**, 5750–5757.

Xiong, J., Fischer, W.M., Inoue, K., Nakahara, M. & Bauer, C.E. (2000) Molecular evidence for the early evolution of photosynthesis. *Science* **289**, 1724–1730.

Chapter 9

Microbiological basis of land use impact on the soil methane sink: molecular and functional analysis

D. B. Nedwell, J. C. Murrell, P. Ineson, D. S. Reay, S. Radajewski, N. McNamara and S. Morris*

Introduction

Methane (CH_4) is the second most significant greenhouse gas (Bouwman 1990) and until recently has been accumulating in the atmosphere at a rate of about 1% per annum. The largest biological sink for atmospheric methane is the soil (see Crutzen 1991), where methane-oxidizing bacteria (MOB) remove methane from the air and oxidize it to CO_2. However, there is increasing evidence that changes in land use disturb this soil CH_4 sink (Ojima *et al.* 1993; Willison *et al.* 1995). Cultivation of fallow land, and addition of fertilizers (Steudler *et al.* 1989; Mosier *et al.* 1991; Hutsch *et al.* 1993) decrease the soil CH_4 sink, presumably because of changes in the MOB community. Forest soils seem to have high CH_4 oxidation capacity (Hütsch *et al.* 1994; Goulding *et al.* 1996), varying with the species of tree, and it has been argued that conversion of forest to agriculture causes a 60% reduction in the CH_4 oxidation capacity of soils (Dobbie *et al.* 1996).

The reason for the reduction in the soil CH_4 oxidation capacity is unclear, although it is central to understanding the global CH_4 cycle. It has been argued (Steudler *et al.* 1989; Hütsch *et al.* 1993; Schnell & King 1994; Willison *et al.* 1995; King & Schnell 1998) that addition of NH_4^+ to soil as fertilizers or manure decreases the rate of soil CH_4 uptake, because of the inhibitory action of NH_4^+ on methane monooxygenase, a key enzyme for CH_4 oxidation in methanotrophs. However, environmental factors other than NH_4^+ concentration such as water stress or ionic strength of the soil water also diminish the activity of the soil MOB (Nesbit & Breitenbeck 1992; Crill *et al.* 1994; Kightley *et al.* 1995; King & Schnell 1998), so that the inhibition of methane oxidation following tilling and fertilization may not result specifically from NH_4^+. Alternatively, addition of NH_4^+ may select for nitrifying (ammonium-oxidizing) bacteria (AOB) in fertilized soils at the expense of the MOB. Residual CH_4-oxidizing capacity remaining even in fertilized soils has been attributed to the selection by fertilizer addition of AOB, which also have an ability to oxidize CH_4 but have a lower affinity than MOB for CH_4 and do not use it as a carbon

* Correspondence author: e-mail nedwd@essex.ac.uk

Figure 9.1 The Gisburn experiment, showing its location in Cumbria, and the arrangement of block and plots within the site.

source (Hyman & Wood 1983; Jones & Morita 1983). Therefore, a variety of different mechanisms may contribute to decrease the CH_4 sink in soils, either directly through inhibitory effects of changed environmental factors on the MOB, through changes in the MOB community, or indirectly through selection of AOB at the expense of the MOB. Understanding the effect of land management on the soil CH_4 sink therefore requires elucidation of the range of the critical factors from these possibilities.

This study was undertaken at the Gisburn experiment of Forest Enterprises Bowland Forest in Lancashire, UK (Brown 1992). This is a replicated, randomized plot experiment established in 1955. It has triplicated half acre (0.2 ha) plots of sessile oak (*Quercus petraea* (Mattuschka) Liebl.), alder (*Alnus glutinosa* (L.) Gaertn.), Norway spruce (*Picea abies* (L.) Karst), Scots pine (*Pinus sylvestris* (L.)), and unforested control plots containing a species-poor rough pasture of *Festuca–Agrostis* with *Nardus stricta, Deschampsia* and *Juncus* spp. Plots of each treatment are distributed randomly (Figure 9.1). The experiment has never been fertilized.

The study aimed to investigate the influence of forestation on the soil methane sink. Molecular biology techniques were used to examine differences between MOB and AOB communities in the different plots. The expression of the genetic potential of these microbial communities in terms of metabolic capacities for CH_4 or NH_4^+ oxidation within the soils was investigated by regular measurements of CH_4

Table 9.1 Soil water chemistry ($\mu g\,L^{-1}$) for soils under the five land-use treatments at Gisburn.

Soil	Alder	Oak	Norway spruce	Scots pine	Grass control
Nitrate	4370 ± 1290	60 ± 5	30 ± 3	70 ± 36	910 ± 830
Ammonium	70 ± 41	60 ± 18	70 ± 30	150 ± 64	30 ± 2
Sulfate	1720 ± 1290	2370 ± 5	1910 ± 3	2400 ± 36	1540 ± 830

Values represent means for the three experimental blocks (A horizon) from 12 monthly sampling visits, starting in July 1999. Errors are standard error about the mean ($n = 36$).

oxidation potential (MOP) or NH_4^+ oxidation potential (AOP), which are relative measures of the size and activity of the MOB and AOB populations present. Such metabolic potentials will in turn be regulated, and become apparent as realized *in situ* rates of activity, by physicochemical environmental variables such as temperature, pH, redox state or nutrient concentrations. A programme of regular sampling in the different forested and control plots was used to reveal site-to-site differences and seasonal changes in these variables.

Soil characteristics

Soil characteristics were measured by taking quadruplicate soil cores from each treatment plot in each of the three experimental blocks. The soils were all acidic upland soils with pH values between 3.8 and 5.0. MOP and AOP of the top 5 cm of the A horizon was measured bimonthly. Soil waters were collected on a monthly basis from all plots within this study for one year (June 1999 to July 2000). Porous cup lysimeters were placed in the A horizons and attached to a 2 L glass bottle and the system maintained under vacuum. The glass bottle was stored in a trench in the ground to keep the sample cool and away from sunlight. Samples were filtered within 24 hours of collection and stored at 2–3°C until analysis. Table 9.1 shows the annual values of concentrations of ions in soil water under the four tree species and in the grassed control plots. A major difference between the plots was that the concentrations of nitrate, but not ammonium, in the A horizon were significantly higher under alder than under the other tree species. This presumably resulted from the ability of alder roots to fix nitrogen. The soil in block 1 of the experiment had higher organic contents than the other two blocks (Moffat & Boswell 1990).

To investigate the edaphic and climatic factors controlling fluxes at the field site, more detailed measurements of the chemistry of soil solution and local climate were monitored within block 2. Local climate and soil physical characteristics were measured within block 2 using Delta-T Automatic Weather Stations (AWS; Type WSO1, Delta-T, Cambridge, UK) configured to monitor air temperature and rainfall for the site as a whole and soil temperatures (5, 10, 15 cm depth), water table depths and volumetric soil moistures for each of the tree species. The temperature and moisture probes were placed close to the trace gas monitoring sites to determine climatic factors driving methane fluxes.

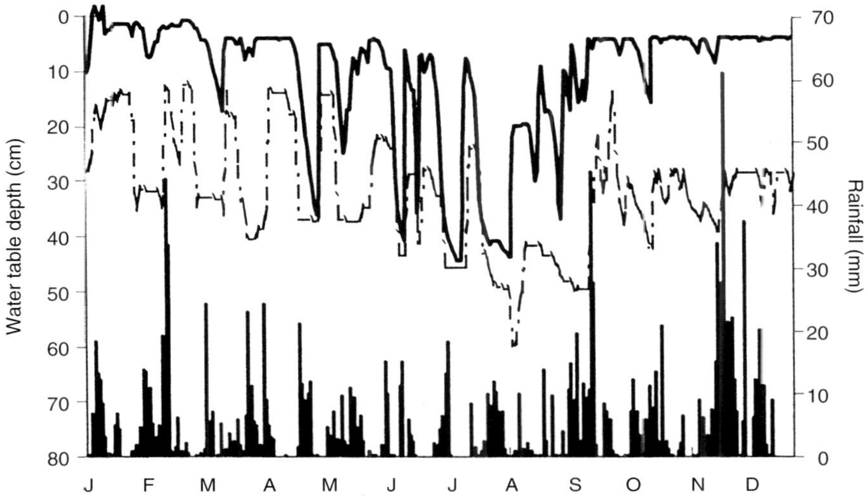

Figure 9.2 Water table fluctuation and response to rainfall events. Black vertical bars represent rainfall events and the heavy and light lines are grass and oak plots, respectively.

Continuous measurements showed that the response of the water table to rainfall was very rapid in forested plots at Gisburn (Figure 9.2) but limited under the unforested grassed control plots. The water table in the grassed control plots remained within 20 cm of the surface for most of the time, but evapotranspiration in the forested plots was evidently sufficient to increase the water table depth, and thus maintain a deeper aerated and oxidized surface layer of soil throughout the year. There were only marginal differences in water tables between the plots for different tree species. In addition to the automatic water table monitors there were three 'manual' dipwells placed in each plot of block 2 to determine intraplot differences in water table. Major intraplot differences were only found for alder in the summer, with variations of up to 50 cm across the plot (data not shown). Otherwise, the single-point continuous monitors provided data for the whole single-species plot.

Molecular analysis of MOB communities

Aerobic methane-oxidizing bacteria (methanotrophs) can use methane as a sole source of carbon and energy. Methanotrophs are widely distributed in the natural environment where they form a major part of the global methane cycle. As many methanotrophs are difficult to isolate into culture a variety of culture-independent molecular biological techniques has been developed to characterize natural methanotroph populations. In this component of the study, we used molecular biological techniques to identify the MOB present in forest soils at the Gisburn Forest Experiment and determine whether the population structure of MOB was affected by the land use regime.

Extant methanotrophs can be classified into two distinct groups on the basis of 16S rRNA phylogeny, morphology, the pathways of carbon assimilation that they use, and membrane type and composition. Type I methanotrophs belong to the γ-subclass of the *Proteobacteria* and presently include seven validly described genera. Type II methanotrophs belong to the α-subclass of the *Proteobacteria* and currently include three genera (Bowman *et al.* 1993; Dedysh *et al.* 2000; Murrell & Radajewski 2000; Wise *et al.* 2001).

In methanotrophs, there are two forms of the enzyme methane monooxygenase (MMO), which is responsible for the first step in the oxidation of methane. A membrane-bound, particulate enzyme (pMMO) has been reported in all methanotrophs except for one recently described genus (*Methylocella* (Dedysh *et al.* 2000)), whereas a cytoplasmic, soluble form of the enzyme (sMMO) is only found in certain strains (Murrell *et al.* 2000). In addition to methanotrophs, pure cultures of ammonia-oxidizing nitrifying bacteria (nitrifiers) can also oxidize methane. However, nitrifiers exhibit a lower affinity for methane than methanotrophs and cannot use methane directly as a carbon source. The enzyme involved in ammonia oxidation in nitrifiers (ammonia monooxygenase; AMO) exhibits sufficient homology to pMMO to suggest that these enzymes may be evolutionarily related (Holmes *et al.* 1995). Furthermore, the phylogeny of the genes that encode the active-site subunits of pMMO and AMO (*pmoA* and *amoA*, respectively) in extant strains reflect that obtained with 16S rRNA genes (Holmes *et al.* 1995).

The restricted phylogenetic distribution of methanotrophs has enabled the design of apparently group-specific PCR primers that target the 16S rDNA (Brusseau *et al.* 1994; Bourne *et al.* 2000; Eller *et al.* 2001). However, when applied to environmental samples, they also appear to detect less related sequences from bacteria which may or may not be able to oxidize methane (McDonald *et al.* 1996; Henckel *et al.* 1999). An alternative approach has been to target the 'functional' genes that code for pMMO and AMO. Conserved domains within the amino acid sequences of the active-site subunits of pMMO and AMO have enabled the design of several sets of PCR amplification primers (Holmes *et al.* 1995; Costello & Lidstrom 1999). We chose the primer set A189/A682 (Holmes *et al.* 1995) to provide an overall assessment of the methane-oxidizing bacterial population in forest soils, as it has previously detected both *pmoA* and *amoA* directly in environmental samples (McDonald & Murrell 1997; Henckel *et al.* 1999; Holmes *et al.* 1999).

Diversity of methane-oxidizing bacteria

Libraries of 100 *pmoA*/*amoA* clones generated from DNA extracted from the 0–5 cm depth layer from all plots in block 2 identified 12 (alder), four (oak), 11 (Norway spruce), 14 (Scots pine) and 10 (grass control) operational taxonomic units (OTUs) based on unique restriction enzyme digest patterns. Each clone was sequenced and following phylogenetic analysis, these sequences were assigned to one of four broad phylogenetic groups, A, B, C and D (Figure 9.3a). The proportion of clones within each of the broad phylogenetic groups is indicated in Figure 9.3b. Only one clone,

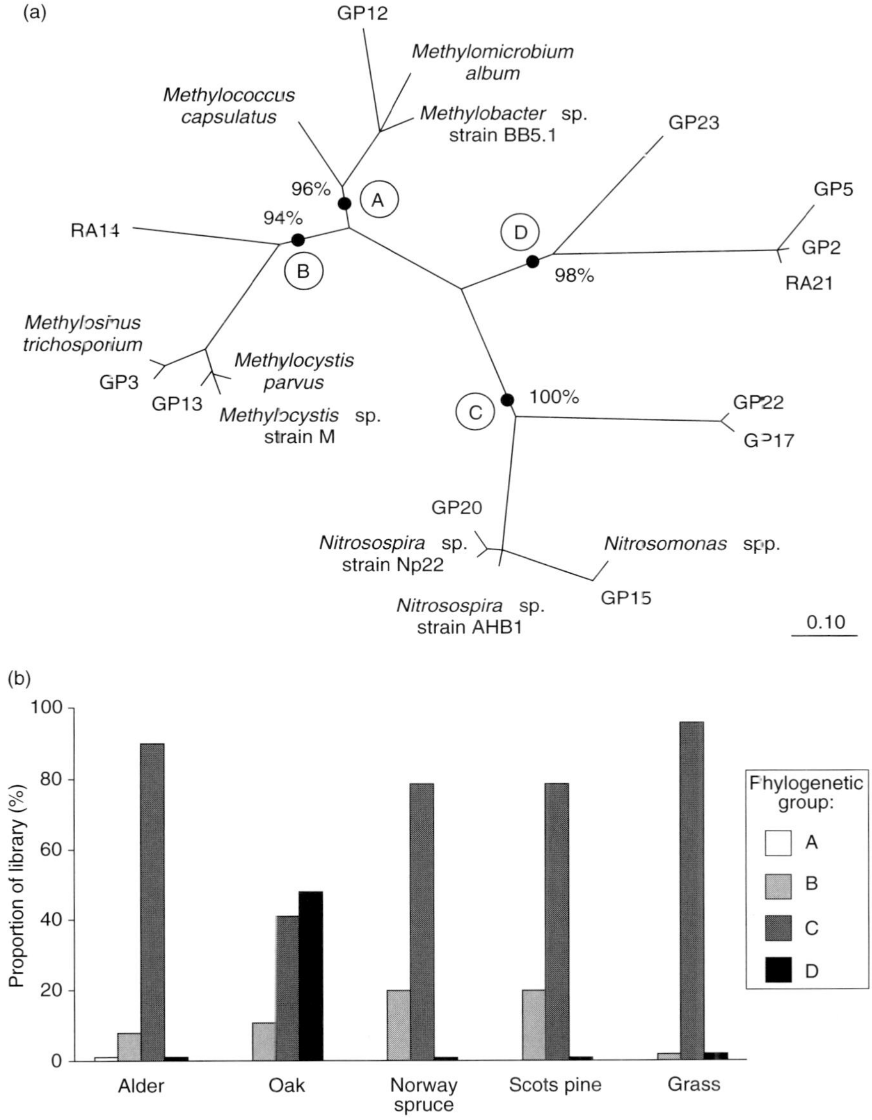

Figure 9.3 (a) Phylogeny of PmoA/AmoA clones (GP prefix) retrieved from Gisburn forest soils (0–5 cm depth) collected in April 1998. Bootstrap values at the nodes of phylogenetic groups A, B, C or D are shown. Scale bar = 0.1 inferred amino acid substitution per amino acid position. *Nitrosomonas* spp. includes 14 highly related OTUs. (b) Distribution of clones in phylogenetic groups A, B, C or D.

which was related to the PmoA of the methanotrophs *Methylobacter* and *Methylomicrobium* (GP12), was recovered in group A. The group B sequences were related to the PmoA of the methanotrophs *Methylosinus* (GP3) or *Methylocystis* (GP13). Clones related to *amoA* of the nitrifier *Nitrosomonas* spp. dominated phylogenetic group C (41%–96% of each library). Group C also included one sequence related to the AmoA of *Nitrosospira* (GP20) and two distinct clones from the Scots pine soil that exhibited sequence homology to both AmoA and PmoA (GP17 and GP22). Group D sequences (GP2, GP5 and GP23) were related to the recently described environmental clone RA21 (Holmes *et al.* 1999).

The proportion of sequences in the clone libraries cannot be used to accurately predict the relative abundance of MOB in the forest soils due to biases inherent in PCR amplification. However, as the soil samples were processed identically, and the only major difference between the plots is the pattern of land use, large differences in the proportion of clones within any particular OTU might reflect a shift in the population of these bacteria in the soil. The predominance of clones in phylogenetic group C (see Figure 9.3), in all soil samples was unexpected because *Nitrosomonas*-like *amoA* sequences have either not been detected in, or are minor components of, other molecular surveys using the general A189/A682 primer set and DNA extracted from soil (McDonald & Murrell 1997; Holmes *et al.* 1999; Henckel *et al.* 2000). Detection of type II methanotroph PmoA sequences (group B) as relatively major components of the alder, oak, Scots pine and Norway spruce libraries was in line with these previous studies. Clones within group D have only been rarely reported (Holmes *et al.* 1999; Henckel *et al.* 2000), but their detection in this study suggested that an unknown bacterium with the potential to oxidize methane may be relatively abundant in the surface of the oak forest soil.

To determine the significance of the PmoA and AmoA homologues, it will be necessary to correlate their presence with methane oxidation and nitrification potentials. No clear relationships were identified between the methane oxidation potentials and the molecular profiles for the five soils examined. For example, type II methanotroph sequences were present in the 0–5 cm layer of each of the four forest soils, even though there was no significant methane oxidation in the alder plot soil. Similarly, there was considerable oxidation of methane in the control soil from block 2 but a low proportion of methanotroph-like sequences. These observations suggest that factors other than water potential might be controlling the oxidation of methane, even under optimized conditions. These may include inhibition of methane oxidation by inorganic soil nutrients, or the involvement of methanotrophs such as *Methylomicrobium album*, or the moderate acidophile *Methylocella palustris*, in which a *pmoA* gene cannot be detected with the PCR primer set (A189/A682) we used (Holmes *et al.* 1995; Dedysh *et al.* 2000).

Identification of active methylotrophs and methanotrophs

To address some of the anomalies identified in the functional gene libraries, we assessed the applicability of stable-isotope enriched compounds (e.g. 99% $^{13}CH_4$) for

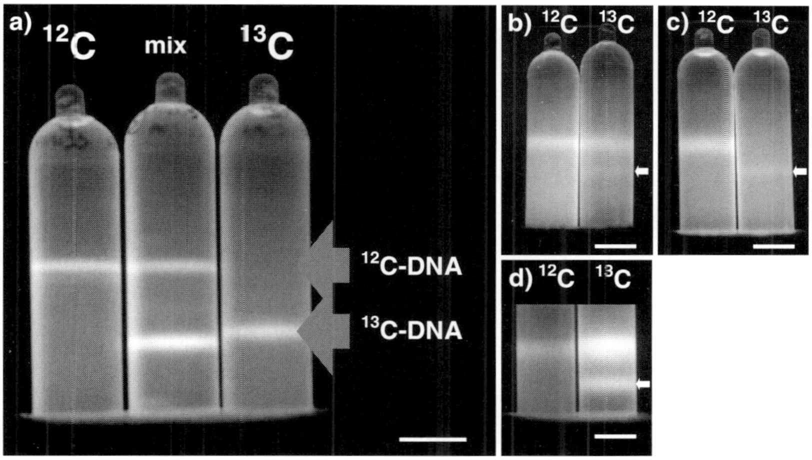

Figure 9.4 Equilibrium centrifugation of isotopically labelled DNA in CsCl/ethidium bromide density gradients. (a) Pure fractions and a mixture of the DNA extracted from a pure culture of a methylotroph, *Methylobacterium extorquens*, grown on either $^{12}CH_3OH$ or $^{13}CH_3OH$ as the sole carbon source. (b,c) DNA extraction from soil that had utilized $^{12}CH_3OH$ or $^{13}CH_3OH$, showing the position of the ^{13}C-DNA fraction (arrow) after the primary (b) and purification (c) ultracentrifugation steps. (d) DNA extraction from an oak soil slurry that had utilized $^{12}CH_4$ or $^{13}CH_4$, showing the position of the ^{13}C-DNA fraction (arrow) that was collected. Scale bar = 1 cm.

identification of microorganisms that are actively involved in methane oxidation. It was demonstrated that the DNA of pure cultures of methanotrophs grown on either $^{12}CH_3OH$ or $^{13}CH_3OH$ can be separated from each other using CsCl/ethidium bromide density gradient centrifugation (Figure 9.4a), enabling characterization of the microbial community involved in substrate metabolism independently of cultivation. We named this technique 'Stable Isotope Probing' (SIP) (Radajewski *et al.* 2000).

Methylotrophs are microorganisms that can use reduced one-carbon compounds, such as methane and methanol, as a sole source of carbon and energy. We used SIP to identify the active aerobic methanol and methane utilizing bacteria in oak soil (0–5 cm depth), and characterized the community using both phylogenetic (16S rRNA) and functional (*pmoA*) gene probes. 'Heavy' DNA was recovered from oak soil (0–5 cm depth) exposed to methanol (0.5% v/w) (Figure 9.4b,c). Despite repeated attempts, no 'heavy' DNA band could be observed after oak soil was exposed to $^{13}CH_4$ (8% v/v), even though CH_4 was rapidly oxidized. However, when dilute minimal medium (containing N and P and trace elements, but no added organic carbon) was added to this soil and the slurry was exposed to $^{13}CH_4$, a distinct ^{13}C-DNA fraction was observed (Figure 9.4d). This illustrated that there was only DNA synthesis after N and P were added and the bacteria were actively growing. Presumably the MOB were initially growth-limited by lack of nutrients in the soil and addi-

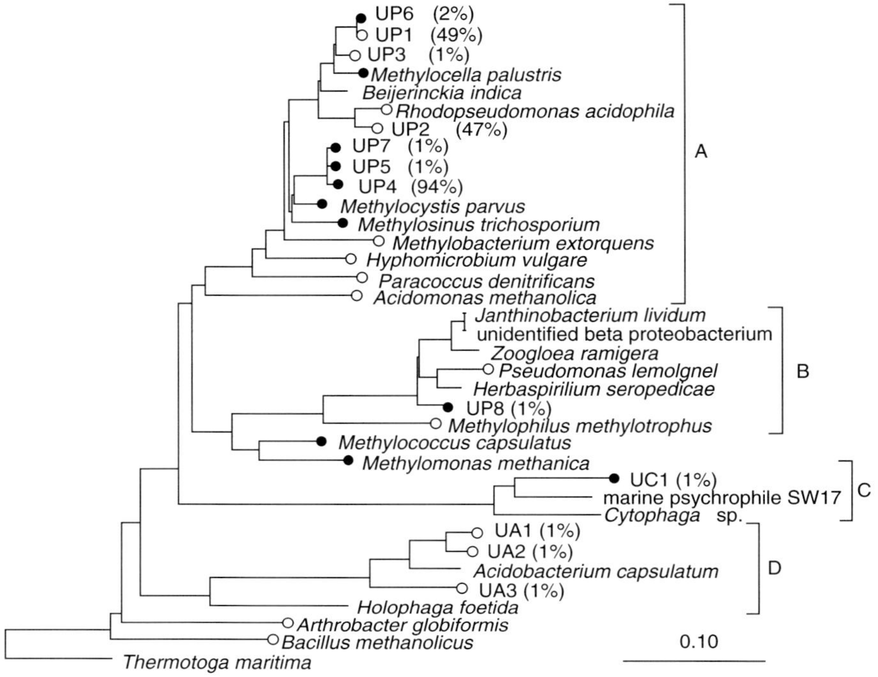

Figure 9.5 Phylogeny of bacterial 16S rDNA sequences amplified from ^{13}C-DNA. Uncultivated *Proteobacteria* (UP) clones UP1–UP3 and uncultivated *Acidobacterium* (UA) clones UA1–UA3 were recovered from a CH_3OH (0.5% v/w) SIP experiment. Clones UP4–UP8 and an uncultivated *Cytophagales* clone (UC1) were recovered from a CH_4 (8% v/v) slurry SIP experiment. Numbers in brackets indicate the percentage of each library represented by that clone type. Sequences were recovered in one of four phylogenetic groups; A, α-*Proteobacteria*; B, β-*Proteobacteria*; C, *Cytophagales*; D, *Acidobacterium*. Known genera or strains of CH_3OH (open circle) and CH_4 (filled circle) utilizing bacteria are shown. Scale bar = 0.1 inferred nucleotide substitution per nucleotide position.

tion of nutrient permitted active growth and DNA synthesis. Sequences from two distinct groups of microorganisms were retrieved from the methanol experiments, which included members of the α-*Proteobacteria* and the *Acidobacterium* division (Figure 9.5). Except for one sequence (UC1), all clones from the methane slurry were identified as members of the *Proteobacteria*, most of which were closely related to the extant methanotrophs *Methylocystis* and *Methylocella*.

Analysis of PmoA clones recovered from the methane slurry SIP experiment identified a cluster of *Methylocystis*-related sequences, which reflected the observations from the 16S rDNA analysis. Unexpectedly, a strong *pmoA*/*amoA* PCR product was obtained from the ^{13}C-DNA from the methanol SIP experiment. Sequence analysis identified that nearly all these clones were related to the AmoA of *Nitrosomonas europaea*. Analysis of the ^{13}C-DNA with nitrifier specific 16S rDNA primers

(G. Webster and J. Prosser, University of Aberdeen) confirmed the presence of nitrifier-related sequences, suggesting AOB may form a minor component of this DNA fraction. However, *amoA*-specific primers failed to generate a product, raising the possibility that an unusual form of the gene *amoA* may be present in certain bacteria within the oak forest soil.

Information about the closest cultivated representatives to dominant OTUs was used to formulate isolation strategies for the bacteria that were active under conditions approaching those occurring *in situ*. Enrichments were established for the isolation of methanol utilizers similar to UP1, using conditions suitable for the isolation of *Methylocella palustris* and *Beijerinckia indica*. Analysis of DNA extracted from enrichments in an acidic (pH 5.3) medium without a nitrogen source identified 16S rDNA sequences highly similar to sequences retrieved directly from the ^{13}C-DNA in the methanol SIP experiment. As the pH of the Gisburn forest soils is acidic (pH 4.0), these results support the hypothesis that acidophilic methylotrophs were detected in the SIP experiments.

Metabolic potentials

A programme of regular measurements of MOP and AOP was undertaken at Gisburn from November 1998 to November 1999 with soil samples from the triplicated trial plots. Triplicate soil samples were taken from the 0–5 cm layer of the A horizon of each trial plot. Soils were sieved to remove stones and roots, and to obtain an homogenized soil sample.The water content of soils can have a major influence on their ability to oxidize CH_4 because of diffusion limitation of both CH_4 and O_2 transport (Whalen *et al.* 1990). The water content of each soil sample was adjusted to 35% of water-holding capacity with a pressure plate, which preliminary work (Reay *et al.* 2001) had shown to be the optimum water content for measurements of CH_4 oxidation. Aliquots (10 g dry weight) of soil were placed in serum bottles (120 ml volume) sealed with a butyl rubber septum and CH_4 was added to the headspace to give a concentration of 50 000 p.p.m.v. (5% v/v), which was a saturating amount of CH_4. The serum bottles were placed on a roller bed (70 r.p.m.) immersed in a water bath adjusted to hold the bottles at the optimum temperature for methane oxidation. The roller beds prevented diffusion limitation of CH_4 oxidation rates.

There was considerable variation in MOP between the replicate plots even under the same tree species. Experimental block 1 tended to have lower MOP values than those measured for the other two blocks. The data are shown in Figure 9.6a as the annual mean values of MOP in all plots. MOP was generally greater in oak soil and in the grassed control plots, than in the Scots pine and Norway spruce soil. MOP in alder soils was significantly less (ANOVAR $P < 0.05$) than in oak or in grass controls: any potential for CH_4 oxidation was essentially absent from the alder soil where rates were not significantly different from zero. It was somewhat surprising, given the generally water-saturated nature of their soil for much of the year and the limited extent of their surface oxic zone, that MOP was as high in the grassed control plots as in the oak soils. Field measurements showed (Table 9.2) that there were net emissions of

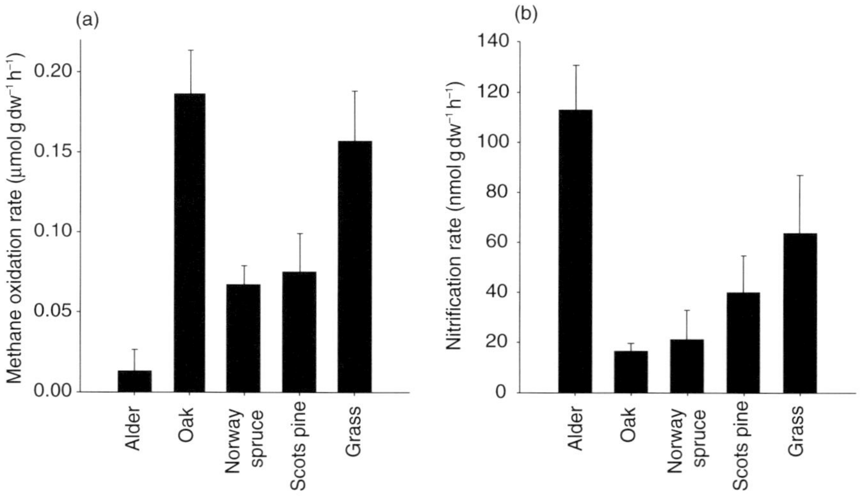

Figure 9.6 Annual averages of (a) methane oxidation potential and (b) ammonium oxidation potential in the soils from different experimental plots at Gisburn. Data show mean values (with standard errors, $n = 21$) from seven bimonthly samplings.

Table 9.2 Seasonal and annual average net fluxes of CH_4 between soil and air in the different plots at Gisburn.

	Grass	Oak	Pine	Alder	Norway
July–Sept	37.2 ± 28.8	−2.5 ± 21.9	10.0 ± 14.6	−8.4 ± 8.2	7.8 ± 25.6
Oct–Dec	79.7 ± 65.7	−8.6 ± 17.5	−6.6 ± 7.6	−2.9 ± 12.9	−8.3 ± 9.3
Jan–Mar	90.1 ± 88.2	5.9 ± 17.2	−6.7 ± 5.6	−5.3 ± 7.1	−7.5 ± 3.5
April–Jul	85.0 ± 89.0	1.5 ± 18.3	−15.5 ± 9.6	−10.7 ± 9.3	−13.0 ± 9.7
Total	72.3 ± 67.9	−0.7 ± 18.8	−4.2 ± 9.4	−6.8 ± 9.2	−4.8 ± 12.2

Rates are $\mu g\ CH_4\ m^{-2}\ h^{-1}$ (±SE, $n = 6$).

CH_4 from the control plot soils, reflecting active subsurface formation of CH_4 within their water-saturated anoxic zone. This contrasted with the forested soils where there was no detectable formation of CH_4 and no CH_4 emissions from their very much more drained and oxic soils. It may be concluded that the greater availability of CH_4 within the control plots maintained a high MOP despite the rather limited depth of the surface oxic layer within which CH_4 oxidation could occur. Even in completely water-saturated soils the roots of higher plants may conduct significant amounts of oxygen downwards, and maintain high MOPs even within apparently anoxic soil (Watson *et al.* 1997).

Ammonium oxidation potentials (AOP) were measured similarly with further aliquots of soils to which 20 ml of ammonium chloride solution was added to give a saturating concentration (5 mM final concentration) of ammonium, and an amount of ^{15}N-ammonium equivalent to 10% of total ammonium concentration was also added. The soil samples were then incubated as described for MOP. At the end of the experiment nitrate was extracted by vigorous shaking in 500 ml UHP water and passed through an ion exchange resin (Amberlite IRA–400, 16–50 mesh) to selectively recover ^{15}N-nitrate from residual ^{15}N-ammonium. The conversion of ^{15}N-ammonium to ^{15}N-nitrate was determined from changes in the $^{14/15}$N ratios of nitrate (measured by mass spectrometry) and hence the AOP.

In contrast to MOP, the AOP in the alder soil was significantly (ANOVAR, $P < 0.05$) greater than in the other plots, indicating the presence of a large and active nitrifying community which was much reduced in the other soils. AOP in oak soil was the smallest (Figure 9.6b). Apart from the grass control plots, there appeared to be an inverse relationship between MOP and AOP. This suggested that a significant capacity for CH_4 oxidation was not maintained in the soil by the presence even of a very active nitrifying community, in contrast to the suggestion (Hyman & Wood 1983; Jones & Morita 1983) that low affinity of AMO for CH_4 might result in significant CH_4 oxidation by nitrifying bacteria.

None of the trial plots in the experiment had received any artificial fertilizer, so it was not possible to correlate application of artificial fertilizers with changes in the soil methane sink. However, the alder soil provided a natural experiment where the soil nitrogen was significantly higher than the other soils, and any significant capacity for CH_4 oxidation had disappeared. The reasons for this were examined experimentally.

Soil chemistry manipulation experiments were used to examine the effect of nitrate, nitrite, ammonium, sodium chloride and polyethylene glycol (PEG—a compound that alters soil water potential without changing ionic concentration) on CH_4 oxidation. Solutions of a range of concentrations of these solutes were added to aliquots of oak soil and the MOP in the resultant soil measured. Where nitrate concentrations in the low nitrate soil (oak) were increased to levels equivalent to those in the high nitrate alder soil, a clear reduction in CH_4 oxidation capacity was observed. Similarly, additions of nitrite, ammonium and sodium chloride also reduced CH_4 oxidation capacity but at different concentrations. Methane oxidation was completely inhibited with initial nitrate, nitrite, ammonium and sodium chloride concentrations of ~500, 1000, 5000 and 10 000 μM respectively. Addition of PEG inhibited MOP at the same concentration as for sodium chloride, indicating that the inhibitory effect of added solutes on MOP was apparently through a non-specific effect due to increased soil water potentials, rather than to increased ionic concentration. These data indicate that inhibition of MOP was not attributable to direct inhibition of MMO in methane oxidizers by ammonia as the inhibition was even stronger by nitrate. Inhibition by ammonia is often related to pH as it changes the equilibrium between NH_4^+ and the NH_3 which is the toxic component. However, the soil pH in our experiments was constant. Benstead and King (2001) demonstrated in Maine forest soil that addition of nitrate inhibited methane oxidation more strongly than addition of sulfuric acid, which eliminated any indirect effect of pH on ammonium in favour of a direct effect of nitrate on methane oxidation. Suggestions (e.g. Schnell & King 1994) that nitrite, an intermediate in ammonium oxidation to nitrate, is a more potent inhibitor of MMO were also discounted by our experiments which showed that it was less effective than nitrate in inhibiting methane oxidation. Clearly, the high concentration of nitrate in alder soil would, on the basis of the results on MOP in oak soil, account for inhibition of methane oxidation.

Large changes in nitrogen species occurred during several of these experiments, particularly in nitrite and ammonium addition experiments. Nitrous oxide production rates and the appearance of increased nitrate concentrations indicated rapid nitrification of ammonium and nitrite. Much of the ammonium initially added was also rapidly bound to soil colloids, so limiting the direct effect of ammonium addition on CH_4 oxidation.

Field studies

Prior to the main field study, intact soil cores were collected from oak and alder plots from Gisburn forest and returned to a constant temperature room maintained at 20°C in the dark. Soil moisture status was standardized to field capacity with deionized water and a Perkin–Elmer 8500 gas chromatograph was used for continuous monitoring of CH_4 concentrations in the airstreams passed over the surface of the soil cores. These experiments therefore examined pseudo *in situ* rates of CH_4 exchange, rather than potentials. Net CH_4 oxidation was found to be significantly lower ($P < 0.05$) for the cores taken from beneath alder ($-10.7 \pm 3.1\,\mu g\,CH_4\,m^{-2}\,h^{-1}$) than for those from oak ($-28.7 \pm 1.4\,\mu g\,CH_4\,m^{-2}\,h^{-1}$). It should be noted that although these data show a significant difference in net CH_4 flux between these plots, it cannot be stated whether these observed differences were a result of changes in methanotrophy or methanogenesis.

The major field study to compare CH_4 flux rates from the soils beneath different tree species utilized the full randomized block design available at Gisburn. One measuring point was randomly assigned in each replicate plot across all three blocks. Additional sampling points were placed in each plot of block 2, adjacent to the automatic weather stations in order to closely relate physical and climatic factors to *in situ* CH_4 fluxes. Methane fluxes were determined on 23 occasions between July 1999 and July 2000 using a closed chamber technique (Bradford *et al.* 2001). The chamber was sealed with a Perspex lid and surrounded by a rubber sleeve to ensure the system was gas-tight. Headspace gas was analysed using a Perkin–Elmer Autosystem XL Gas Chromatograph fitted with a flame ionization detector. Methane flux was calculated from the start and end concentrations of methane in the headspace (approx. 2 L) of the chamber.

Net CH_4 flux rates, averaged annually and across each season for the period of study, are shown in Table 9.2. The flux rates for grassland were much higher than any of the tree plots for the majority of the study period, with net outputs of CH_4 compared to net uptakes of CH_4 in forested plots. These differences were not statistically significant because of the large variances in the grassed control plots. These high variances were attributed to the large interblock variability in net CH_4 flux rates observed throughout the study.

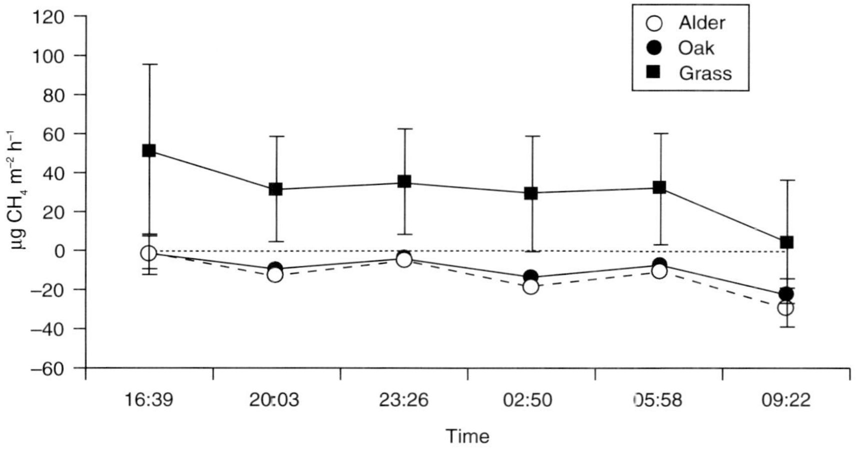

Figure 9.7 *In situ* CH_4 exchange rates measured with a mobile laboratory over 24 hours in Gisburn block 2 during August 1999, for control, oak and alder plots (mean ± SE, $n = 3$).

A mobile laboratory was also used to monitor field CH_4 fluxes from oak, alder and grassland within block 2 over short sampling periods. Samples were taken via PTFE tubing attached to 20 cm diameter galvanized steel chambers inserted 5 cm into the ground. The GC was configured as in the laboratory experiment, but with nine sequentially scanned ports to allow for the greater sample numbers. All data analyses and statistical comparisons were performed using SAS (SAS Institute 1988) with repeated-measures ANOVA used to assess the impacts of treatment additions on trace gas fluxes. Figure 9.7 shows data for a 24 hour monitoring period in August 1999, showing mean CH_4 fluxes from grass, oak and alder of 30.4 ± 31.2; −9.7 ± 3.4 and −12.9 ± 4.6 μg CH_4 m^{-2} h^{-1}, respectively. Flux rates measured using continuous flow monitoring were similar to those measured using static chambers. As with the static chambers, variability of the rates for grassland plots was again high. There were no significant differences detected between the treatments, although grassland again was a net producer during this measurement period.

Conclusions

Our data demonstrated that, as shown in other studies, change in land use significantly influenced the size of the soil methane sink. At Gisburn the conversion from grassed acidic upland soils to forestry had a major effect on the soil environment. Uptake of CH_4 from the air across the soil–air interface was greatly stimulated in the forested plots compared to the grassed control plots because of the deeper water table in forested plots; with the exception of alder plots where CH_4 was not removed from the air by the soil MOB. In contrast to the forested plots, the control plot soils always exhibited net output of CH_4 to the atmosphere.

Interestingly, an MOP broadly similar in magnitude to those in the forested plots remained in the soil of the grassed control plots, despite their more water saturated nature. However, the *in situ* measurements showed that, despite this high MOP, there was net output of methane to the atmosphere from the control plots, presumably because methanogenesis was greater in these more water logged soils. Clearly, the considerable metabolic potential for CH_4 oxidation in the soil of the control plots was limited by environmental variables, and was not expressed *in situ*.

There was a broadly inverse relationship between the magnitude of MOP and AOP in the various soils, but the presence of even a considerable AOP in alder soil was not sufficient to maintain any significant MOP. Therefore there was no evidence that this active nitrifying community was capable of oxidizing methane by low affinity of *amo* for methane: an observation also reported in Maine forest soil (Schnell & King 1994).

The inhibition of CH_4 oxidation in alder soil was not the result of inhibition of MMO by ammonium, but rather seemed to be the consequence of high nitrate concentrations in the alder soil. Experiments confirmed that additions of nitrate to oak soil, which had the greatest MOP, completely inhibited CH_4 oxidation at nitrate concentrations equivalent to that in alder soil, but the mechanism for this inhibition is not known. However, it is evident that, whatever the mechanism, increase of soil solute concentrations, either by artificial fertilization or by other mechanisms such as N_2 fixation, is likely to diminish the soil CH_4 sink relative to undisturbed soils.

There was no evidence for qualitative changes in the phylogenetic composition of the MOB and AOB communities of soils under the different land-use regimes. 16S rDNA sequences characteristic of both CH_4 and ammonium oxidizers were present in all soils, even in soils where CH_4 or ammonium oxidation potentials were not measurable. However, there may be differences in the amounts or numbers of the various phylogenetic groups present in the different soils, which could not be quantified with the techniques used, and which remains to be examined in future work. Stable isotope probing with ^{13}C-labelled methanol demonstrated the presence of microorganisms capable of oxidizing C_1 compounds such as methanol which had not previously been recognized as methylotrophs, although the bacteria identified as active methanotrophs with $^{13}CH_4$ *in situ* were predominantly of extant methanotrophs

Acknowledgements

This work was funded by the Natural Environment Research Council's thematic programme EDGE (Ecological Dynamics and Genes).

References

Benstead, J. & King, G.M. (2001) The effect of soil acidification on atmospheric methane uptake by a Maine forest soil. *FEMS Microbiology Ecology* **34**, 207–212.

Bourne, D.G., Holmes, A.J., Iversen, N. & Murrell, J.C. (2000) Fluorescent oligonucleotide rDNA probes for specific detection of methane oxidizing bacteria. *FEMS Microbiology Ecology* **31**, 29–38.

Bouwman, A.F. (1990) Introduction. In: *Soils and*

the Greenhouse Effect (ed. A. F. Bouwman), pp. 25–32. John Wiley, Chichester.
Bowman, J.P., Sly, L.I., Nicholas, P.D. & Hayward, A.C. (1993) Revised taxonomy of the methanotrophs: description of *Methylobacter* gen. nov., emendation of *Methylococcus*, validation of *Methylosinus* species and a proposal that the family *Methylococcaceae* includes only group I methanotrophs. *International Journal of Systematic Bacteriology* **43**, 735–753.
Bradford, M.A., Ineson, P., Wookey, P.A. & Lappin-Scott, H.M. (2001) The effects of acid nitrogen and acid sulphur deposition on CH_4 oxidation in a forest soil: a laboratory study. *Soil Biology and Biochemistry* **33**, 1695–1702.
Brown, A.F.H. (1992) Functioning of mixed-species stands at Gisburn, NW England. In: *The Ecology of Mixed Species Stands of Trees* (eds M. G. R. Cannell, D. C. Malcolm & P. A. Robertson), pp 125–150. Blackwell, Oxford.
Brusseau, G.A., Bulygina, E.S. & Hanson, R.S. (1994) Phylogenetic analysis and develpment of probes for differentiating methylotrophic bacteria. *Applied and Environmental Microbiology* **60**, 626–636.
Costello, A.M. & Lidstrom, M.E. (1999) Molecular characterization of functional and phylogenetic genes from natural populations of methanotrophs in lake sediments. *Applied and Environmental Microbiology* **65**, 5066–5074.
Crill, P.M., Martikainen, P.J., Nykanan, P. & Silvola, J. (1994) Temperature and N fertilization effects on methane oxidation in a drained peatland soil. *Soil Biology and Biochemistry* **26**, 1331–1339.
Crutzen, P.J. (1991) Methane's sinks and sources. *Nature* **350**, 380–381.
Dedysh, S.N., Liesack, W., Khmelenina, V.N. *et al.* (2000) *Methylocella palustris* gen. nov., sp. nov., a new methane-oxidizing acidophilic bacterium from peat bogs, representing a novel subtype of serine-pathway methanotrophs. *International Journal of Systematic and Evolutionary Microbiology* **50**, 955–969.
Dobbie, K.E., Smith, K.A., Prieme, A. *et al.* (1996) Effect of land use on the rate of methane uptake by surface soils in northern Europe. *Atmospheric Environment* **30**, 1005–1011.
Eller, G., Stubner, S. & Frenzel, P. (2001) Group-specific 16S rRNA targeted probes for the detection of type I and type II methanotrophs by fluorescence *in situ* hybridisation. *FEMS Microbiology Letters* **198**, 91–97.
Goulding, K.W.T., Willison, T.W., Webster, C.P. & Powlson, D.S. (1996) Methane fluxes in aerobic soils. *Environmental Monitoring and Assessment* **42**, 175–187.
Henckel, T., Friedrich, M. & Conrad, R. (1999) Molecular analyses of the methane-oxidizing microbial community in rice field soil by targeting the genes of the 16S rRNA, particulate methane monooxygenase, and methanol dehydrogenase. *Applied and Environmental Microbiology* **65**, 1980–1990.
Henckel, T., Jäckel, U., Schnell, S. & Conrad, R. (2000) Molecular analyses of novel methanotrophic communities in forest soils that oxidize atmospheric methane. *Applied and Environmental Microbiology* **66**, 1801–1808.
Holmes, A.J., Costello, A., Lidstrom, M.E. & Murrell, J.C. (1995) Evidence that particulate methane monooxygenase and ammonia monooxygenase may be evolutionarily related. *FEMS Microbiology Letters* **132**, 203–208.
Holmes, A.J., Roslev, P., McDonald, I.R. *et al.* (1999) Characterization of methanotrophic bacterial populations in soils showing atmospheric methane uptake. *Applied and Environmental Microbiology* **65**, 3312–3318.
Hütsch, B.W., Webster, C.P. & Powlson, D.S. (1993) Long term effects of nitrogen fertilization on methane oxidation in soil from the Broadbalk wheat experiment. *Soil Biology and Biochemistry* **25**, 1307–1315.
Hütsch, B.W., Webster, C.P. & Powlson, D.S. (1994) Methane oxidation in soil as affected by land use, soil pH and N fertilization. *Soil Biology and Biochemistry* **26**, 1613–1622.
Hyman, M.R. & Wood, P.M. (1983) Methane oxidation by *Nitrosomonas europaea*. *Biochemistry Journal* **212**, 31–37.
Jones, R.D. & Morita, R.Y. (1983) Methane oxidation by *Nitrosococcus oceanus* and *Nitrosomonas europaea*. *Applied and Environmental Microbiology* **45**, 401–410.

Kightley, D., Nedwell, D.B. & Cooper, M. (1995) Capacity for methane oxidation in landfill cover soils measured in laboratory-scale microcosms. *Applied and Environmental Microbiology* **61**, 592–601.

King, G.M. & Schnell, S. (1998) Effects of ammonium and non-ammonium salts additions on methane oxidation by *Methylosinus trichosporium* OB3b and Maine forest soils. *Applied and Environmental Microbiology* **64**, 253–257.

McDonald, I.R. & Murrell, J.C. (1997) The particulate methane monooxygenase gene *pmoA* and its use as a functional gene probe for methanotrophs. *FEMS Microbiology Letters* **156**, 205–210.

McDonald, I.R., Hall, G.H., Pickup, R.W. & Murrell, J.C. (1996) Methane oxidation potential and preliminary analysis of methanotrophs in blanket bog peat using molecular ecology techniques. *FEMS Microbiology Ecology* **21**, 197–211.

Moffat, A.J. & Boswell, R.C. (1990) Effect of tree species and species mixtures on soil properties at Gisburn Forest, Yorkshire. *Soil Use and Management* **6**, 46–51.

Mosier, A., Schimel, D., Valentine, D., Bronson, K. & Parton, W. (1991) Methane and nitrous oxide fluxes in native fertilized and cultivated grass-lands. *Nature* 350, 330–332.

Murrell, J.C. & Radajewski, S. (2000) Cultivation-independent techniques for studying methanotroph ecology. *Research in Microbiology* **151**, 807–814.

Murrell, J.C., Gilbert, B. & McDonald, I.R. (2000) Molecular biology and regulation of methane monooxygenase. *Archives of Microbiology* **173**, 325–332.

Nesbit, S.P. & Breitenbeck, G.A. (1992) A laboratory study of factors influencing methane uptake by soils. *Agriculture, Ecosystems and Environment* **41**, 39–54.

Ojima, D.S., Valentine, D.W., Mosier, A.R., Parton, W.J. & Schimel, D.S. (1993) Effect of land use change on methane oxidation in temperate forest and grassland soils. *Chemosphere* **26**, 675–685.

Radajewski, S., Ineson, P., Parekh, N.R. & Murrell, J.C. (2000) Stable-isotope probing as a tool in microbial ecology. *Nature* **403**, 646–649.

Reay, D.S., Nedwell, D.B. & McNamara, N. (2001) Physical determinants of methane oxidation capacity in a temperate soil. *Water Air and Soil Pollution: Focus* **1**, 401–414.

SAS Institute (1988) *SAS/STAT™ Users Guide*, release 6.03 edition. SAS Institute, Cary, NC.

Schnell, S. & King, G.M. (1994) Mechanistic analysis of ammonium inhibition of atmospheric methane consumption in forest soils. *Applied and Environmental Microbiology* **60**, 3514–3521.

Steudler, P.A., Bowden, R.D., Melillo, J.M. & Aber, J.D. (1989) Influence of nitrogen fertilisation on methane uptake in temperate forest soils. *Nature* **341**, 314–316.

Watson, A., Stephen, K.D., Nedwell, D.B. & Arah, J.R.M. (1997) Oxidation of methane in peat: kinetics of CH_4 and O_2 removal and the role of plant roots. *Soil Biology and Biochemistry* **29**, 1257–1267.

Whalen, S.C., Reeburgh, W.S. & Sandbeck, K.A. (1990) Rapid methane oxidation in a landfill cover soil. *Applied and Environmental Microbiology* **56**, 3405–3411.

Willison, T.W., Webster, C.P., Goulding, K.W.T. & Powlson, D.S. (1995) Methane oxidation in temperate soils: effects of land use and the chemical form of nitrogen fertilizer. *Chemosphere* **30**, 539–546.

Wise, M.G., McArthur, J.V. & Shimkets, L.J. (2001) *Methylosarcina fibrata* gen. nov., sp. nov. and *Methylosarcina quisquiliarum* sp. nov., novel type I methanotrophs. *International Journal of Systematic and Evolutionary Microbiology* **51**, 611–621.

Chapter 10

Molecular genetic analysis of the ammonia-oxidizing bacterial community in a defined hypereutrophic freshwater lake

*C. B. Whitby, R. A. Meade, G. Hall, J. R. Saunders, R. W. Pickup and A. J. McCarthy**

Background

Chemolithoautotrophic ammonia-oxidizing bacteria (AOB) play a crucial role in freshwater nitrification, a key stage in the natural cycling of nitrogen. Nitrification is a two-stage process in which ammonia is oxidized to nitrite by ammonia-oxidizing bacteria and nitrite is oxidized to nitrate by nitrite-oxidizing bacteria. Nitrification has a significant ecological impact because nitrates are more easily leached from the soil matrix than ammonia. Moreover, nitrate can be used as an electron acceptor for denitrification. Both of these mechanisms contribute to fertilizer loss from agricultural soils and pollution of water supplies with nitrates (Hall 1986). In addition, AOB activity is important in wastewater treatment to achieve the necessary reduction in the ammonia content of sewage prior to discharge, reducing the risk of oxygen depletion, preventing eutrophication and protecting aquatic life (Hall 1986). Therefore, understanding the dynamics of nitrifier populations in both terrestrial and aquatic environments has been the subject of considerable research effort (e.g. Wagner *et al.* 1995; Pommerening-Röser *et al.* 1996; Stephen *et al.* 1996; Kowalchuk *et al.* 1997; Ward *et al.* 1997; Hastings *et al.* 1998; Phillips *et al.* 1999; Whitby *et al.* 1999; Bano & Hollibaugh 2000).

In view of the difficulty in culturing AOB in the laboratory, microbial ecologists have been faced with the challenge of correlating ammonia-oxidizer population diversity with environmental parameters. There are two major groups of ammonia-oxidizers within the β-*Proteobacteria* (the nitrosospiras and the nitrosomonads) with only one described species, *Nitrosococcus oceanus*, outside this phylum in the gamma subclass of the *Proteobacteria* (Head *et al.* 1993). For a long time, the physiology and biochemistry of ammonia oxidation was based on laboratory experiments using the nitrosomonad *Nitrosomonas europaea*, probably because it is the most easily isolated and cultured ammonia-oxidizer species (Prosser 1989). Now that it is

* Correspondence author: aj55m@liverpool.ac.uk

possible to determine the composition of ammonia-oxidizing communities by direct analysis of environmental DNA, the importance and greater ubiquity of nitrosospiras has been revealed (Hiorns *et al.* 1995; Stephen *et al.* 1996; Kowalchuk *et al.* 1997, 1999; Hastings *et al.* 1998; Bano & Hollibaugh 2000). The application of 16S rDNA sequence analysis to investigate ammonia-oxidizer population diversity in various environments is now widespread and has allowed the 16S rRNA sequence data base to be considerably expanded (Stephen *et al.* 1996; Hastings *et al.* 1998; Speksnijder *et al.* 1998; Whitby *et al.* 1999; Bano & Hollibaugh 2000; Kowalchuk *et al.* 2000; Purkhold *et al.* 2000).

More recently, the range of molecular tools to infer the presence of nitrifiers in environmental samples has been extended by the development of techniques specific for the ammonia monooxygenase structural gene *amoA,* which occurs in all chemoautotrophic AOB (McTavish *et al.* 1993; Sinigalliano *et al.* 1995; Rotthauwe *et al.* 1997; Stephen *et al.* 1998; Mendum *et al.* 1999; Purkhold *et al.* 2000). The *amoA* gene has provided an alternative marker for molecular ecological studies on both nitrosospiras and nitrosomonads and has been used to identify the presence of species that could not be detected using the more conventional phylogenetic target, the 16S rRNA gene (Sinigalliano *et al.* 1995; Hastings *et al.* 1997; Rotthauwe *et al.* 1997; Stephen *et al.* 1998; Purkhold *et al.* 2000). Purkhold *et al.* (2000) have recently reevaluated existing 16S rRNA and *amoA* gene probes and amplification primers, as well as conducting a thorough phylogenetic analysis of all available 16S rRNA and *amoA* ammonia-oxidizer sequences to provide a framework for future molecular ecological studies on these organisms.

The correlation of species identification and detection with metabolic function in the environment is vital in microbial ecology. A number of approaches have been applied to AOB, of which fluorescent *in situ* hybridization (FISH) using 16S ribosomal RNA as the target is the most common. Actively metabolizing bacterial cells will contain high numbers of ribosomes, the RNA of which will bind to labelled oligonucleotide probes that are introduced into permeabilized, fixed cells (Amman *et al.* 1995). The technique is not universally applicable, and although well suited to the *in situ* analysis of the large AOB populations that colonize wastewater treatment plants (Wagner *et al.* 1995), high background fluorescence in combination with low abundance of AOB in freshwater limits its application. Similarly, recovery of *amoA* messenger RNA and its reverse transcription into cDNA as a direct method for identifying active AOB species is likely to suffer from problems of insufficient abundance for detection. Radajewski *et al.* (2000) have developed a technique for analysing the active components of microbial populations. The method relies on the organism utilizing an isotopically enriched substrate, followed by separation of labelled and unlabelled DNA by ultracentrifugation. Although there is now the potential for widespread application of this stable isotope probing in microbial ecology, the requirement for a period of laboratory incubation must be borne in mind when interpreting data.

The molecular ecology of AOB has been comprehensively reviewed by Kowalchuk and Stephen (2001).

(a) (b)

Figure 10.1 Aerial photographs of (a) Esthwaite Water and Priest Pot and (b) Priest Pot.

Priest Pot—the study site

Priest Pot is a small (1 ha), shallow (3.5 m max. depth) hypereutrophic freshwater lake situated at the northern end of Esthwaite Water in the English Lake District, UK (Figure 10.1). The lake has been studied over many years and the combination of shallow water and sheltered location combine to provide steep chemical gradients that are sites of intense biological activity. Priest Pot has high levels of essential nutrients compared to other freshwater lakes in the catchment area. (e.g. nitrate $\geq 1\,mg\,L^{-1}$; ammonia $\geq 1\,mg\,L^{-1}$; inorganic carbon $\leq 4\,mmol\,L^{-1}$; total phosphorus $600\,\mu g\,L^{-1}$; chlorophyll a $300\,\mu g\,L^{-1}$ (Finlay *et al.* 1997). A summary of the interactions between the physical parameters, chemical fluxes and biological activity in Priest Pot is shown in Figure 10.2.

The increasing temperature of the surface water during early summer facilitates the establishment of a density gradient in the water column. As the elevated temperatures persist throughout the summer, the oxygen gradient moves higher in the water column and becomes more defined (Figure 10.3a). The sediment and deeper water can experience anaerobic conditions throughout the stratified period. Later in the year the temperature of the surface water decreases and the water column becomes mixed, leading to the collapse of stratification (Figure 10.3b). The basic limnology of Priest Pot has been described by Finlay *et al.* (1997).

The stratification that occurs in Priest Pot is unusual in that the oxy/thermocline occurs over a very narrow depth range (ca. 50–75 cm). Larger lakes such as Windermere stratify, but the oxy/thermocline occurs over several metres and the water column is unstable. In general, temperate freshwater lakes undergo stratification in summer to the extent that populations of AOB become concentrated at the oxy/thermocline where oxygen and ammonia gradients combine to produce conditions favourable for their proliferation (Hall 1986; Hastings *et al.* 1998; Whitby *et al.* 1999, 2001a). Therefore, Priest Pot is an excellent site in which to study the influence of changing environmental factors, both in time and space, on microbial community structure.

The concentrations of ammonia, nitrite and nitrate were monitored in the surface sediment and water column of Priest Pot during the seasonal cycle (1998–1999). Residual ammonia, nitrite and nitrate levels in lakewater were low ($<2\,\mu gN\,ml^{-1}$)

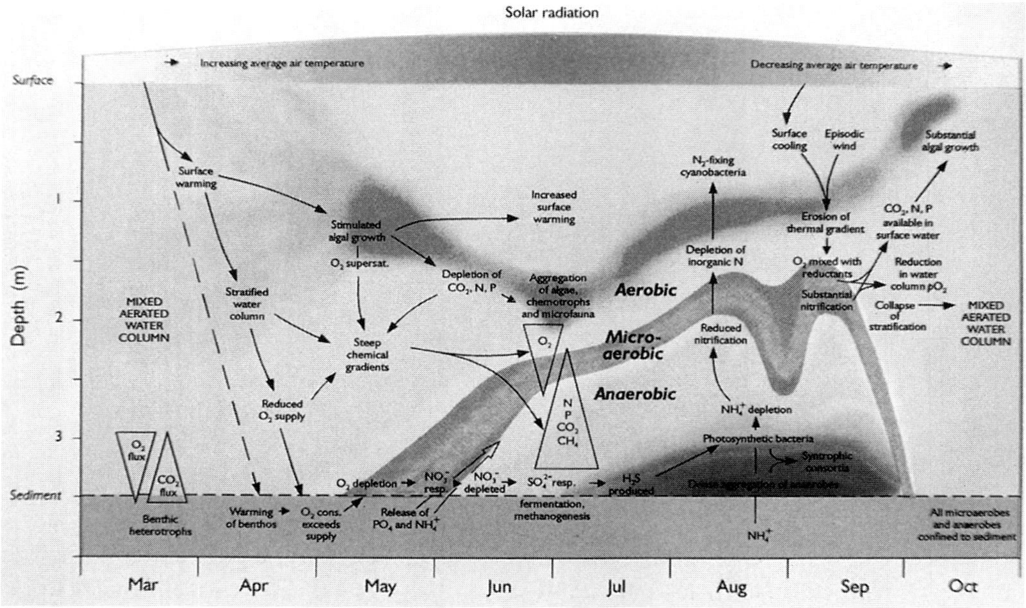

Figure 10.2 An illustration of some reciprocal interactions involving physical, chemical and microbiological variables in Priest Pot, based on data obtained in 1996. (Reproduced with permission from Finlay *et al.* 1997.)

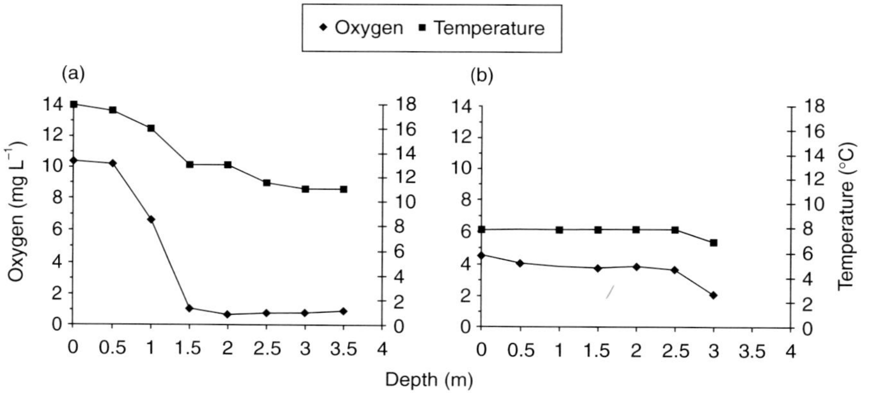

Figure 10.3 Graphs of temperature and oxygen concentrations in Priest Pot on (a) 6 July 1998 and (b) 5 October 1998.

throughout the year. Sediment was by contrast relatively rich in ammonia and nitrate, but not nitrite, which as an intermediate in both nitrification and denitrification, is unlikely to accumulate. The highest concentration of ammonia (250 μgN g dry weight^{-1}) was recorded in the summer months, when levels of ammonia in the surface sediment were high and showed the greatest monthly variation. Populations

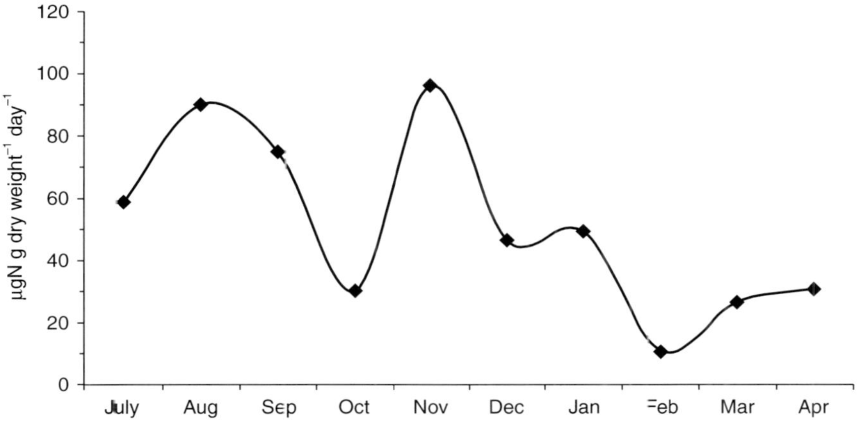

Figure 10.4 Nitrification potentials of Priest Pot sediment sampled at monthly intervals: July 1998 to April 1999; surface sediment 0–0.5 cm.

of microorganisms fluctuate in hypereutrophic Priest Pot and exert considerable effects on nutrient concentrations such that it is difficult to make correlations or draw conclusions from point measurements of inorganic nitrogen species.

Measurement of nitrification potential provides an indication of nitrifier activity and is done by incubating sediment samples with ammonia and determining nitrite/nitrate production over a defined period. Nitrification potentials for Priest Pot sediments are presented in Figure 10.4. Rates were variable during the summer when the lake had stratified, and peaked in November during overturn prior to a decrease over the winter months. The nitrification potentials measured in Priest Pot sediments are similar to those reported for profundal sediment samples of Esthwaite Water (Hastings *et al.* 1998), a larger eutrophic lake nearby (see Figure 10.1a).

AOB are chemoautotrophic and therefore cannot be enumerated in environmental samples by conventional microbiological plating procedures. A most probable number (MPN) technique, in which serially diluted samples are incubated in tubes of media and scored positive if nitrite/nitrate are detected after incubation is an accepted standard enumeration technique (Rowe *et al.* 1977). This approach generated estimates of 10^4–10^7 AOB g dry weight^{-1} of sediment, and only 10^1–10^4 AOB per 50 ml lakewater in the water column (Figure 10.5). The order of magnitude difference between sediment and lakewater AOB population sizes correlates with the higher concentrations of ammonia (substrate) detected in the sediment (see above) and the fact that nitrification potentials (see Figure 10.4) could only be obtained for sediment samples; lakewater nitrification potential was found to be below the threshold of detection. In addition, the MPN data for November support the nitrification potential data for that month. The MPN estimates for AOB can be compared to those obtained from other freshwater lakes of lower trophic status within the catchment area since the same methodologies were applied. Sediment counts are greater in Priest Pot (maximum 10^7 cells g dry weight^{-1}) than Windermere, a eutrophic/mesotrophic lake (maximum 10^6 cells g dry weight^{-1}), and Buttermere

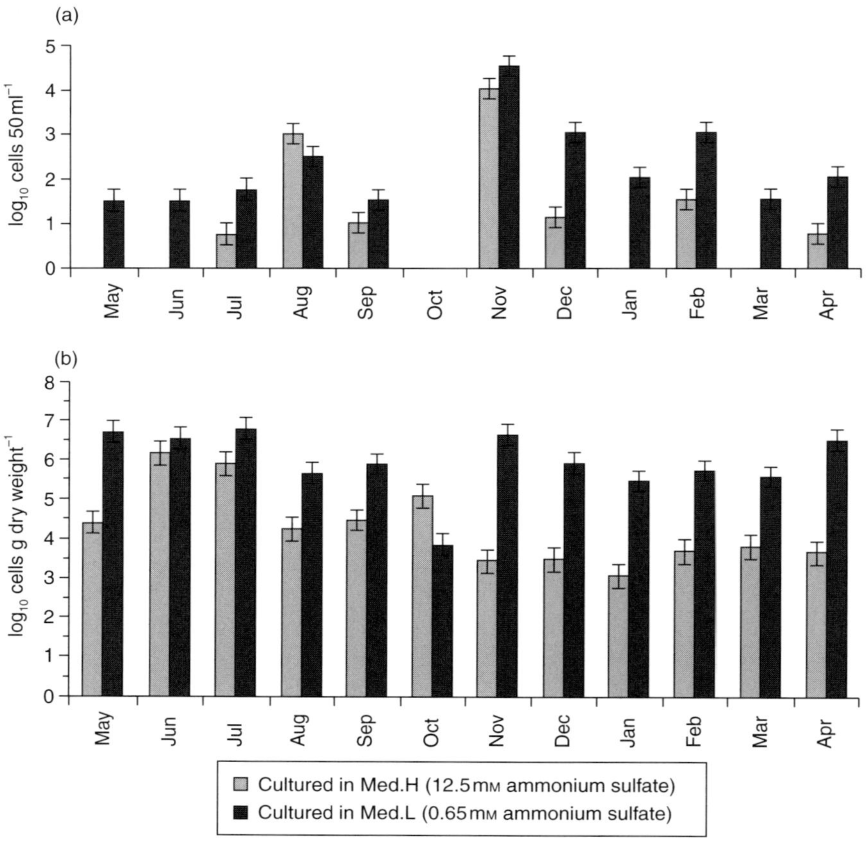

Figure 10.5 MPN values for AOB in Priest Pot. (a) Lakewater 1.5 m. (b) Surface sediment 0–0.5 cm. Bars illustrate 95% confidence limits.

(maximum 10^3 cells g dry weight^{-1}), an oligotrophic lake. However, while nitrification is directly detectable in Priest Pot water column samples taken for MPN determinations, there is an absolute requirement for 100-fold concentration of lakewater by tangential flow filtration to enable nitrification to be detected in Windermere and Buttermere water column samples. On the basis of MPN determinations therefore, both Priest Pot sediment and water column contain a high population of AOB in keeping with its classification as a hypereutrophic lake.

Genetic diversity of ammonia-oxidizing bacteria in Priest Pot

A parallel approach using *amoA* (ammonia monooxygenase gene) and 16S rRNA gene-specific primers and Temporal Temperature Gradient Gel Electrophoresis (TTGE) facilitated the analysis of temporal and spatial changes in the ammonia-oxidizer community structure in Priest Pot. The vast majority of ecologically significant AOB reside in the β-*Proteobacteria* division, and in addition to the recognition of two major genera (Head *et al.* 1993; Teske *et al.* 1994), 16S rDNA sequence analysis has enabled a number of clusters to be delineated (Stephen *et al.* 1996; Purkhold *et al.* 2000). We targeted three groups within this classification by applying 16S rDNA amplification (PCR) primers for all AOB (Kowalchuk *et al.* 1997), nitrosospiras (Hiorns *et al.* 1995) and the traditional *Nsm. europaea–eutropha* nitrosomonad lineage (Hiorns *et al.* 1995). In pristine freshwater environments, it is necessary to perform an initial amplification of total eubacterial 16S rDNA to detect AOB after the second round of specific amplification ('nested' PCR) (Hiorns *et al.* 1995; Hastings *et al.* 1998; Whitby *et al.* 1999, 2001a). 'Total' DNA extracted from Priest Pot lakewater and sediment samples during the 1998/99 season were analysed in this way and despite using a 'nested' PCR approach, no *Nsm. europaea–eutropha* 16S rDNA PCR products were recovered. By contrast, *Nitrosospira* spp. 16S rDNA was amplified from all samples of lakewater and sediment throughout the year, and their fidelity confirmed by obtaining hybridization signals with an internal ammonia-oxidizer specific oligonucleotide probe (AAO258; Hiorns *et al.* 1995). Although not a quantitative analysis *per se*, the consistent detection of the strongest hybridization signals with DNA amplified from sediments correlates with the nitrification potential and MPN data that demonstrated the presence of a much larger population of AOB in sediment compared to the water column. This scenario—ubiquity of nitrosospiras, failure to directly detect *Nsm. europaea–eutropha* and the concentration of AOB in the sediment—has been described for the adjoining lake, Esthwaite Water (Hastings *et al.* 1998).

Further analysis of the Priest Pot AOB community was pursued using the specific PCR primers (CTO189f-GC and CTO654r) designed by Kowalchuk *et al.* (1997) to amplify the 16S rDNA of all β-proteobacterial AOB. Amplified environmental DNA was analysed by Temporal Temperature Gradient Electrophoresis (TTGE) in which DNA fragments of identical size generated by PCR and unresolved by conventional gel electrophoresis, are separated to provide a genotypic fingerprint. These can be compared to observe the gross effects of environmental parameters on population composition, and further processed by determining the sequence of DNA bands excised from the gel matrix. The TTGE profiles obtained demonstrated that the AOB community structures in the sediment and lakewater were distinct from one another and that generally the diversity within the sediment ammonia-oxidizer populations did not fluctuate significantly throughout the year. The DNA fingerprints obtained from the lakewater sampled at different depths during the winter months were similar, but did fluctuate during the summer (August/September) when stratification clearly exerted effects on the AOB population within, above and below the oxycline. This is a clear demonstration of spatial segregation of a microbial population in a lake in response to changing environmental conditions, and is identifiable here because Priest Pot is such a well-defined ecosystem. These features are illustrated in an example of a TTGE gel presented in Figure 10.6, where in the August sample in particular the AOB banding pattern shows a strong shift between 1.0 and 1.5 m depth.

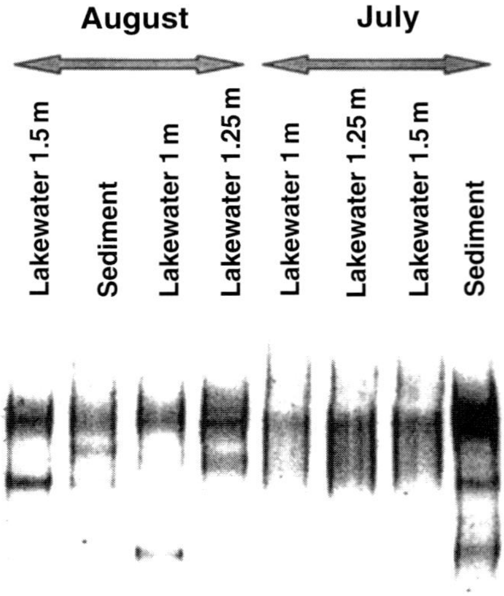

Figure 10.6 TTGE gel of 16S rDNA amplified from sediment and lakewater sampled in July and August.

Sequences for a total of 43 16S rDNA fragments (224 bp), excised from TTGE gels and re-amplified, were determined. Their phylogenetic analysis is presented in Figure 10.7. Representatives of five of the seven AOB 16S rDNA sequence clusters described by Stephen *et al.* (1996) were detected—three equated with *Nitrosospira* spp. and two with *Nitrosomonas* spp. Nitrosospiras of clusters one and five were not found, in keeping with the previous assignment of these two groups to marine rather than freshwater environments (Stephen *et al.* 1996). Again, *Nsm. europaea–eutropha* lineage sequences were only recovered from sediment and lakewater enrichments and not from environmental DNA (Figure 10.7). Most of the sequences obtained from sediment and lakewater that had not been enriched in laboratory media were located within four *Nitrosospira* clusters (Figure 10.7); three of these groups do not contain described species (Purkhold *et al.* 2000). This emphasizes the poor relationship between named, cultured species and AOB genotypes that predominate in the environment. The distribution pattern of these genotypes did not correlate with seasonal occurrence nor were they partitioned between water column and sediment.

Analysis of bacterial community diversity in the environment can be given added significance if more than one gene target is used. This requires the identification of a unique gene that shows sufficient variation to provide information that can be phylogenetically interpreted. For AOB, the gene encoding the α-subunit of ammonia monooxygenase (*amoA*) that contains the active site is such a gene (McTavish *et al.* 1993), because general congruence between phylogenetic trees of 16S rDNA and *amoA* sequences has been demonstrated (Rotthauwe *et al.* 1997; Purkhold *et al.* 2000). Gene fragments of *amoA* (428 bp) were amplified directly from all of the lakewater and sediment samples using AOB-specific primers (Rotthauwe *et al.* 1997;

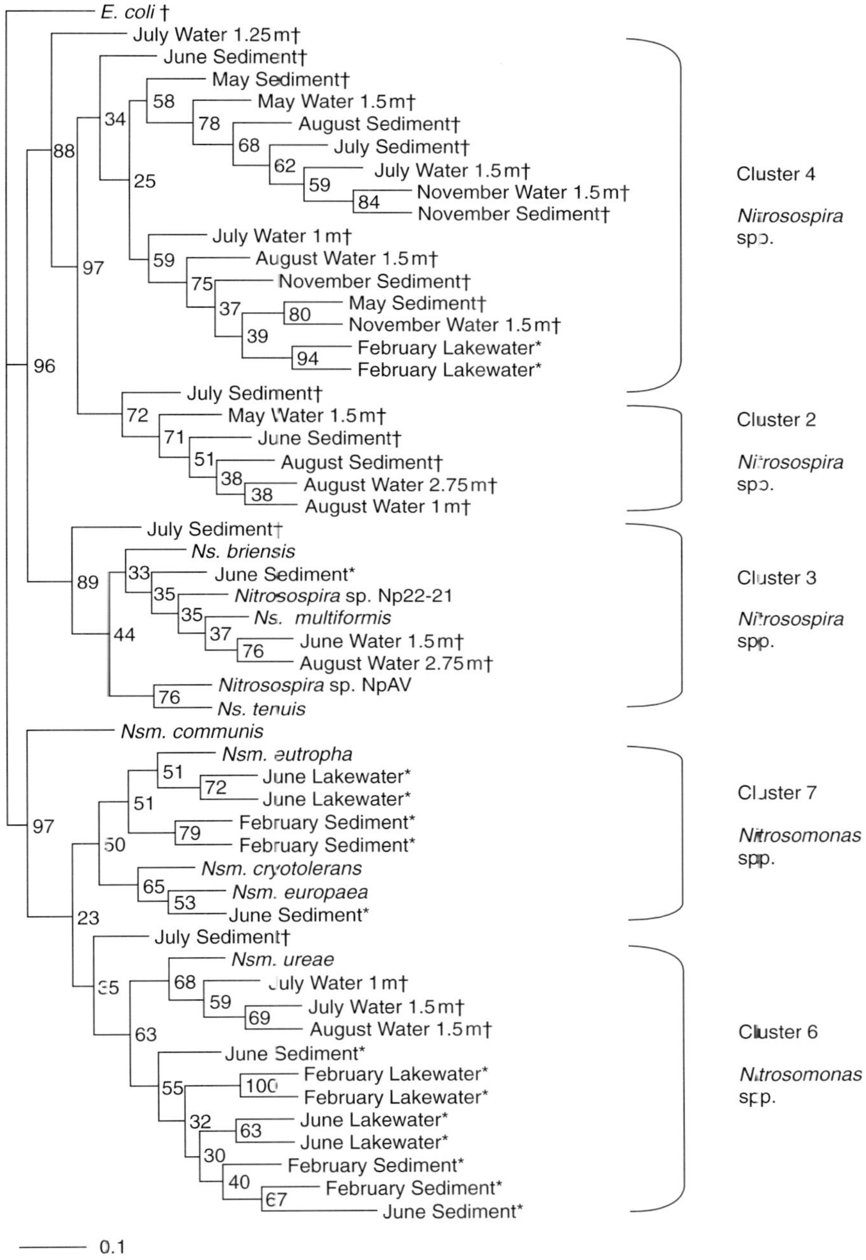

Figure 10.7 Phylogenetic tree showing the position within the β-*Proteobacteria* of partial 16S rRNA sequences (224 bp) recovered from Priest Pot. Scale bar indicates 0.1 substitutions per base. *Escherichia coli* was used as an outgroup. * Sequences are from enrichment cultures; † sequences are from environmental DNA. Numbers at nodes are % bootstrap values.

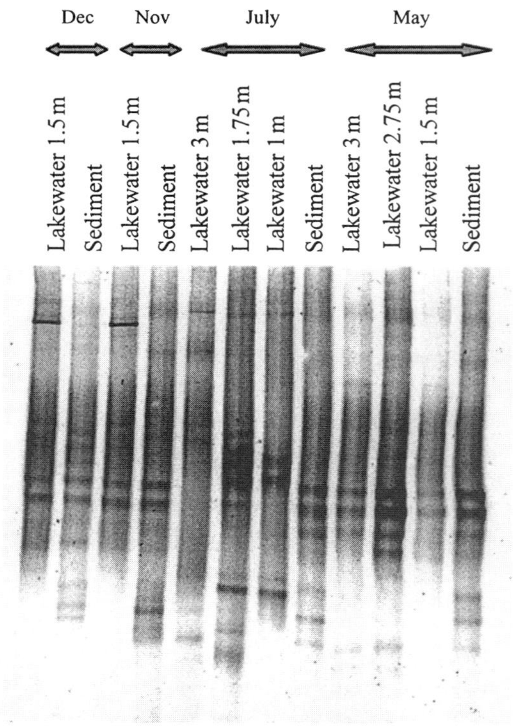

Figure 10.8 TTGE gel of the *amoA* gene amplified from DNA extracted from sediment and lakewater samples.

Stephen *et al.* 1998). There was no requirement for a preliminary round of non-specific amplification to obtain amplified DNA of the correct size, and of sufficient concentration to be visible on ethidium bromide-stained agarose gels. Amplified environmental DNA was analysed by TTGE using the *amoA* primers designed by Rotthauwe *et al.* (1997), modified by Stephen *et al.* (1998) and contained a GC-clamp on the forward primer. These *amoA* PCR amplification products were analysed in a polyacrylamide gel using the method of Muyzer *et al.* (1993) at a temperature gradient of 47–59°C for 16 hours. TTGE profiles of this amplified *amoA* DNA contained a larger number of distinct bands than those obtained with 16S rDNA, potentially providing a better basis for diversity analysis (Figure 10.8). Again, sediment and water column samples gave TGGE fingerprints that were distinct from one another; AOB diversity in the sediment was relatively constant throughout the seasonal cycle, and water column AOB diversity showed variation with both depth and season. For *amoA* sequence analysis, the dominant bands were excised from the TTGE gels and combined with *amoA* sequences from clones to generate a total of 42 sequences (200 bp). The resulting phylogenetic tree is presented in Figure 10.9.

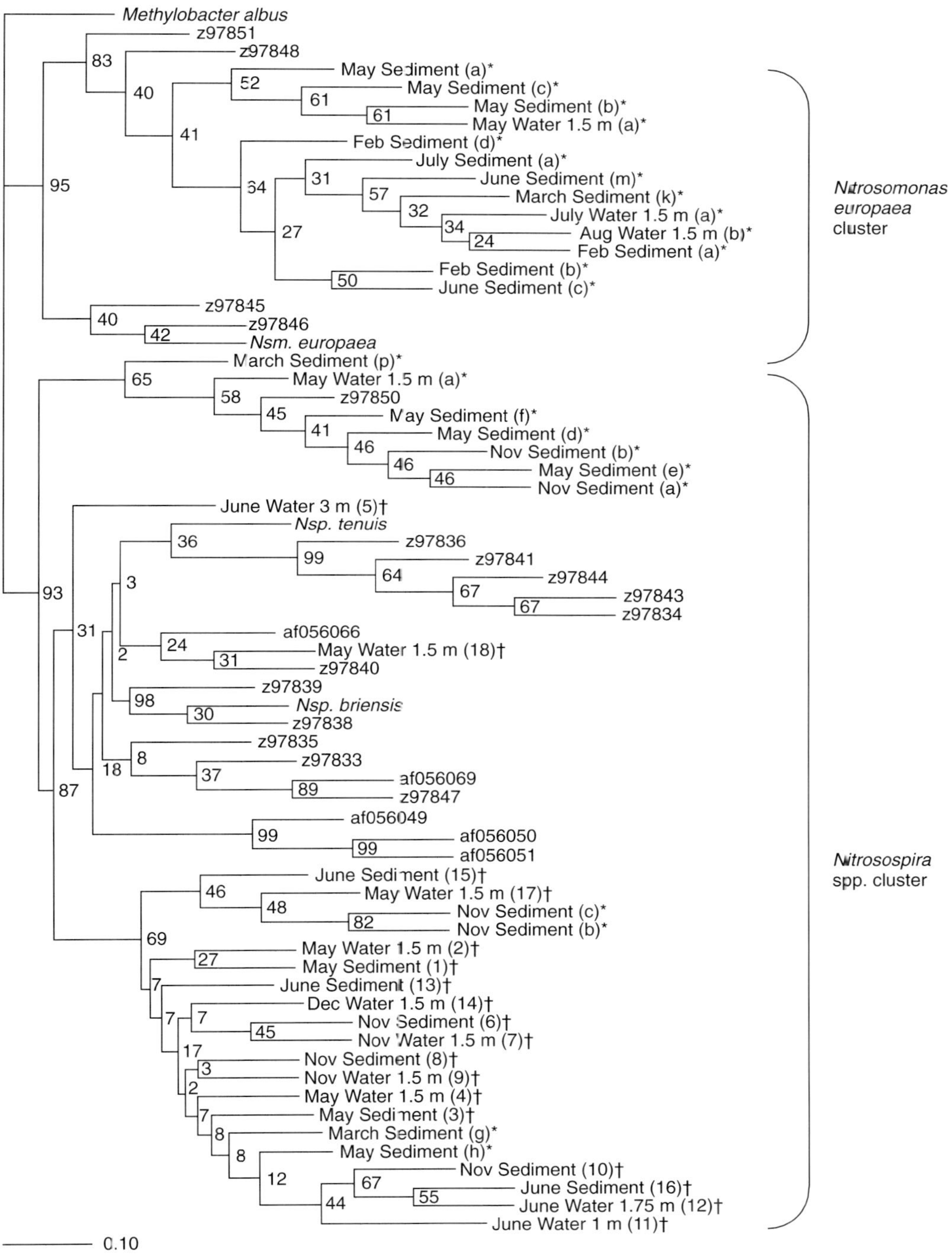

Figure 10.9 Phylogenetic tree showing the position within members of the class β-*Proteobacteria* of partial *amoA* gene sequences (200 bp) recovered from Priest Pot. Sequences from named strains were obtained from GenBank and EMBL databases. Scale bar indicates 0.1 substitutions per base. The *pMMOA* sequence of *Methylobacter albus* was used as an outgroup. * Sequences are from clone libraries; † sequences are from bands excised from TTGE gels.

Again, the clustering of sequences was not correlated with the sample location (sediment or water column) or date (summer stratification versus winter overturn). Unlike the 16S rDNA sequences discussed above, fine-scale resolution of the phylogenetic tree to enable identification of sequences with previously described clusters is not possible. Examination of the data does however show that the sequence clusters did not contain named species and that randomly selected clones revealed greater diversity within the Priest Pot AOB than could be discerned simply by sequencing DNA bands from TTGE gels. While TTGE is a powerful technique for producing genotypic fingerprints that enable microbial population shifts to be detected, and community dynamics to be monitored, it does not provide a complete picture of the diversity in the native microflora.

Analysis of ammonia-oxidizer populations in enrichment cultures

There is now increasing evidence that ammonia concentration in the environment can exert a strong influence on AOB community structure (Hastings *et al.* 1997; Phillips *et al.* 1999). The evidence points to tolerance of high ammonia as the key factor. Partial sequence analysis of the ammonia-sensitive and ammonia-tolerant strains of *Nitrosomonas* spp. isolated from the River Elbe (Stehr *et al.* 1995) and from sewage (Suwa *et al.* 1997) revealed segregation in relation to ammonium sensitivity. Both studies demonstrated that ammonium-sensitive strains were related to *Nitrosomonas oligotropha* and *Nitrosomonas ureae*, while ammonium-tolerant strains were related to *Nitrosomonas europaea* and *Nitrosomonas eutropha*. For the freshwater environment, we have demonstrated that both oligotrophic and eutrophic lakes contain AOB populations that respond differently to ammonia concentration in enrichment cultures (Hastings *et al.* 1998; Whitby *et al.* 1999). Essentially, sediment sample AOB enrichments nitrify at different concentrations, while water column samples nitrify only at low ammonia concentrations in enrichment culture. Subculture of this enriched water column AOB population into high ammonia medium does not produce nitrification, while sediment enrichments can be subcultured to enrich under any conditions. Priest Pot, with its high biological productivity and classification as a hypereutrophic lake, provided an excellent opportunity to further test the hypothesis that environmental ammonia concentration exerts a profound effect on AOB community structure.

Lakewater (1.5 m depth) and sediment (0–0.5 cm depth) sampled from Priest Pot in July and February, representative of summer and winter respectively, were inoculated into media containing high (12.5 mM), intermediate (5 mM) and low (0.65 mM) ammonium sulfate concentrations. These enrichment cultures were monitored for nitrification for up to 61 days by measuring the accumulation of nitrate and nitrite. Enrichment cultures inoculated from lakewater samples only exhibited nitrification at the low and intermediate ammonium sulfate concentrations, whilst sediment samples nitrified across the whole concentration range. As found in nearby Esthwaite Water, sediment enrichments could be further subcultured to give nitrification across the concentration range, whilst lakewater enrichments never nitrified upon subculture into the high (12.5 mM) ammonium sulfate medium. Therefore, despite Priest Pot's status as a hypereutrophic lake, the partitioning of AOB into ammonium-tolerant and -sensitive with sediment and lakewater respectively, followed the pattern established for other lakes of lower nutrient status (Hastings *et al.* 1998). DNA was recovered from the enrichment cultures by PCR amplification with identification of 16S rDNA bands by TTGE. This analysis revealed that *Nitrosospira* spp. and *Nitrosomonas europaea–eutropha* were omnipresent. Amplified DNA was also hybridized against oligonucleotide probes for different AOB clusters (Stephen *et al.* 1998) and these data were used together with sequence analysis of bands excised from TTGE gels to follow the occurrence of different AOB during the enrichment periods. *Nsm. europaea–eutropha*, which could never be detected in Priest Pot by direct amplification of environmental DNA, increased during enrichment of both lakewater and sediment and under all ammonia concentration conditions. Again, the identification of this AOB lineage as a successful competitor in laboratory enrichment cultures, but a minor player in the freshwater environment (Hiorns *et al.* 1995; Hastings *et al.* 1997; Whitby *et al.* 1999), is confirmed. Nitrosospiras predominated prior to enrichment, and TTGE analysis showed that their diversity decreased during enrichment under all conditions. In all cases, *Nitrosospira* spp. with 16S rDNA sequences that place them in a group identified as cluster 2 soil clones by Stephen *et al.* (1996) persisted alongside *Nsm. europaea–eutropha* as enrichment progressed. Our direct analysis of Priest Pot DNA demonstrated that the sediment generally contained greater AOB diversity than the water column, and when sediment enrichments were subcultured into fresh media containing a higher or lower ammonium sulfate concentration, other DNA bands reappeared periodically. This shows that laboratory enrichment selects a stable predominant AOB community that can be perturbed to allow growth of strains that are not part of that dominant population or are merely surviving. Such population dynamics revealed by molecular biological analysis of AOB DNA in simple laboratory enrichment cultures must be only a fraction of that which occurs in the lakewater environment itself.

These experiments based on PCR amplification of 16S rDNA can show how diversity is affected by environmental parameters and gradients, but generate data that can only be strictly interpreted in qualitative terms. Avoidance of PCR by directly labelling of 16S rRNA and enumerating the cells that bind specific oligonucleotides is one approach to determining the relative abundance of different AOB. It had been our initial intention to apply this technique, fluorescent *in situ* hybridization (FISH), to lakewater and sediment by reasoning that Priest Pot had sufficiently high biological activity to support AOB populations of a directly detectable size. This proved untenable—not only were there insufficient AOB in the microbial population, but problems of high background fluorescence in the sample biomass and low rRNA content of AOB giving a weak signal proved insurmountable. The technique was however usefully developed for application to the enrichment cultures described above. We used three fluorescently labelled (FITC) oligonucleotide 16S rRNA probes that target AOB in general (AAO258), *Nitrosospira* spp. (Ns85) and *Nsm. europaea–eutropha* (Nm75) (Hiorns *et al.* 1995). Enrichment cultures of sediment in high ammonia media were dominated by *Nsm. eutropha–europaea*, but subculturing into low ammonia medium encouraged the growth of nitrosospiras with reversion to *Nsm. europaea–eutropha* predominance again when these enrichments were reintroduced into high ammonia medium. The circle was completed by performing competition experiments in which media were inoculated with equivalent numbers (8×10^5 cells) of individual *Nitrosospira* spp. (the NpAv strain or *Nsp. multiformis*) in combination with either *Nsm. europaea* or *Nsm. eutropha*. These co-cultures were analysed by FISH using the probes applied to enrichment cultures described above. In high ammonium (12.5 mM) media, the nitrosomonad always predominated but in the low ammonium (0.65 mM) medium inoculated with *Nitrosospira* sp. NpAv and *Nsm. europaea*, the nitrosospira predominated (Figure 10.10). This is all further evidence in support of the view that nitrosomonads of the *Nsm. europaea–eutropha* lineage predominate over nitrosospiras in nitrogen-rich conditions, laboratory or environmental (Wagner *et al.* 1995; Mobarry *et al.* 1996; Hastings *et al.* 1997; Juretschko *et al.* 1998; Schramm *et al.* 1998).

Active versus inactive ammonia-oxidizer populations

Despite the now widespread application of molecular biological methods to analyse bacterial communities in nature, increasing attention is being given to the use of these methods to generate quantitative data that relate to metabolic activity *in situ*. This ability to link metabolic activity with taxonomic identity and species abundance is vital in microbial ecology. While fluorescent *in situ* probing, discussed above, has been successfully applied to analyse ammonia-oxidizing bacterial populations (e.g. Wagner *et al.* 1995; Schramm *et al.* 1998), the technique is not without problems. Probes need to be designed from known (cultivated species) sequences, background and non-specific signals can interfere, and the numbers of the indigenous targets may be below detection thresholds. There is a direct relationship between the number of ribosomes per bacterial cell and metabolic activity, such that isolation of RNA from environmental samples followed by quantification of specific ribosomal RNAs can be used as an indicator that target species are active in that habitat. This can be achieved by determining relative hybridization ratios to specific and general bacterial rRNA (e.g. Purdy *et al.* 1997), but for AOB, the numbers of cells present are usually insufficient for detection by direct hybridization to RNA extracts. We introduced an amplification step by using ammonia-oxidizer specific 16S rDNA primers in a PCR reaction with cDNA template generated from reverse transcriptase treatment of RNA (RT-PCR) extracted from sediment, lakewater and enrichment cultures. Only the latter were reproducibly amenable to this analysis, and RT-PCR amplified DNA fragments could be separated by TTGE and identified by sequencing. While the correlation of ribosome number with growth rate, and therefore metabolic activity, is valid for heterotrophic bacteria such as *E. coli* (Gausing 1977), it has been shown that slow-growing chemoautotrophic bacteria such as AOB can maintain relatively high numbers of ribosomes under starvation conditions (Natuscka *et al.* 1999). An alternative approach is to target messenger RNA of functional genes in a RT-PCR analysis,

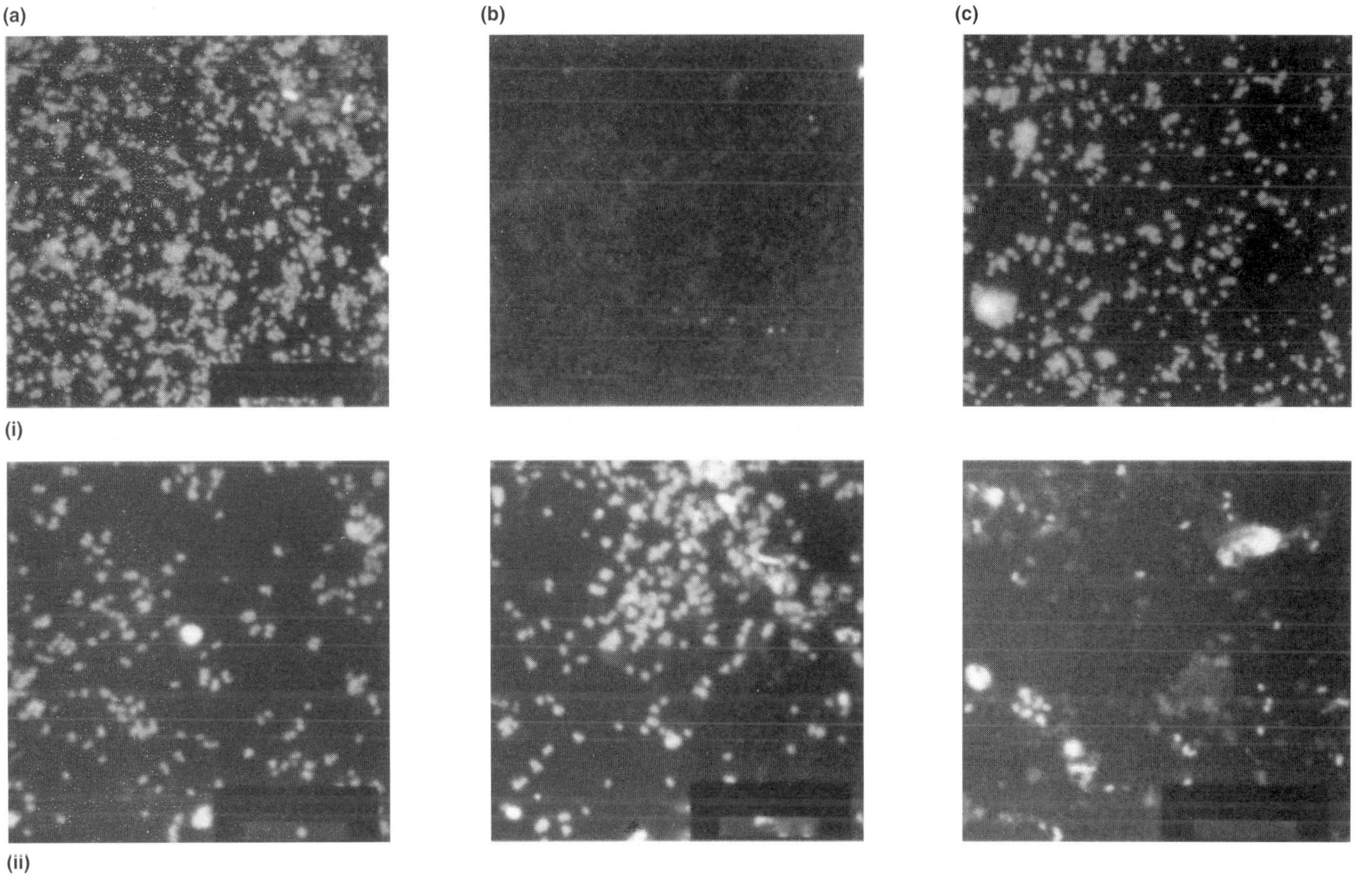

Figure 10.10 Fluorescence *in situ* hybridization of (i) *Nitrosospira* NpAv/*Nitrosomonas europaea* co-culture inoculated to 12.5 mM ammonium sulfate medium; (ii) *Nitrosospira multiformis*/*Nitrosomonas europaea* co-culture inoculated to 0.65 mM ammonium sulfate medium. 16S rDNA in cells visualized by hybridization to oligonucleotide probes AAO258 (a), NS85′ (b) and NM75′ (c).

and although the implication of metabolic activity is therefore unequivocal, the instability and low abundance of mRNA is a serious constraint.

An entirely novel approach to the use of molecular biology techniques to demonstrate the activity of a bacterial species has been described recently by Radajewski *et al.* (2000). The method relies on the incorporation of ^{13}C into the DNA of bacteria that are assimilating a ^{13}C-enriched substrate, such that subsequent density gradient ultracentrifugation can be used to separate 'heavy' from 'light' DNA. Specific PCR amplification of ^{13}C-DNA therefore yields products that can only have originated from metabolically active species and in this way Radajewski *et al.* (2000) were able to unequivocally identify active methylotrophs from an oak forest soil enrichment supplemented with ^{13}C methanol as the sole carbon source. Although difficult to envisage the application of this technique directly in the environment, its application to the analysis of enrichments or samples subjected to incubation is an important development. Following this lead, we incubated high ammonium sediment enrichment cultures (12.5 mmol L^{-1} $(NH_4)_2SO_4$) for 5 days in the presence of ^{13}C-labelled CO_2 with the aim of recovering 'heavy' DNA from active CO_2-assimilating species. This was successful, and mass spectrometry analysis confirmed the separation of the 'heavy' ^{13}C-labelled DNA fraction from the ^{12}C-unlabelled DNA fraction. Ammonia-oxidizer 16S rDNA amplified from these gradients was then characterized by TTGE and sequence analysis to identify the metabolically active components of the population. The TTGE fingerprints revealed one band isolated from the 'heavy' ^{13}C-DNA with 97% sequence homology to *Nsm. europaea–eutropha* and one band obtained from the 'light' ^{12}C-DNA fraction with 98% sequence homology to the *Nitrosospira* spp. recovered in cluster 3 by Stephen *et al.* (1996). Again, the predominance of nitrosomonad (*Nsm. europaea–eutropha*) activity in enrichment cultures is identified, on this occasion by DNA analysis that is directly targeted at actively metabolizing cells (Whitby *et al.* 2001b). This use of stable isotope incorporation into ammonia-oxidizer DNA circumvents the problems associated with RNA analysis and appears to provide a more reliable approach to equating species presence and activity.

Perspective

For AOB, whose taxonomy and ecology are very poorly described by conventional microbiology, the influence of environmental parameters on community structure is only beginning to be determined, and is almost dependent on direct analysis of environmental DNA. This has revealed the presence of previously unknown sequences, which are inferred to be novel, as yet uncultured members of the β-subgroup *Proteobacteria* ammonia-oxidizers (Stephen *et al.* 1996, 1998, 1999; Kowalchuk *et al.* 1997; Speksnijder *et al.* 1998; McCaig *et al.* 1999; Mendum *et al.* 1999, Phillips *et al.* 1999; Whitby *et al.* 1999). From their origin in the distinction of only two major genera of β-*Proteobacteria* AOB (Head *et al.* 1993), 16S rDNA phylogenetic trees have expanded to provide taxonomic frameworks within which clusters (species?) can be recognized (Stephen *et al.* 1996; Purkhold *et al.* 2000).

Although the trees are often based on partial sequences of ca. 300–400 bp, sequence similarities between clusters that are high and subclusters that are arbitrarily defined, reliable identification of unknowns can generally be achieved. From an ecological viewpoint, application of the classification shows that similar or closely related AOB strains occur in very different environments. There have also been attempts to obtain cultured representatives of all AOB 16S rDNA sequence clusters (Utaker *et al.* 1995) and this remains an important priority. The congruence between the 16S rDNA and ammonia monooxygenase (*amoA*) sequence phylogenies has been demonstrated to be insufficient for the fine-scale resolution between and within clusters (Rotthauwe *et al.* 1997; Mendum *et al.* 1999; Horz *et al.* 2000; Purkhold *et al.* 2000).

The extensive use of molecular tools in microbial ecology is now invaluable in addressing the question of which bacterial species are the most important in different habitats under the prevailing environmental conditions. The influence of environment on AOB genotypic diversity was the important driver in this project with the objective of determining whether a unique or unusual AOB community occupied this small, segregated hypereutrophic lake. The AOB sequences recovered (both 16S rDNA and *amoA*), their segregation between sediment and water column, and their seasonal distribution were mostly as described for larger lakes that sustain much less biological activity. There were similarly strong selective effects of substrate (ammonium) concentration on AOB enrichment from different compartments of the lake. This detailed analysis of the AOB populations in Priest Pot therefore demonstrates that the site is a valid model for studying freshwater lake nitrification.

Acknowledgements

This research was funded by the Natural Environment Research Council (NERC) EDGE Programme, UK.

References

Amman, R.I., Ludwig, W. & Schleifer, K-H. (1995) Phylogenetic identification and *in situ* detection of individual microbial cells without cultivation. *Microbiological Reviews* **59**, 143–169.

Bano, N. & Hollibaugh, J.T. (2000) Diversity and distribution of DNA sequences with affinity to ammonia-oxidizing bacteria of the beta subdivision of the class *Proteobacteria* in the Arctic Ocean. *Applied and Environmental Microbiology* **66**,1960–1969.

Finlay, B.J., Maberly, S.C. & Cooper, J.I. (1997) Microbial diversity and ecosystem function. *OIKOS* **80**, 209–213.

Gausing, K. (1977) Regulation of ribosome production in *Escherichia coli*: synthesis and stability of ribosomal RNA at different growth rates. *Journal of Molecular Biology* **115**, 335–354.

Hall, G.H. (1986) Nitrification in Lakes. In: *Nitrification* (ed. J. I. Prosser), pp. 127–156. IRL Press, Oxford.

Hastings, R.C., Ceccherini, M.T., Miclaus, N. *et al.* (1997) Direct molecular biological analysis of ammonia-oxidizing bacteria populations in cultivated soil plots treated with swine manure. *FEMS Microbiology Ecology* **23**, 45–54.

Hastings, R.C., Saunders, J.R., Hall, G.H., Pickup, R.W. & McCarthy, A.J. (1998) Application of molecular biological techniques to a seasonal study of ammonia oxidation in a eutrophic freshwater lake. *Applied and Environmental Microbiology* **64**, 3674–3682.

Head, I.M., Hiorns, W. D., Embley, T.M., McCarthy, A.J. & Saunders, J.R. (1993) The phylogeny of autotrophic ammonia-oxidizing bacteria as determined by analysis of 16S ribosomal RNA gene sequences. *Journal of General Microbiology* **139**, 1147–1153.

Hiorns, W.D., Hastings, R.C., Head, I.M. *et al.* (1995) Amplification of 16S ribosomal RNA genes of the autotrophic ammonia-oxidizing bacteria demonstrates the ubiquity of nitrosospiras in the environment. *Microbiology* **141**, 2793–2800.

Horz, H-P, Rotthauwe, J-H., Lukow, T. & Liesack, W. (2000) Identification of major sub-groups of ammonia-oxidizing bacteria in environmental samples by T-RFLP analysis of *amoA* PCR products. *Journal of Microbiological Methods* **39**, 197–204.

Juretschko, S., Timmermann, G., Schmid, M. *et al.* (1998) Combined molecular and conventional analyses of nitrifying bacterium diversity in activated sludge: *Nitrosococcus mobilis* and *Nitrosospira*-like bacteria as dominant populations. *Applied and Environmental Microbiology* **64**, 3042–3051.

Kowalchuk, G.A. & Stephen, J.R. (2001) Ammonia-oxidizing bacteria: a model for molecular microbial ecology. *Annual Review of Microbiology* **55**, 485–530.

Kowalchuk, G.A., Stephen, J.R., DeBoer, W. *et al.* (1997) Analysis of ammonia-oxidizing bacteria of the β-subdivision of the class Proteobacteria in coastal sand dunes by denaturing gradient gel electrophoresis and sequencing of PCR-amplified 16S ribosomal DNA fragments. *Applied and Environmental Microbiology* **63**, 1489–1497.

Kowalchuk, G.A., Naoumenko, Z.S., Derikx, P. *et al.* (1999) Molecular analysis of ammonia-oxidizing bacteria of the β-subdivision of the class Proteobacteria in compost and composted materials. *Applied and Environmental Microbiology* **65**, 396–403.

Kowalchuk, G.A., Stienstra, A.W., Hans, G. *et al.* (2000) Changes in community structure of ammonia-oxidizing bacteria during secondary succession of calcareous grasslands. *Environmental Microbiology* **2**, 99–110.

McCaig A.E., Phillips, C.J., Stephen, J.R. *et al.* (1999) Nitrogen cycling and community structure of Proteobacterial β-subgroup ammonia-oxidizing bacteria within polluted marine fish farm sediments. *Applied and Environmental Microbiology* **65**, 213–220.

McTavish, H., Fuchs, J.A. & Hooper, A.B. (1993) Sequence of the gene encoding for ammonia monooxygenase in *Nitrosomonas europaea. Journal of Bacteriology* **175**, 2436–2444.

Mendum, T.A., Sockett, R.E. & Hirsch, P.R. (1999) Use of molecular and isotopic techniques to monitor the response of autotrophic ammonia-oxidizing populations of the β-subdivision of the class *Proteobacteria* in aerobic soils to nitrogen fertilizer. *Applied and Environmental Microbiology* **65**, 4155–4162.

Mobarry, B.K., Wagner, M., Urbain, V., Rittmann, B.E. & Stahl, D.A. (1996) Phylogenetic probes for analysing abundance and spatial organisation of nitrifying bacteria. *Applied and Environmental Microbiology* **62**, 2156–2162.

Muyzer, G., DeWaal, E.C. & Uitterlinden, A.G. (1993) Profiling of complex microbial populations by denaturing gradient gel electrophoresis analysis of polymerase chain reaction-amplified genes coding for 16S rRNA. *Applied and Environmental Microbiology* **59**, 695–700.

Natuscka, L., Nielsen, O., Andreasen, K. *et al.* (1999) Combination of fluorescent *in situ* hybridization and microautoradiography—a new tool for structure–function analyses in microbial ecology. *Applied and Environmental Microbiology* **65**, 1289–1297.

Phillips, C.J., Smith, Z., Embley, T.M. & Prosser, J.I. (1999) Phylogenetic differences between particle-associated and planktonic ammonia-oxidizing bacteria of the β-subdivision of the class Proteobacteria in the Northwestern

Mediterranean Sea. *Applied and Environmental Microbiology* **65**, 779–786.

Pommerening-Röser, A., Rath, G. & Koops, H-P. (1996) Phylogenetic diversity within the genus *Nitrosomonas. Systematic and Applied Microbiology* **19**, 344–351.

Prosser, J.I. (1989) Autotrophic nitrification in bacteria. *Advances in Microbial Physiology* **30**, 125–181.

Purdy, K.J., Nedwell, D.B., Embley, T.M. & Takii, S. (1997) Use of 16S rRNA-targeted oligonucleotide probes to investigate the occurrence and selection of sulfate-reducing bacteria in response to nutrient addition to sediment slurry microcosms from a Japanese estuary. *Systematic and Applied Microbiology* **24**, 221–234.

Purkhold, U., Pommerening-Roser, A., Juretschko, S. *et al.* (2000) Phylogeny of all recognized species of ammonia oxidizer based on comparative 16S rRNA and *amoA* sequence analysis: implications for molecular surveys. *Applied and Environmental Microbiology* **66**, 5368–5382.

Radajewski, S., Ineson, P., Parekh, N. & Murrell, C. (2000) Stable-isotope probing as a tool in microbial ecology. *Nature* **403**, 646–649.

Rotthauwe, J-H., Witzel, K.-P. & Liesack, W. (1997) The ammonia monooxygenase structural gene *amoA* as a functional marker: molecular fine scale analysis of the natural ammonia-oxidizing populations. *Applied and Environmental Microbiology* **63**, 4704–4712.

Rowe, R., Todd, R. & Waide, J. (1977) Microtechnique for most probable number analysis. *Applied and Environmental Microbiology* **33**, 134–141.

Schramm, A., De Beer, D., Wagner, M. & Amann, R. (1998) Identification and activities *in situ* of *Nitrosospira* and *Nitrospira* spp. as dominant populations in a nitrifying fluidized bed reactor. *Applied and Environmental Microbiology* **64**, 3480–3485.

Sinigalliano, C.D., Kuhn, D.N. & Jones, R.D. (1995) Amplification of the *amoA* gene from diverse species of ammonia-oxidizing bacteria and from an indigenous bacterial population from seawater. *Applied and Environmental Microbiology* **61**, 2702–2706.

Speksnijder, A., Kowalchuk, G.A., Roest, K. & Laanbroek, H. (1998) Recovery of a *Nitrosomonas*-like 16S rDNA sequence group from freshwater habitats. *Systematic and Applied Microbiology* **21**, 321–330.

Stehr, G., Bottcher, B., Dittberner, P., Rath, G. & Koops, H-P. (1995) The ammonia-oxidizing nitrifying populations of the River Elbe estuary. *FEMS Microbiology Ecology* **17**, 177–186.

Stephen, J.R., McCaig, A.E., Smith, Z., Prosser, J.I. & Embley, T.M. (1996) Molecular diversity of soil and marine 16S rRNA gene sequences related to β-subgroup ammonia-oxidizing bacteria. *Applied and Environmental Microbiology* **62**, 4147–4154.

Stephen, J.R., Kowalchuk, G.A., Bruns, M-A. *et al.* (1998) Analysis of β-subgroup Proteobacterial ammonia-oxidizer populations in soil by denaturing gradient gel electrophoresis analysis and hierarchal phylogenetic probing. *Applied and Environmental Microbiology* **64**, 2958–2965.

Stephen, J.R., Chang, Y-J., Macnaughton, S.J. *et al.* (1999) Effect of toxic metals on indigenous soil β-subgroup Proteobacterium ammonia-oxidizer community structure and protection against toxicity by inoculated metal-resistant bacteria. *Applied and Environmental Microbiology* **65**, 95–101.

Suwa, Y., Sumino, T. & Noto, K. (1997) Phylogenetic relationships of activated sludge isolates of ammonia-oxidizers with different sensitivities to ammonium sulphate. *Journal of General and Applied Microbiology* **43**, 373–379.

Teske, A., Alm, E., Regan, J.M. *et al.* (1994) Evolutionary relationships among ammonia- and nitrite-oxidizing bacteria. *Journal of Bacteriology* **176**, 6623–6630.

Utaker, J.B., Bakken, L., Jiang, Q. & Nes, J. (1995) Phylogenetic analysis of seven new isolates of ammonia-oxidizing bacteria based on 16S rRNA gene sequences. *Systematic and Applied Microbiology* **18**, 549–559.

Wagner, M., Rath, G., Amann, R., Koops, H-P. & Schleifer, K-H. (1995) *In situ* identification of ammonia-oxidizing bacteria. *Systematic and Applied Microbiology* **18**, 251–264.

Ward, B.B., Voytek, M.A. & Witzel, K-P. (1997) Phylogenetic diversity of natural populations of ammonia-oxidizers investigated by specific PCR amplification. *Microbial Ecology* **3**, 87–96.

Whitby, C.B., Saunders, J.R., Rodriguez, J., Pickup, R.W. & McCarthy, A.J. (1999) Phylogenetic differentiation of two closely related *Nitrosomonas* spp. that inhabit different sediment environments in an oligotrophic freshwater lake. *Applied and Environmental Microbiology* **65**, 4855–4862.

Whitby, C.B., Saunders, J.R., Pickup, R.W. & McCarthy, A.J. (2001a) A comparison of ammonia-oxidizer populations in eutrophic and oligotrophic basins of a large freshwater lake. *Antonie van Leuwenhoek* **79**, 179–188.

Whitby, C.B., Hall, G., Pickup, R.W. *et al.* (2001b) ^{13}C incorporation into DNA as a means of identifying the active components of ammonia-oxidizer populations. *Letters in Applied Microbiology* **32**, 398–401.

Chapter 11

The influence of selection pressures on species diversity, functional gene diversity and activity of ammonia-oxidizing bacteria

J. I. Prosser, T. M. Embley and G. Webster*

Introduction

A major objective of ecological research is the establishment of a theoretical and conceptual framework that both explains and predicts ecosystem behaviour through an understanding of the biological communities and environmental characteristics controlling ecosystem processes. An important element of this understanding is knowledge of the interactions between environmental factors and species diversity. For example, do changes in environmental factors influence diversity and heterogeneity, are species and functional diversity linked and do changes in species and functional diversity influence the rates of ecosystem processes? These questions are of particular significance when considering the influence of pollution and other anthropogenic processes on ecosystem behaviour, and the extent to which this may be predicted through knowledge of the functional characteristics of individual species within the community.

Microorganisms are essential components of all ecosystems. They drive biogeochemical cycles, degrade organic matter and organic pollutants, act as plant pathogens and symbionts, and are essential for the growth and activity of all higher organisms. Despite this, our understanding of the links between the structure of microbial communities and their role in natural ecosystems lags significantly behind that of higher plants and animals. Our ignorance stems, largely, from a lack of appreciation of their role because of their microscopic size. This lack of appreciation persists despite the demonstration that 50% of global biological carbon and 90% of global biological nitrogen are contained within prokaryotic organisms (Whitman *et al.* 1998). In addition, reliable analysis of natural microbial communities has been severely limited by the lack of appropriate methods.

The analysis of microbial populations has traditionally relied on cultivation-based techniques. Typically, species composition is determined by enrichment of microorganisms on laboratory media, isolation of pure cultures and identification

*Correspondence author: j.prosser@abdn.ac.uk

of these cultures using morphological, physiological and other phenotypic characteristics. Even with modern, automated techniques this process is tedious and time consuming, but the major limitation is reliance on laboratory cultivation. Several studies have provided compelling evidence that only a small proportion of the natural microbial communities can be cultivated. For example, total bacterial counts are typically one or two orders of magnitude greater than culturable counts, suggesting that 90–99% of the population is not detected on laboratory media, due to inability to design culture media and conditions appropriate for all organisms, entry of cells into dormant states and competition during incubation. Consequently, our understanding of microbial ecology is based on the physiology of relatively few organisms that have been selected under laboratory conditions, which are considerably different to those experienced in the natural environment.

Characterization of genes representative of members of microbial populations removes many of the above methodological limitations. This approach involves extraction and purification of environmental DNA, followed by amplification and analysis of genes that provide information on the identity, function and activity of microbial communities (Giovannoni *et al.* 1990). These techniques, although in their infancy, have demonstrated considerable diversity at a scale previously unsuspected and have led to the discovery of major groups of bacteria with high relative abundance in both terrestrial and aquatic environments. These groups have no or few cultured representatives and it is therefore not possible to define their role in ecosystem function.

Currently, molecular techniques provide one realistic approach to the characterization of *in situ* microbial diversity and assessment of environmental impact on community composition and structure. They provide us with information on the patterns and phenomena that we need in order to explain why particular groups are important in natural microbial populations. In addition, they provide a means of testing the appropriateness and limitations of established approaches and alternative techniques are available for cross-checking and validation. As a consequence, they are providing the beginnings of a theoretical framework for microbial communities, which have largely been restricted to higher organisms. Prokaryotes, however, are more numerous, more diverse and more active in the environment. Mechanisms influencing their ecology operate at a much smaller spatial scale and ecological and evolutionary timescales for microorganisms are much closer than for higher organisms. It is therefore not clear whether established ecological theory is capable of describing microbial populations, potentially limiting its ability to describe the majority of organisms responsible for ecosystem function.

The impact of molecular techniques on microbial ecology will be illustrated in this article by studies of ammonia-oxidizing bacteria in the soil, although many of the findings apply equally to other microbial groups of importance in natural ecosystems. Ammonia oxidizers are ideal model organisms for such studies. They carry out a biogeochemical process of importance on both local and global scales and molecular techniques have provided information on population ecology that was unattainable by traditional approaches.

Community analysis through characterization of 16S rRNA genes and functional genes

16S rRNA genes have provided the basis for prokaryotic taxonomy for many years. All prokaryotes possess ribosomes and the 16S subunit is sufficiently large (approximately 1500 bases) to distinguish different members of the prokaryotic kingdom. Certain regions of the 16S rRNA gene sequence are conserved in all prokaryotes, while others vary to different extents. Conserved regions facilitate alignment of sequences when comparing 16S rRNA genes from different organisms. Variable regions enable distinction between groups and provide hierarchical grouping of sequences to species, and often subspecies levels. The discriminatory power of 16S rRNA gene sequences varies between established prokaryote groups but greater discrimination, and finer-scale information, can be achieved by extending analysis into spacer regions between genes encoding the 16S and 23S subunits.

Similarities in sequence can be used to design specific primers for PCR amplification of 16S rRNA genes of particular prokaryotic groups from mixed community DNA. This provides the basis for analysis of diversity in natural populations. Following extraction and purification of environmental DNA, PCR amplification of 16S rRNA genes using specific primers generates a population of 16S rRNA genes, which may then be analysed by several techniques. PCR products may be cloned, members of the clone library sequenced and sequences subjected to phylogenetic analysis. Alternatively, more rapid analysis may be achieved using fingerprinting techniques. For example, denaturing and temperature gradient gel electrophoresis (DGGE, TGGE; Muyzer & Smalla 1998) separate products on the basis of differences in sequence through chemical or heat denaturation, respectively. Bands on gels may be excised and sequenced, while semiquantitative analysis of relative abundance of different groups may be achieved through comparisons of band intensities (McCaig *et al.* 1999, 2001). Other fingerprinting techniques, e.g. terminal restriction length polymorphism (T-RFLP, Liu *et al.* 1997), involve restriction of amplification products. Sequence information can also be used to design and construct fluorescently labelled oligonucleotide probes specific for particular microbial groups for *in situ* detection by fluorescence *in situ* hybridization (FISH, DeLong *et al.* 1989), while quantification of gene copies may be achieved by quantitative PCR methods, including competitive PCR (Jansson & Leser 1996) and real-time PCR (Heid *et al.* 1996).

The analysis of 16S ribosomal DNA (rDNA) sequences provides information on diversity, enables quantification of phylogenetic relationships and, through comparison with database sequences, enables identification of community members. This approach therefore provides more rapid, reliable and informative analysis of natural communities than traditional techniques, but its major advantage is the lack of requirement for laboratory enrichment, isolation and identification of pure cultures of community members. These benefits and advantages must be balanced against potential limitations. PCR products can be influenced by methods used to extract and purify environmental DNA and by PCR bias, primer bias, cloning, formation of chimeras and sequencing errors, and cells may contain multiple rRNA operons. The extent of these limitations will only become apparent as studies increase in number, but it is possible to test for and control many of these limitations and to validate them using alternative molecular methods.

Ammonia-oxidizing bacteria and nitrification

Nitrification is the oxidation of ammonia to nitrate, via nitrite, and is carried out mainly by two groups of autotrophic bacteria, defined on the basis of their abilities to oxidize ammonia to nitrite (ammonia oxidizers) and nitrite to nitrate (nitrite oxidizers), both groups obtaining organic carbon by fixation of carbon dioxide. Ammonia oxidizers are physiologically versatile and are also involved in the degradation of organic carbon, nitrous oxide production, denitrification, urea hydrolysis and methane oxidation. They therefore play a key role in global nitrogen cycling and are responsible for greenhouse gas production and degradation of organic pollutants. Their activities lead to significant losses of ammonia-based fertilizers through conversion to nitrate, which is then leached, polluting groundwaters, or denitrified to nitrogen gas. The process of soil nitrification is relatively easy to measure but analysis of nitrifier populations has been limited by difficulties in isolating and identifying natural populations. Their low growth rates, faster growth of co-enriched heterotrophs and poor growth on solid media make isolation difficult. Identification is traditionally based on a limited range of morphological and other phenotypic characteristics making taxonomy and identification difficult. Together these prevent meaningful analysis of the diversity of natural populations, although some attempts have been made using immunological methods (Belser & Schmidt 1978).

Taxonomy of ammonia-oxidizing bacteria

Until recently, ammonia oxidizers were placed in five genera, *Nitrosomonas*, *Nitrosospira*, *Nitrosovibrio*, *Nitrosolobus* and *Nitrosococcus*, defined by cell morphology, arrangement of intracytoplasmic membranes and a limited number of physiological characteristics, with relationships confirmed by DNA : DNA hybridization. Analysis of 16S rRNA gene sequences (Head *et al.* 1993) places *Nitrosococcus oceanus* in the γ-*Proteobacteria* and suggests two genera within the β-*Proteobacteria*, *Nitrosomonas* and *Nitrosospira* (including *Nitrosovibrio* and *Nitrosolobus*). This organization has been supported by further sequencing of both 16S rRNA and functional genes (see Purkhold *et al.* 2000).

In addition to autotrophs, a wide phylogenetic range of heterotrophic bacteria and fungi are capable of nitrification, producing nitrate directly from organic N substrates through secondary metabolic processes (Killham 1986). Heterotrophic nitrifiers gain no obvious benefit from this process and exhibit much lower specific nitrification activities than autotrophs. Their contributions to nitrification are generally considered to be low, although they may be important in environments where autotrophic activity is limited, e.g. in acid soils. Anaerobic ammonium oxidation has also been characterized in wastewater treatment processes, involving oxidation of ammonium by nitrite, resulting in production of nitrogen gas (Jetten *et al.* 1998).

Primer and probe design

Information on 16S rDNA sequences of pure cultures of ammonia oxidizers (Head *et al.* 1993) led to the design of primers for the amplification of ammonia-oxidizer 16S rRNA genes in enrichment cultures and environmental DNA and the design of probes for FISH. The usefulness of such primers and probes depends on their ability to select for the target group only. However, in the absence of detailed knowledge of natural ammonia-oxidizer populations, it is important to design primers with reduced selectivity, to avoid missing groups with 16S rDNA sequence closely related to, but not within those of pure cultures. Primers used in early studies (e.g. those designed by McCaig *et al.* 1994) satisfied these criteria, amplifying ammonia-oxidizer 16S rRNA genes but also those of closely related heterotrophs. Sequence data from subsequent studies of pure cultures, enrichment cultures and environmental clone libraries have enabled the design of primers with greater specificity, the aim being to amplify all ammonia-oxidizer sequences while excluding non-ammonia oxidizers. The majority have targeted autotrophic ammonia oxidizers belonging to the β-subgroup of the *Proteobacteria*, which appear to be most common in natural populations. This design process evolves continually, as more sequence data become available, but a recent survey (Purkhold *et al.* 2000) indicates that the primers of McCaig *et al.* (1994) amplify all known β-proteobacterial ammonia oxidizers, while primers designed by Kowalchuk *et al.* (1997) (CTO primers) show greatest specificity but are not all inclusive. 16S rRNA probes have also been developed for FISH analysis of specific groups of ammonia oxidizers (Utåker & Nes 1996; Schramm *et al.* 1999). The accumulation of sequence data for continued development and refinement of primers and probes is typical for other microbial groups and provides an evolving and expanding picture of microbial diversity, linking phylogenies to the ecology and habitat of microbial populations.

The key enzyme in ammonia oxidation is ammonia monooxygenase, which catalyses conversion of ammonia to hydroxylamine. The *amoA* gene, encoding one subunit of ammonia monooxygenase, has been sequenced in a number of pure cultures of ammonia oxidizers, particularly in nitrosomonads, where phylogenetic groups appear to follow those for 16S rRNA genes (Purkhold *et al.* 2000). Primers have been designed that are specific for *amo*A genes of β-proteobacterial ammonia oxidizers, while excluding closely related *pmo*A genes, encoding methane monooxygenase in methane-oxidizing bacteria (Rotthauwe *et al.* 1997). Other functional genes, including hydroxylamine oxidoreductase and nitrite reductase (Bruns *et al.* 1998), have been investigated for analysis of natural populations.

Diversity of natural populations of ammonia-oxidizing bacteria

PCR-based techniques focused initially on amplification of 16S rRNA genes (e.g. McCaig *et al.* 1994; Stephen *et al.* 1996) and, more recently, on *amo*A genes (e.g. Rotthauwe *et al.* 1997). Target sequences are pure cultures, enrichment cultures and DNA extracted directly from environmental samples. PCR products are analysed by sequencing or by fingerprinting techniques, such as DGGE, TGGE or

T-RFLP (Horz *et al.* 2000). The most important benefit is the ability to detect and classify ammonia oxidizers without the requirement for laboratory cultivation. This removes the well-accepted potential for selection of particular organisms under laboratory conditions but is particularly important for ammonia oxidizers, which grow slowly, are difficult to isolate in pure culture and for whom identification is difficult.

The diversity of ammonia-oxidizer populations in a range of natural environments has been assessed by characterization of 16S rRNA genes. Figure 11.1 contains sequences amplified from pure cultures, from cultures enriched from soil and seawater and from clone libraries constructed from DNA extracted directly from terrestrial and aquatic environments (Stephen *et al.* 1996). Sequence analysis of representatives of clone libraries of PCR amplification products showed considerable diversity. Duplicate sequences were not obtained from amplified extracted DNA and were never identical to the pure culture sequences included at that time. The study confirmed separation of β-*Proteobacteria* into two genera but also indicated considerable novel diversity within these genera. Sequence analysis tentatively indicated subdivision into seven clusters, four within *Nitrosospira* and three within *Nitrosomonas*. This provided a hypothesis for further testing, through acquisition of further sequence data, and the basis for ecological studies. Clusters were related to the environments from which sequences were obtained, clusters 2, 3 and 4 from terrestrial environments and clusters 1 and 5 dominated by sequences from marine environments. Thus, terrestrial and marine '*Nitrosospira*-like' clone sequences are phylogenetically distinct and this approach provided the first evidence of *Nitrosospira* in marine environments, where they are now found to be common. Clusters 1 and 5 have no pure culture or enrichment culture representatives, making it difficult to predict their potential ecological roles. Remaining clusters contained sequences from several environments. *Nitrosomonas europaea*, originally isolated from soil and the type culture for ammonia oxidizers, falls within cluster 7 but is only distantly related to the majority of other sequences obtained from terrestrial environments. This highlights the potential dangers of studies based on laboratory

---

Figure 11.1 (*facing page*) Relationships of environmental partial 16S rDNA sequences to partial sequences from reference β-proteobacterial ammonia oxidizers, based upon 1099 bases of aligned 16S rDNA sequence. Environmental sequences were obtained from samples of soil, sediment and sewage. Scale bar represents 10% estimated substitutions calculated using the Jukes and Cantor correction. Abbreviations are exemplified by the following examples: EnvA2–4, clone number 4 from polluted sediment A 16S rDNA library, sediment sample replicate 2; C1aW3, marine enrichment from non-polluted sample C, sample replicate number 1, enrichment replicate a, medium W, 3rd generation enrichment; pH7C/54, clone 54 from 16S rDNA library from replicate soil sample C, pH 7.0 soil; enrich ZD5, pH 7.0 soil enrichment sequence; enrich ZB3, pH 4.2 soil enrichment; AEM–5, sequence from enriched surface water samples; Sewage, sequence from a wastewater treatment plant.

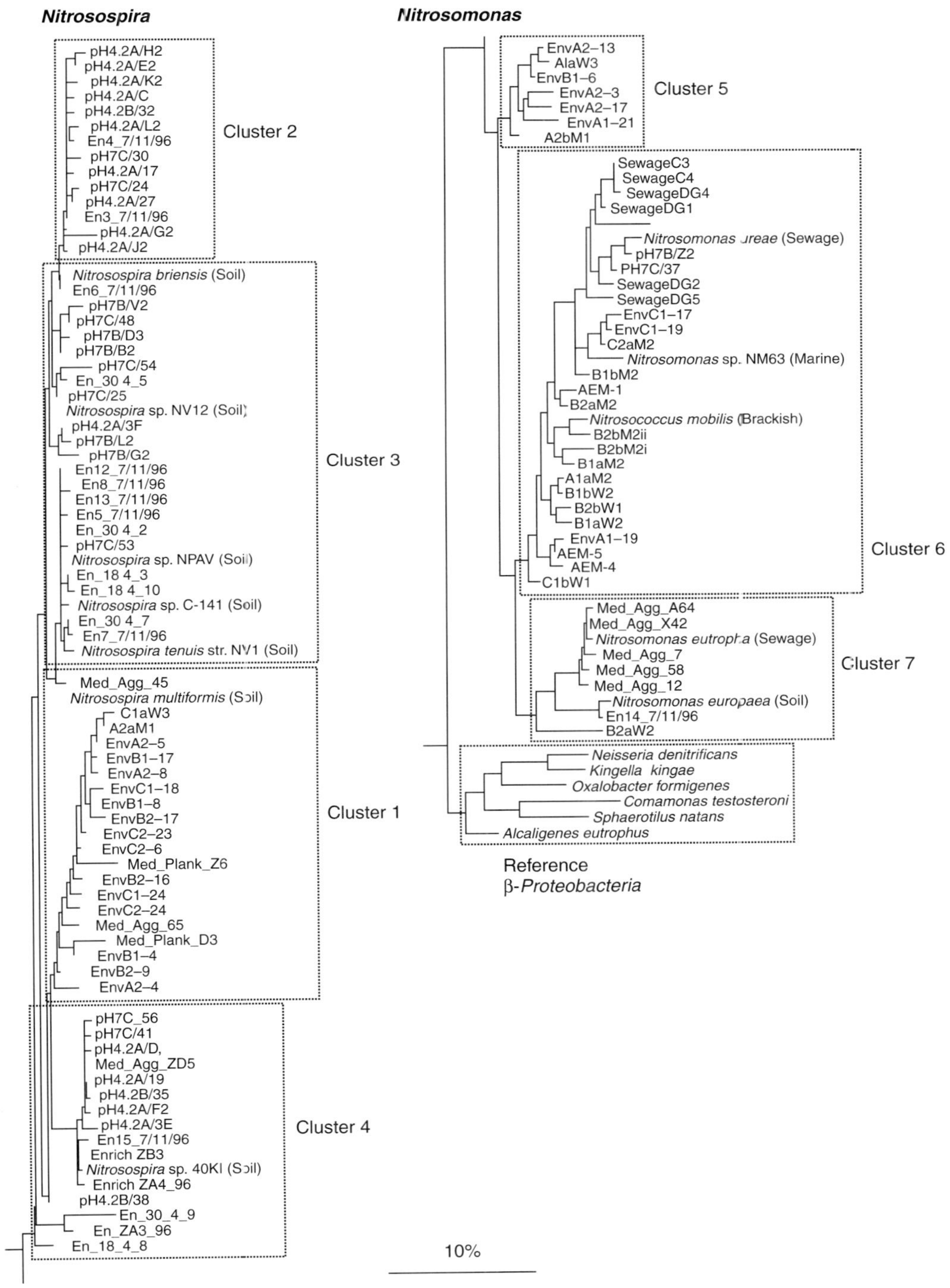
Nitrosospira
pH4.2A/H2
pH4.2A/E2
pH4.2A/K2
pH4.2A/C
pH4.2B/32
pH4.2A/L2
En4_7/11/96
pH7C/30
pH4.2A/17
pH7C/24
pH4.2A/27
En3_7/11/96
pH4.2A/G2
pH4.2A/J2
Cluster 2
Nitrosospira briensis (Soil)
En6_7/11/96
pH7B/V2
pH7C/48
pH7B/D3
pH7B/B2
pH7C/54
En_30 4_5
pH7C/25
Nitrosospira sp. NV12 (Soil)
pH4.2A/3F
pH7B/L2
pH7B/G2
En12_7/11/96
En8_7/11/96
En13_7/11/96
En5_7/11/96
En_30 4_2
pH7C/53
Nitrosospira sp. NPAV (Soil)
En_18 4_3
En_18 4_10
Nitrosospira sp. C-141 (Soil)
En_30 4_7
En7_7/11/96
Nitrosospira tenuis str. NV1 (Soil)
Cluster 3
Med_Agg_45
Nitrosospira multiformis (Soil)
C1aW3
A2aM1
EnvA2-5
EnvB1-17
EnvA2-8
EnvC1-18
EnvB1-8
EnvB2-17
EnvC2-23
EnvC2-6
Med_Plank_Z6
EnvB2-16
EnvC1-24
EnvC2-24
Med_Agg_65
Med_Plank_D3
EnvB1-4
EnvB2-9
EnvA2-4
Cluster 1
pH7C_56
pH7C/41
pH4.2A/D,
Med_Agg_ZD5
pH4.2A/19
pH4.2B/35
pH4.2A/F2
pH4.2A/3E
En15_7/11/96
Enrich ZB3
Nitrosospira sp. 40KI (Soil)
Enrich ZA4_96
pH4.2B/38
En_30_4_9
En_ZA3_96
En_18_4_8
Cluster 4
Nitrosomonas
EnvA2-13
AlaW3
EnvB1-6
EnvA2-3
EnvA2-17
EnvA1-21
A2bM1
Cluster 5
SewageC3
SewageC4
SewageDG4
SewageDG1
Nitrosomonas ureae (Sewage)
pH7B/Z2
PH7C/37
SewageDG2
SewageDG5
EnvC1-17
EnvC1-19
C2aM2
Nitrosomonas sp. NM63 (Marine)
B1bM2
AEM-1
B2aM2
Nitrosococcus mobilis (Brackish)
B2bM2ii
B2bM2i
B1aM2
A1aM2
B1bW2
B2bW1
B1aW2
EnvA1-19
AEM-5
AEM-4
C1bW1
Cluster 6
Med_Agg_A64
Med_Agg_X42
Nitrosomonas eutropha (Sewage)
Med_Agg_7
Med_Agg_58
Med_Agg_12
Nitrosomonas europaea (Soil)
En14_7/11/96
B2aW2
Cluster 7
Neisseria denitrificans
Kingella kingae
Oxalobacter formigenes
Comamonas testosteroni
Sphaerotilus natans
Alcaligenes eutrophus
Reference
β-Proteobacteria
10%

cultivation, in focusing on strains that may not be representative of natural populations.

Interpretation of these data requires consideration of a number of factors. Although duplicate sequences were not obtained, differences were often limited to 1–2 bases and may have arisen through sequencing or cloning errors. However, comparisons were done over regions of approximately 300 bp only and greater lengths may reveal greater diversity. Indeed, full-length sequencing provided support for subdivision into clusters (Stephen *et al.* 1996). Although identity with pure cultures was never observed, relatively few ammonia oxidizers have been purified and even fewer sequenced. The number of environments represented in Figure 11.1 is relatively small, but more sequence data are now available and the general conclusions remain valid. Additional clusters or subclusters have been characterized and, particularly for *Nitrosomonas*, 16S rDNA and *amoA* trees are congruent (Purkhold *et al.* 2000).

Selection in laboratory cultures

Evidence for selection during laboratory cultivation arises from comparisons between sequences amplified from DNA extracted directly from environmental samples with those from enrichment cultures or from cultures used in most probable number (MPN) counts. Smith *et al.* (2001) compared 16S rDNA cloned from DNA extracted from arable soils and marine sediments with sequences from enrichment cultures obtained from the same sites. Soil enrichment cultures were dominated by *Nitrosospira* cluster 3 sequences but clone sequences were distributed more evenly between *Nitrosospira* clusters 2, 3 and 4. Although this suggests laboratory selection of *Nitrosospira* cluster 3, these differences may also be due to bias associated with the molecular techniques. However, a higher proportion of enrichment cultures than clones had identical sequences, suggesting growth of strains adapted to the medium and cultivation conditions employed. Cluster 3 sequences also represent the dominant pure culture isolates of *Nitrosospira* species cited in the literature, and this group may be selected by use of similar isolation conditions by different groups. Although laboratory selection is likely, pure and enrichment cultures are essential for characterization of the physiology, and hence ecological function, of community members. An important role for molecular techniques is, therefore, to identify which laboratory cultures are representative of the natural community. In this study, 16% of enrichment culture sequences were identical to those in environmental clones, suggesting that a significant proportion of the natural population was culturable. For marine samples, *Nitrosomonas* sequences dominated enrichments, while *Nitrosospira* sequences were most abundant among environmental clones, but with less evidence for laboratory selection. Only one of 40 marine enrichment sequences was found among clones, indicating poorer culturability of marine strains represented in the clone library, under the conditions employed.

Inferences drawn from such studies depend on the biases associated with molecular techniques, the discriminatory ability of the 16S rDNA region analysed and

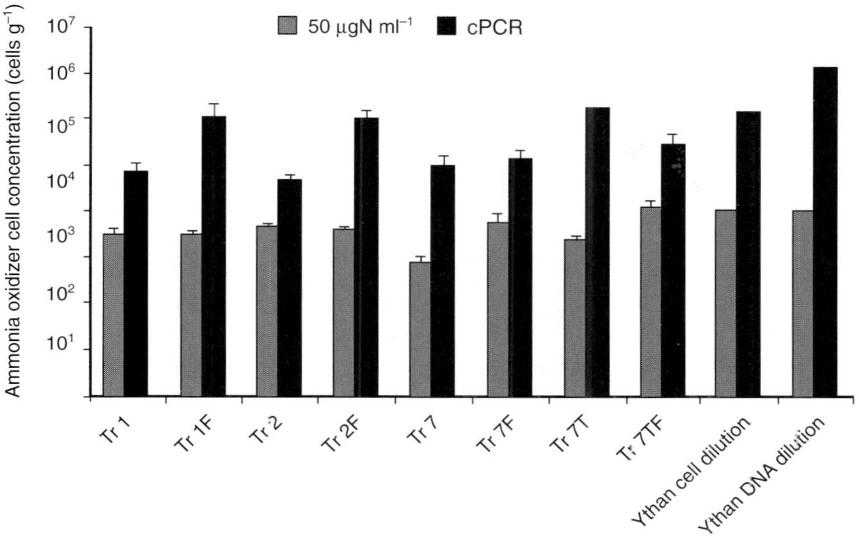

Figure 11.2 Ammonia-oxidizer cell concentrations in triplicate samples from LTER soils and Ythan Estuary sediments estimated by the Most Probable Number method using medium containing 50 μg NH_4^+-N mL^{-1} and by cPCR. Results are shown as means of triplicate samples ± standard error. Tr 1 and Tr 2 are cultivated soils while Tr 7 is a successional grassland soil. The suffix F refers to a fertilized treatment while Tr 2 and the suffix T refers to soils that have been tilled. (Reproduced with permission from Phillips *et al.* 2000b.)

the cultivation conditions employed. Smith *et al.* (2001) employed a relatively restricted range of cultivation conditions, but still found that 7% of the sequence diversity in clones was observed also in laboratory enrichment culture, which is higher than most estimated values for culturability within natural populations. Additional, independent experimental approaches are required to determine conclusively which picture of diversity is correct but molecular techniques provide the potential for this, through use of FISH or analysis of nucleic acids using different primer sets or primers for different genes.

Enumeration of ammonia oxidizers is traditionally carried out using cultivation-based MPN techniques. Problems of laboratory selection are exacerbated by the slow growth of ammonia oxidizers, requiring incubation for several weeks. Quantification of ammonia oxidizers by competitive PCR (cPCR) provides evidence for such selection, and for consequent underestimation of populations. For example, cell concentrations determined by cPCR quantification of β-proteobacterial ammonia-oxidizing bacteria in agricultural soils and estuarine sediments were between one and three orders of magnitude greater than MPN counts (Figure 11.2; Phillips *et al.* 2000b). Molecular analysis of MPN cultures provides further evidence of such selection (see below). Mendum *et al.* (1999) used cPCR of 16S rRNA and *amoA* genes to assess the response of ammonia-oxidizer populations to addition of ammonia nitrate fertilizer. Samples taken 3 days after fertilizer addition showed increases in nitrification rate but no detectable population growth. After 6 weeks, however, when nitrification rates had decreased, numbers of ammonia oxidizers were greater.

Influence of soil pH

Autotrophic nitrification is frequently reported in acid soils (De Boer *et al.* 1988), but the minimum pH for growth in liquid batch culture is 6.5 (Allison & Prosser 1991). Possible explanations for this contradiction are surface growth (Allison & Prosser 1993), aggregate formation (De Boer *et al.* 1991), ureolysis (De Boer & Laanbroek 1989; Allison & Prosser 1991; Burton & Prosser 2001) and heterotrophic nitrification (Killham 1986). A further explanation is the existence of strains adapted to low pH environments, which could not be tested using traditional methodologies. Stephen *et al.* (1996, 1998) used 16S rDNA-based methods to analyse ammonia-oxidizer communities in soil plots maintained for 36 years at a range of pH values. Initial analysis involved PCR amplification of 16S rDNA using β-AMO primers from soils of low (4.5) and neutral (7.2) pH, with subsequent sequencing and phylogenetic analysis of clones. Representatives of clusters 2, 3, 4 and 6 were found, with some evidence for greater proportions of cluster 2 in acid and cluster 3 in neutral soils. To provide greater replication and analysis of a greater number of pH values, Stephen *et al.* (1998) subjected PCR products from CTO primers to DGGE analysis combined with oligonucleotide probing, to distinguish clusters with similar banding patterns. Quantification of band intensities normalized against a general probe indicated differences in relative abundance of different clusters associated with soil pH (Figure 11.3). The relative signal strength for the *Nitrosospira* cluster 3 probe decreased with decreasing soil pH (6.6–3.9), while that of *Nitrosospira* cluster 2 increased. Signals from *Nitrosospira* cluster 4 were greatest at pH 5.5, while *Nitrosomonas* cluster 6a signals did not vary significantly with pH. Kowalchuk *et al.* (2000a) reported similar selection for *Nitrosospira* clusters 2 and 4 in acid successional grassland soils. These results indicate significant pH-associated differences in *Nitrosospira* populations but, in themselves, provide no information on the physiological characteristics within different clusters that may be responsible. Hastings *et al.* (2000) also detected *Nitrosospira* in acid forest soils (pH 4.1) and demonstrated its survival during simulated drought conditions. *Nitrosomonas* was less abundant but their data indicated that it responded more rapidly on rewetting of soil.

Soil management regimes

Soil management and fertilization might be expected to influence ammonia-oxidizer community structure and composition. In particular, liming and ammonia fertilization will encourage ammonia-oxidizer growth, providing the potential for selection of strains better adapted to the prevailing conditions. When looking for

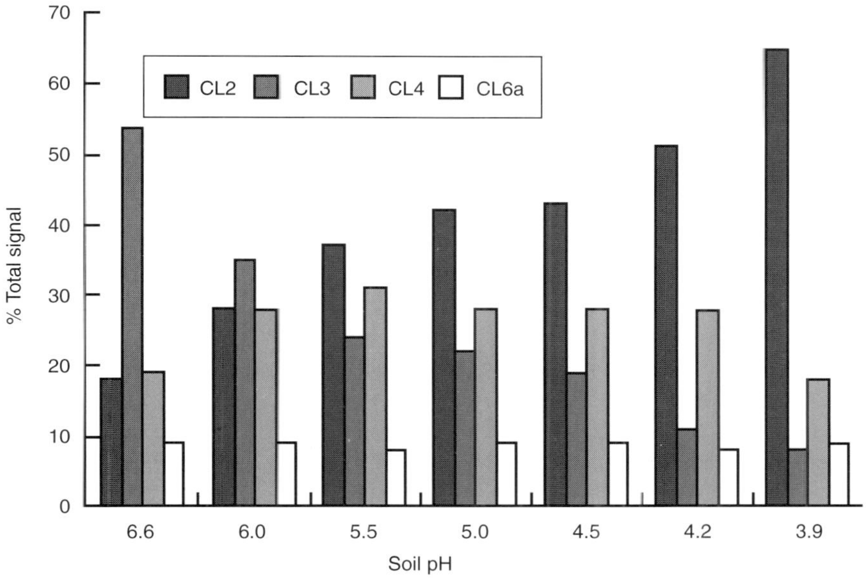

Figure 11.3 Quantification of hybridization signals from DGGE gels of 16S rRNA genes amplified using ammonia-oxidizer specific primers, from DNA extracted from soil maintained at a range of pH values. Quantification was achieved by phosphoimaging and profile analysis of each lane. The percentage total signal is the hybridization signal for each sample, expressed as a normalized percentage of the signal for the cluster standard. Least significant difference (LSD) values for clusters 2, 3, 4 and 6a were 2.13, 2.36, 1.50 and 1.15 respectively. (Reproduced with permission from Stephen *et al.* 1998.)

such effects, however, it is important to consider the likely potential timescale required for changes in community structure, which must involve growth or death of components of the population sufficient to cause detectable changes in relative abundance, as differences in relative abundance of different groups of clones or by fingerprinting methods such as DGGE. For the latter, which detects differences in relative abundance of different clusters, it is also possible that significant changes may occur within clusters, which might have significant effects on functional diversity of ammonia oxidizers.

Bruns *et al.* (1999) and Phillips *et al.* (2000a) examined the influence of a number of management regimes on nitrification and ammonia-oxidizer populations in Long-Term Ecological Research (LTER) sites in Michigan. In addition to assessing 16S rRNA gene diversity in extracted DNA, they obtained MPN counts of ammonia oxidizers. MPN counts suffer from the limitations of cultivation-based techniques but analysis of positive cultures at the lowest dilutions potentially provides information on the dominant populations and the method can also be used to assess selectivity of growth media, by comparison of molecular sequence data on positive cultures with those amplified from extracted DNA. Bruns *et al.* (1999) demonstrated a reduction in MPN counts in media containing high ammonium concentration (2 mg NH_4^+-N mL^{-1}), particularly in soils that had never been tilled. This suggests that a proportion of viable cells was sensitive to high ammonium concentrations and that this proportion is influenced by soil management regimes. Phillips *et al.* (2000a) also observed a decrease in MPN counts at high ammonium concentrations and carried out DGGE analysis of 16S rRNA genes amplified both from MPN tubes showing growth and from extracted DNA. DGGE bands were found in MPN tubes that were not detected in extracted DNA, suggesting selection during laboratory growth. Several of these bands were amplified from medium containing the highest ammonium concentration (1 mg NH_4^+-N mL^{-1}), indicating that selection may be linked to ammonium concentrations higher than those experienced in the soil. MPN sequences were dominated by *Nitrosomonas* (some with close similarity to *N. europaea*), often to the exclusion of *Nitrosospira* sequences, which dominated environmental sequences. Counts obtained using cPCR were from 10- to 100-fold greater than MPN counts, as also found by Kowalchuk *et al.* (2000b). The results therefore indicate that MPN counts both significantly underestimate the size of ammonia-oxidizer populations and enumerate only those best adapted to growth under laboratory conditions, with ammonium concentration an important selective factor.

Bruns *et al.* (1999) observed differences in ammonia-oxidizer communities between different treatments. Populations in tilled soils were less diverse than those from never-tilled soils and only *Nitrosospira* cluster 3 sequences were detected in tilled soils. By contrast, soils which had never been tilled yielded sequences representing three clusters, *Nitrosospira* clusters 3 and 4 and *Nitrosomonas* cluster 6, possibly reflecting greater physical heterogeneity in these soils. There were also indications that N fertilization led to increased relative abundance of representatives of *Nitrosospira* cluster 3. Phillips *et al.* (2000a) found no correlation between nitrification rates and ammonia-oxidizer DGGE banding patterns in cultivated, successional and native deciduous forest soils, suggesting that differences resulting from soil management in these sites required the greater precision of the cloning/sequencing approach, rather than DGGE. Kowalchuk *et al.* (2000b) also assessed the influence of management regimes on ammonia-oxidizer populations in chalk grassland soils in which N fertilization had been discontinued at different times. DGGE analysis and sequencing detected only *Nitrosospira* cluster 3 and 4 sequences, with a decrease in the relative abundance of the former with increasing time since fertilization. Analysis of sequences from MPN counts indicated that this change in diversity was driven by ammonium supply, rather than changes in plant cover and other factors.

Heterogeneity

Heterogeneity in soil physicochemical properties means that bacterial communities in close proximity may experience different environmental conditions and selection pressures, potentially leading to similar heterogeneity in microbial communities. To assess heterogeneity in soil ammonia-oxidizer populations, Webster *et al.* (2002)

analysed 16S rRNA genes and *amoA* genes in improved grassland soils (regular addition of N fertilizer) and unimproved soils (no fertilization). Both sites were grazed by sheep, which provided the major source of N input in unimproved sites. DGGE analysis did not detect selection for particular clusters in either site, as found by Phillips *et al.* (2000a). However, DGGE patterns from unimproved soils indicated greater heterogeneity, with representatives of a greater number of ammonia-oxidizer clusters than improved soils. In addition, 0.5 g-subsamples of soil from unimproved soils demonstrated greater small-scale heterogeneity than those from improved soils. DGGE banding patterns from unimproved soil samples varied significantly between replicates samples, in terms of both the type and number of clusters represented, while banding patterns from replicate samples of improved soil were more homogeneous. Controls demonstrated that this was not due to PCR error. Heterogeneity in soil pH and ammonium concentration in 0.5 g samples were also greater for unimproved samples, suggesting that soil properties may drive heterogeneity in ammonia-oxidizer populations.

Influence of high urea/ammonia concentration

Heterogeneity in ammonium supply is likely to be greater in unimproved soils, where the major N input is sheep urination, supplying higher and more localized urea concentrations, than in sites receiving regular N input. Consequences for nitrification and for ammonia-oxidizer populations were investigated in microcosms amended with synthetic sheep urine providing 1 mg urea-N g^{-1} soil. Incubation of microcosms for 12 weeks reduced homogeneity of ammonia-oxidizer populations, usually with selection for a single cluster. Sequencing of DGGE bands indicated that the final dominant band was representative of *Nitrosospira* cluster 3. Phylogenetic analysis of sequences of these bands led to a re-evaluation of cluster 3, with subdivision into clusters 3a and 3b, the latter containing sequences of the selected bands. Urea was hydrolysed in microcosms to ammonium within 10 hours, and was then oxidized to nitrite and nitrate after a lag period ranging from 0 to approximately 4 weeks. The length of this lag period was correlated with the relative abundance of different clusters of ammonia oxidizers initially present in the soil, being lowest in soils initially containing high relative abundance of *Nitrosospira* cluster 3b or *Nitrosomonas* cluster 7 sequences. Those dominated by *Nitrosospira* cluster 3a or other *Nitrosospira* clusters exhibited long lag periods, while initially mixed populations gave intermediate lag periods. Laboratory growth experiments showed these lag periods to be due to differences in tolerance to high ammonium concentrations. Pure and enrichment culture representatives of *Nitrosospira* cluster 3b and *Nitrosomonas* cluster 7 grew on medium containing 1 mg NH_4^+ mL^{-1}, while representatives of clusters 2, 3a and 4 were inhibited. These data therefore demonstrate a direct link between bacterial community structure and ecosystem processes. The presence of particular groups of ammonia oxidizers determined whether N input as sheep urine was converted rapidly to nitrate, which determined the rate of loss N loss through leaching, and its availability for plant growth. The study also demonstrates that the diversity and community of bacterial populations in the soil significantly influence ecosystems processes and that this influence can be explained on the basis of the physiological characteristics of different members of the community.

Discussion

The above studies demonstrate the power of molecular techniques in characterizing the diversity and community structure of microbial populations in soil and the influence of environmental factors. They indicate that the diversity of ammonia-oxidizing bacteria, organisms that are central to biogeochemical cycling of nitrogen, is considerable and make feasible characterization of natural populations without lengthy enrichment, isolation and identification procedures. They also indicate the existence and potential environmental importance of groups of ammonia oxidizers that are not yet represented in laboratory culture. We are now beginning to increase our understanding of the mechanisms controlling diversity and, importantly, the relationships between species diversity, physiological diversity and ecosystem process rates. This, however, requires a combination of molecular techniques and traditional techniques for the measurement of process rates and physiological studies on cultivated organisms. The studies described in this article have been based, mainly, on 16S rRNA genes. For ammonia oxidizers they have proved a valuable tool and, because the majority of autotrophic soil ammonia oxidizers belong to a single phylogenetic group, also provide information on ecosystem function. Confirmation of findings has been obtained by analysis of functional genes. Ammonia oxidizers have proved to be excellent model organisms for community studies and the findings are of relevance and importance for other microbial groups. The challenge is to extend these studies to determine the fundamental mechanisms driving diversity, and links with physiological diversity within ammonia oxidizers and with environmental factors and ecosystem process rates. The development of techniques such as STARFISH, which combines 16S rRNA probing and microautoradiography (Ouverney & Fuhrman 1999), and stable isotope probing (Radajewski *et al.* 2000) potentially enable *in situ* physiological studies. However, there is still an important role for laboratory studies of microbial physiology, informed by molecular analysis to ensure that organisms studied in the laboratory are representative of indigenous communities. Molecular studies also provide the basis for genomic studies of ecologically important microbial groups, which will provide more comprehensive information on the genes underlying the biochemistry of ammonia oxidizers, differences in metabolic potential between closely related groups and their relevance for microbial ecology and evolution.

Acknowledgements

This work was funded by the Natural Environment Research Council's thematic programme EDGE (Ecological Dynamics and Genes).

References

Allison, S.M. & Prosser, J.I. (1991) Urease activity in neutrophilic autotrophic ammonia oxidizing bacteria isolated from acid soils. *Soil Biology and Biochemistry* **23**, 45–51.

Allison, S.M. & Prosser, J.I. (1993) Ammonia oxidation at low pH by attached populations of nitrifying bacteria. *Soil Biology and Biochemistry* **25**, 935–941.

Belser, L.W. & Schmidt, E.L. (1978) Serological diversity within a terrestrial ammonia-oxidizing population. *Applied and Environmental Microbiology* **36**, 584–588.

Bruns, M.A., Fries, M.R., Tiedje, J.M. & Paul, E.A. (1998) Functional gene hybridization patterns of terrestrial ammonia-oxidizing bacteria. *Microbial Ecology* **36**, 293–302.

Bruns, M.A., Stephen, J.R., Kowalchuk, G.A. *et al.* (1999) Comparative diversity of ammonia oxidizer 16S rRNA gene sequences in native, tilled, and successional soils. *Applied and Environmental Microbiology* **65**, 2994–3000.

Burton, S.A.Q. & Prosser, J.I. (2001) Autotrophic ammonia oxidation at low pH through urea hydrolysis. *Applied and Environmental Microbiology* **67**, 2952–2957.

De Boer, W. & Laanbroek, H.J. (1989) Ureolytic nitrification at low pH by *Nitrosospira* species. *Archives of Microbiology* **152**, 935–941.

De Boer W., Duyts, H. & Laanbroek H.J. (1988) Autotrophic nitrification in a fertilised acid heath soil. *Soil Biology and Biochemistry* **20**, 845–850.

De Boer, W., Klein Gunnewiek, P.J.A., Veenhuis, M. *et al.* (1991) Nitrification at low pH by aggregated chemolithotrophic bacteria. *Applied and Environmental Microbiology* **57**, 3600–3604.

DeLong, B.K., Wickham, G.S. & Pace, N.R. (1989) Phylogenetic stains: ribosomal RNA-based probes for the identification of single cells. *Science* **243**, 1360–1363.

Giovannoni, S.J., Britschgi, T.B., Moyer, C.L. & Field, K.G. (1990) Genetic diversity in Sargasso Sea bacterioplankton. *Nature* **345**, 60–63.

Hastings, R.C., Butler, C., Singleton, I. *et al.* (2000) Analysis of ammonia-oxidizing bacteria populations in acid forest soil during conditions of moisture limitation. *Letters in Applied Microbiology* **30** 14–18.

Head, I.M., Hiorns, W.D., Embley, T.M. *et al.* (1993) The phylogeny of autotrophic ammonia-oxidizing bacteria as determined by analysis of 16S ribosomal RNA gene sequences. *Journal of General Microbiology* **139**, 1147–1153.

Heid, C.A., Stevens, J., Livak, K.J. & Williams, P.M. (1996) Real time quantitative PCR. *Genome Research* **6**, 986–994.

Horz, H.P., Rotthauwe, J.H., Lukow, T. & Liesack, W. (2000) Identification of major subgroups of ammonia oxidizing bacteria in environmental samples by T-RFLP analysis of *amo*A PCR products. *Journal of Microbiological Methods* **39**, 197–204.

Jansson, J.K. & Leser, T.D. (1996) Quantitative PCR of environmental samples. *Molecular Microbial Ecology Manual* **2.7.4**, 1–19.

Jetten, M.S.M., Strous, M., van de Pas Schoonen, K.T. *et al.* (1998) The anaerobic oxidation of ammonium. *FEMS Microbiology Reviews* **22**, 421–437.

Killham, K. (1986) Heterotrophic nitrification. In: *Nitrification* (ed. J. I. Prosser), pp. 117–126. IRL Press, Oxford.

Kowalchuk, G.A., Stephen, J.R., De Boer, W. *et al.* (1997) Analysis of ammonia-oxidizing bacteria of the beta subdivision of the class Proteobacteria in coastal sand dunes by denaturing gradient gel electrophoresis and sequencing of PCR-amplified 16S ribosomal DNA fragments. *Applied and Environmental Microbiology* **63**, 1489–1497.

Kowalchuk, G.A., Stienstra, A.W., Helig, G.H.J. *et al.* (2000a) Molecular analysis of ammonia-oxidizing bacteria in soil of successional grasslands of the Drentsche A (The Netherlands). *FEMS Microbiology Ecology* **31**, 207–215.

Kowalchuk, G.A., Stienstra, A.W., Helig, G.H.J. *et al.* (2000b) Changes in the community structure of ammonia oxidizing bacteria during secondary succession of calcareous grasslands. *Environmental Microbiology* **2**, 99–110.

Liu, W.T., Marsh, T.L., Cheng, H. & Forney, L.J. (1997) Characterization of microbial diversity by determining terminal restriction fragment length polymorphisms of genes encoding 16S rRNA. *Applied and Environmental Microbiology* **63**, 4516–4522.

McCaig, A.E., Embley, T.M. & Prosser, J.I. (1994) Molecular analysis of enrichment cultures of marine ammonia oxidizers. *FEMS Microbiology Letters* **120**, 363–368.

McCaig, A.E., Glover, L.A. & Prosser, J.I. (1999) Molecular analysis of eubacterial community structure and diversity in unimproved and improved upland grass pastures. *Applied and Environmental Microbiology* **65**, 1721–1730.

McCaig, A.E., Glover, L.A. & Prosser, J.I. (2001) Numerical analysis of grassland bacterial community structure under different land management regimes using 16S rDNA sequence data and DGGE banding patterns. *Applied and Environmental Microbiology*, **67**, 4554–4559.

Mendum, T.A., Socket, R.E. & Hirsch, P.R. (1999) Use of molecular and isotopic techniques to monitor the response of autotrophic ammonia-oxidizing populations of the β-subdivision of the class *Proteobacteria* in arable soils to nitrogen fertilizer. *Applied and Environmental Microbiology* **65**, 4155–4162.

Muyzer, G. & Smalla, K. (1998) Application of denaturing gradient gel electrophoresis (DGGE) and temperature gradient gel electrophoresis (TGGE) in microbial ecology. *Antonie van Leeuwenhoek* **73**, 127–141.

Ouverney, C.C. & Fuhrman, J.A. (1999) Combined microautoradiography—16S rRNA probe technique for determination of radioisotope uptake by specific microbial cell types *in situ*. *Applied and Environmental Microbiology* **65**, 1746–1752

Phillips, C.J., Harris, D., Dollhopf, S.L. *et al.* (2000a) Effects of agronomic treatments on the structure and function of ammonia oxidizing communities. *Applied and Environmental Microbiology* **66**, 5410–5418.

Phillips, C.J., Paul, E.A. & Prosser, J.I. (2000b) Quantitative analysis of ammonia oxidizing bacteria using competitive PCR. *FEMS Microbiology Ecology* **32**, 167–175.

Purkhold, U., Pommerening-Roser, A., Juretschko, S. *et al.* (2000) Phylogeny of all recognized species of ammonia oxidizers based on comparative 16S rRNA and *amo*A sequence analysis. Implications for molecular diversity surveys. *Applied and Environmental Microbiology* **66**, 5368–5382.

Radajewski, S., Ineson, P., Parekh, N.R. & Murrell, J.C. (2000) Stable-isotope probing as a tool in microbial ecology. *Nature* **403**, 646–649.

Rotthauwe, J-H., Witzel, K-P. & Liesack, W. (1997) The ammonia monooxygenase structural gene *amoA* as a functional marker: molecular fine-scale analysis of natural ammonia-oxidizing populations. *Applied and Environmental Microbiology* **63**, 4704–4712.

Schramm, A., de Beer, D., van den Heuvel, J.C. *et al.* (1999) Microscale distribution of populations and activities of *Nitrosospira* and *Nitrospira* spp. along a macroscale gradient in a nitrifying bioreactor: quantification by *in situ* hybridization and the use of microsensors. *Applied and Environmental Microbiology* **65**, 3690–3696.

Smith, Z., McCaig, A.E., Stephen, J.R. *et al.* (2001) Species diversity of uncultured and cultured populations of soil and marine ammonia oxidizing bacteria. *Microbial Ecology*, **42**, 228–237.

Stephen, J.R., McCaig, A.E., Smith, Z. *et al.* (1996) Molecular diversity of soil and marine 16S rDNA sequences related to β-subgroup ammonia oxidizing bacteria. *Applied and Environmental Microbiology* **62**, 4147–4154.

Stephen, J.R., Kowalchuk, G.A., Bruns, M.A. *et al.* (1998) Analysis of β-subgroup ammonia-oxidizer populations in soil by DGGE analysis and hierarchical phylogenetic probing. *Applied and Environmental Microbiology* **64**, 2958–2965.

Utåker, J.B. & Nes, I.F. (1996) A qualitative evaluation of the published oligonucleotides specific for the 16S rRNA gene sequences of the ammonia oxidizing bacteria. *Systematics and Applied Microbiology* **21**, 72–82.

Webster, G., Embley, T.M. & Prosser, J.I. (2002) Grassland management regimes reduce small-scale heterogeneity and species diversity of β-proteobacterial ammonia oxidizer populations. *Applied and Environmental Microbiology* **68**, 20–30.

Whitman, W.B., Coleman, D.C. & Wiebe, W.J. (1998) Prokaryotes: the unseen majority. *Proceedings of the National Academy of Sciences of the United States of America* **95**, 6578–6583.

Part 3

Host–pathogen interactions

Chapter 12

Mutualistic interactions amongst viruses?

M. Naylor, H. C. J. Godfray, D. W. Pallett, M. Tristem, J. P. Reeves and J. I. Cooper*

Introduction

Scientists working with viral plant pathogens were initially concerned with establishing aetiology and patterns of epidemics but have also been in the forefront of biological theory and technology. For example, tobacco mosaic virus (*sensu lato*) was the primary test bed for an astonishing number of key processes; nucleoprotein structure, RNA genetics, intra- and inter-cellular protein trafficking, pathogenesis, resistance, virus-mediated gene delivery, genetic modification and RNA infectiousness (Scholthoff *et al.* 1999). Additionally, viruses were among the first sources of transgenic genes used to confer resistance or tolerance in transgenic crops (e.g. Fitchen & Beachy 1993; Baulcombe 1994; Palukaitis & Zaitlin 1997). Furthermore, plant viruses are in the vanguard of studies to elucidate, for example, the molecular basis of 'natural' gene expression or the opportunities for moderating gene expression through 'silencing' (e.g. Vance & Vaucheret 2001; Waterhouse *et al.* 2001). The revolution in molecular biology over the last few decades has had as much impact on plant virology as it has had on other areas of virology and microbial biology. The genomes of an increasingly diverse range of viruses have been characterized—with an unknown number of sequences privately lodged within the vaults of commercial companies. The current publicly accessible data include more than 100 plant pathogenic virus genomes (out of a total of 525 virus genomes that have been completely sequenced). Additionally, there are substantial numbers of partial viral sequences and all these resources are the focus of intense research interest for targeted manipulation of genetic structures and also 'gene mining'.

Viral genomes are dynamic populations of mutable molecules, and viruses currently pose numerous classificatory problems that the International Committee on Virus Taxonomy (ICTV) has yet to grapple with fully. Nevertheless, new insights into virus ecology and evolution have been obtained from virus phylogenies inferred from genomic base sequence data. At least temporarily, these approaches have resolved the relationships between some viruses, and confirmed the longstanding

* Correspondence author: MNA@wpo.nerc.ac.uk

view that evolutionary relationships among virus groups are as much reticulate as tree-like. The 'new' molecular tools, and particularly their ready accessibility to non-specialized laboratories in the form of inexpensive high-throughput 'kits', is revolutionizing understanding of epidemiology and ecology. As our part of the NERC-funded EDGE thematic programme we focused on a key enzyme, RNA-directed RNA polymerase (also more simply called polymerase or replicase), and developed a model for co-virus replication and transmission of two viruses that infect carrots and other Apiaceae. The two viruses we studied (carrot red leaf virus luteovirus and carrot mottle virus umbravirus) are economically significant (e.g. Watson *et al.* 1964, 1998; Waterhouse & Murant 1983; Watson & Falk 1994; Gibbs *et al.* 1996a,b) and were suspected to have a mutualistic relationship.

Interrelationships between luteoviruses and umbraviruses

Luteoviruses seem to have had complex origins and currently are assigned by ICTV to three distinguishable types: the genus *Luteovirus* (most similar to Tombusviridae in terms of polymerase sequence), the genus *Enamovirus* and the genus *Polerovirus*. Promiscuous recombination involving genes from these differing gene pools (Mayo & Jolly 1991; Gibbs & Cooper 1995) confounds attempts at pigeonholing luteoviruses and the taxonomy is somewhat tentative. The CtRLV that we used to make molecular detection tools is unassigned by ICTV and the genome has not been sequenced in toto. However, as we will describe in detail elsewhere, sequence comparisons lead us to conclude that CtRLV is more similar to viruses in the *Polerovirus* genus than to those in the *Luteovirus* genus or the *Enamovirus* genus.

Whereas luteoviruses form nucleoprotein virions and are only transmitted by aphids, umbraviruses are transmissible manually, but do not form 'conventional' (nucleoprotein) virions. Indeed, none of the umbraviruses characterized to date has a gene sequence coding for capsid protein. With that knowledge, it is now possible to explain why the umbraviruses are transmitted (by aphids) from plants also containing a luteovirus. The co-replicating luteovirus contributes the protein coat for both viral genomes.

An umbravirus isolate obtained from carrot in the UK and maintained by serial manual transfer was the basis for a substantial amount of biological/epidemiological data and was named carrot mottle. The sequenced virus (from Australia) was named carrot mottle mimic umbravirus (CtMoMV; U57305). When investigating virus prevalence and gene dynamics in the South of England, we used the CtMoMV sequence to provide molecular tools for CtMoV detection.

Luteoviruses are usually considered to be phloem limited but short-term observations (Barker 1989) showed that a number of different viruses in luteovirus-infected plants enhanced rates of luteovirus accumulation plausibly reflecting exploitation of extravascular cells.

This chapter is divided into three parts. To begin, we review the diversity of natural histories that are found amongst the RNA viruses and other entities which parasitize plants. In the second section we derive and explore an epidemiological model to predict the circumstances under which mutualism is stable. In the final section we give an overview of our experimental work, and highlight areas where more research is required to further understand the evolution and ecology of these intriguing viruses.

The diversity of plant virus natural histories

Viruses constitute a class of obligate parasites that is very heterogeneous (Van Regenmortel *et al.* 2000) and the viruses that infect vascular plants have complex interrelationships that span the spectrum from parasitism through commensalism to mutualism (e.g. Liu & Cooper 1994; Pirone & Blanc 1996; Taliansky *et al.* 2000). The ultimate infectious unit in two-thirds of the known viruses that infect vascular plants is a single-stranded nucleic acid (RNA) that directs the synthesis of protein on ribosomes of host cells and does not produce DNA copies (Domingo & Holland 1994). Some viruses (currently suspected to be a small minority) integrate and are retained in the germ line (e.g. Harper *et al.* 1999). Other plant pathogenic viruses have RNA genomes that require copying before translation into proteins on ribosomes.

Viruses with compartmentalized/divided genomes

A diverse range of viruses has genomes composed of two, three or more nucleic acid species, each with a more or less distinct base sequence (e.g. Chao 1992). In some viruses, the different parts of the divided genome occur together within coats (capsids) of virus-coded protein. In others, the constituent parts of a genome are packaged separately in tubular (e.g. tobacco rattle virus) or polyhedral particles. Indeed, a bipartite genome strategy is widespread among viruses harmful in a range of crops and such viruses are also prevalent in wild plants (e.g. Cooper & Harrison 1973). Compartmentalization of genes facilitates genome re-assortment and may provide one means whereby viral phenotypes rapidly adapt to new environmental circumstances. Alternatively, or additionally, multipartite genomes may limit the impacts of copy error on virus integrity and survival (Nee 2000). It is not always easy to recognize multipartite genomes. Some viruses encapsidate RNA molecules that may be virus derived or host derived (e.g. Cooper *et al.* 1994). Independently pathogenic nucleic acid species such as viroids almost certainly have a non-viral origin and replicate autonomously with or without associated disease (e.g. Elena *et al.* 1991; Chela-Flores 1994). They do not encode their own specific proteins and have been observed to be encapsidated in structural proteins synthesized in response to luteovirus infections (Querci *et al.* 1997). In these circumstances, the viroids are temporarily endowed with the ecologically significant aphid transmission properties determined by the luteoviral capsid proteins.

The curious case of pea enation mosaic 'virus'

One pair of dependent mutualist viruses was thought to be a single agent with a bipartite genome and was given the name pea enation mosaic virus (PEMV). Now, this virus is known to contain within a common protein coat, two phenotypically complementary RNAs each capable of independent replication (German & de Zoeten 1975; Demler *et al.* 1994a; de Zoeten & Skaf 2001).

Whereas RNA2 of PEMV (also called PEMV-2) has the sequence organization and plant invasion properties of umbraviruses (Demler *et al.* 1994b), RNA1 of PEMV is distinctive in being generally luteovirus-like and incapable of independent systemic movement within a plant (Demler & de Zoeten 1991). Consequently, PEMV-1 is suspected to derive transmission benefits from properties coded in PEMV-2 and the co-transmission of these two RNAs might be an early stage in an association capable of evolving further into a monopartite genome. Packaging constraints seem not to be critical because satellite RNA seems to be encapsidated with either the RNA1 or the RNA2.

A model of mutualistic viruses

Under what circumstances is the carrot red leaf/carrot mottle virus mutualism described in the introduction stable, i.e. under what conditions will both viruses persist in a system, and not be susceptible to invasion by a single virus that expresses both phenotypes? There are a number of ways in which this problem could be modelled but here we follow Nee (2000) who uses an approach that is jointly rooted in classical epidemiology and modern metapopulation theory. We develop the model in four stages. We begin with (i) a very simple framework of a single (non-compartmentalized) virus and then (ii) briefly describe Nee's model of two nucleic acids, one of which codes for the replication machinery, the other for dispersal capability. We then (iii) modify Nee's framework to explore the carrot virus interaction and (iv) explore whether the system is stable to invasion by a dual-phenotype virus. To facilitate comparisons, we retain where possible Nee's algebraic conventions.

The basic framework

Consider a virus that attacks a species of plants. Let the density of uninfected plants be x and the density of infected plants be y. The vast majority of plant viruses require vectors, typically nematodes or insects that use the plant for food, for transmission, and this fact plus the spatial distribution of the host population may lead to complex epidemiology. Here we ignore these complexities and simply assume that the probability of transmission is a linear function of the product of the densities of infected and uninfected plants, cxy, where c is the transmission parameter. This mass action assumption is the starting point for most explorations of host–pathogen dynamics. Finally, assume that uninfected plants are born at a rate b and die at a rate a while infected plants die at an elevated rate $a + \alpha$ and clear themselves of the virus at rate e (we assume no vertical transmission). These assumptions suggest the following model of the dynamics of the system:

$$\begin{aligned}\frac{\mathrm{d}x}{\mathrm{d}t} &= (b-a)x - cxy + ey \\ \frac{\mathrm{d}y}{\mathrm{d}t} &= cxy - (a+\alpha+e)y\end{aligned} \qquad (12.1)$$

where we have implicitly ignored such things as age structure and seasonality so that we can phrase the model as a simple system of ordinary differential equations. Were infected plants to reproduce, we would need to add a *by* term to the first or second equation, depending upon whether vertical transmission of the virus occurs. Equations (12.1) are the classical SI (Susceptible–Infected) model whose properties have been exhaustively explored by theoretical epidemiologists (Anderson & May 1979). However, following Nee (2000), we can simplify further if we are willing to make the assumption that the virus does not kill the host ($\alpha = 0$) and that plant population dynamics occur at a much slower rate than that of the virus so that births and deaths can be ignored ($b = a = 0$). This gives us:

$$\begin{aligned}\frac{dx}{dt} &= -cxy + ey \\ \frac{dy}{dt} &= cxy - ey\end{aligned} \qquad (12.2)$$

which is another very well understood model, this time the Levin's metapopulation model (Hanski 1999). In its original guise, x and y represented the fraction of habitat patches unoccupied and occupied by a particular species respectively (the close linkage between metapopulation and epidemiological models is discussed by Nee *et al.* 1997). In the absence of fluctuations in host (patch) densities it is useful to define total host density, $h = x + y$.

Although Equations (12.2) have three parameters (c, e, h) the dynamics are determined by just two parameter combinations, h and e/c. The equilibrium density of infected plants, y^* (here and in future the asterisk indicates an equilibrium) is simply $h - e/c$: the greater the transmission rate and the lower the clearance rate, the larger the fraction of host plants that will be infected (Nee 2000). The model also tells us that if $h < e/c$ the density of infected plants is negative, which means that there is a threshold plant density below which the infection cannot maintain itself. Again, the greater the colonization rate and the smaller the clearance rate, the lower the plant density at which the virus can spread.

Nee's mutualism model

Nee used the above framework to explore the dynamics of a system motivated by tobacco rattle virus (TRV). TRV is a two-component virus with an RNA1 encoding a replicase and associated replication machinery while the coat protein is encoded on the smaller RNA2. An infection containing RNA1 can maintain itself but is trapped in the plant unless the RNA2 is present to provide the coat protein. Let x be the density of uninfected plants, y the density of plants infected by RNA1, and z the density of plants infected by both RNA1 and RNA2. The rate at which plants move out of the uninfected class is given by the term cxz where c again is the transmission parameter. Of these plants, a fraction f acquire only RNA1, the remaining $(1 - f)$ acquiring both nucleic acids (recall RNA2 on its own cannot maintain itself in the plant). Plants containing only RNA1 can acquire RNA2 by a second infection, and the numbers moving from the y to z classes are given by the term cyz which again assumes mass action and the same colonization parameter. As before let $h = x + y + z$ be total plant density and assume the clearance rate, e, is the same for the two categories of infection. Nee's model can thus be represented by three equations:

$$\begin{aligned}\frac{dx}{dt} &= -cxz + ey + ez \\ \frac{dy}{dt} &= fcxz - cyz - ey \\ \frac{dz}{dt} &= (1-f)cxz + cyz - ez\end{aligned} \tag{12.3}$$

Adding the extra equation leads to substantially more complex dynamics, which are fully explored in Nee (2000). The most interesting aspect of the model is how the stability of the system is related to plant density. Recall that in the single-virus model analysed in the previous section, the density of infected plants was given by the expression $h - e/c$. As total plant density (h) is reduced, the density of infected plants declines until the virus becomes extinct (when $h = e/c$). For some parameter values, the present model shows the same behaviour: a stately reduction in prevalence as plant densities drop. However, for other parameter values, the densities of infected plants do not decline smoothly to zero: a threshold is reached where a mathematical catastrophe occurs and a small drop in plant density causes the prevalence to drop precipitately from relatively high levels to zero.

A model of the carrot virus system

The CtRLV/ CtMoV system is a subtly different type of mutualism from the TRV-motivated interaction modelled by Nee. In particular, the luteovirus (CtRLV-equivalent to RNA1) is capable of independent transmission, while we expect the presence of the umbravirus (CtMoV-equivalent to RNA2) to influence the colonization and extinction rates. To accommodate this natural history we analyse the following model:

$$\begin{aligned}\frac{dx}{dt} &= -c_y xy - c_z xz + e_y y + e_z z \\ \frac{dy}{dt} &= c_y xy + fc_z xz - (1-f)c_z yz - e_y y \\ \frac{dz}{dt} &= (1-f)c_z z(x+y) - e_z z\end{aligned} \tag{12.4}$$

which is illustrated schematically in Figure 12.1. Again we set $h = x + y + z$ to be total plant density.

The first equation describes changes in the density of uninfected plants. Losses to this class occur through infections originating from y plants containing only CtRLV, and from z plants containing both viruses. We assume that infection is in both cases linear, but that the transmission parameter may be different (c_y and c_z

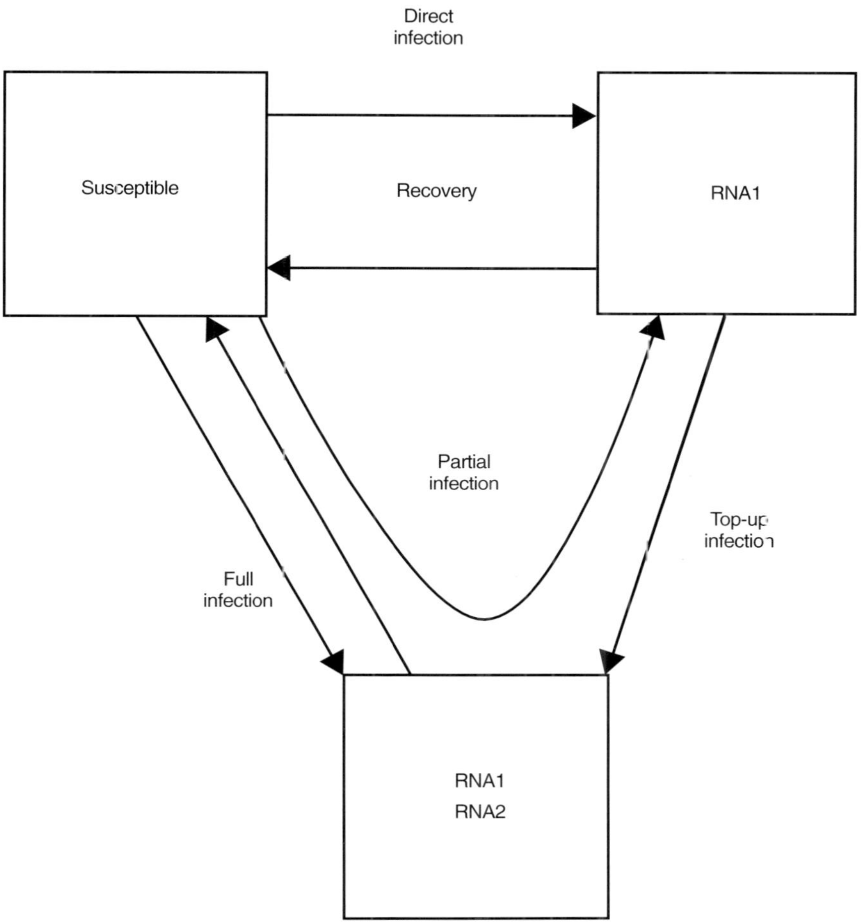

Figure 12.1 Schematic representation of the carrot virus mutualism model. Plants are characterized as uninfected and susceptible (x), containing either the luteovirus (RNA1) and umbravirus (RNA2) (z), or just the luteovirus (y). The different fluxes between the three categories of plants are shown.

respectively). Additions to this class occur when virus clearance occurs, and this we assume can happen at different rates (e_y and e_z) in the two categories of infected plants.

The second equation describes the changes in density of plants infected only with CtRLV. The first term ($c_y xy$) represents infections originating from plants containing only luteovirus while the second term ($fc_z xz$) represents infections originating from plants containing both viruses, but which transmit only the luteovirus (a fraction f). The third term is the loss to the luteovirus-only class due to infections with the umbravirus. We assume that the efficiency of this process ($(1 - f)c_z$) is the same as when susceptible plants are infected (a slightly different assumption to Nee's). The last term is the clearance rate.

Finally, the third equation represents changes in the densities of doubly infected plants. Both uninfected and luteovirus-infected plants move into this category at a rate determined by $(1 - f)c_z$. The final term is the dual-infection specific clearance rate. In deriving these equations we have assumed that the probability of umbravirus-only infections occurring, and being subsequently rescued by luteovirus infection is small. We have never observed umbravirus-only infections in the field (see below).

Before analysing the full system we can note that in the absence of the umbravirus, Equations (12.4) collapse to the single-virus situation (Equations 12.1) with the density of the luteovirus given by $y^+ = h - e_y/c_y$, where we use the plus to identify the equilibrium with no umbravirus in the system.

Now consider the system with both viruses present and solve first for the equilibrium density of z, that is of plants containing both CtRLV and CtMoV:

$$z^* = h - \frac{e_z}{c_z(1-f)} \qquad (12.5)$$

This expression is structurally very similar to that for y^+. The fraction of the total plant population harbouring a joint infection goes up when the colonization rate (c_z) and the efficiency of joint infections $(1 - f)$ are high, but declines as the clearance rate (e_z) goes down.

The conditions under which we expect to find different combinations of the two viruses present are determined solely by host plant population density, h, and two parameter combinations, e_y/c_y and $e_z/c_z(1 - f)$. We can distinguish four different situations, which are also illustrated in Figure 12.2.

1 $e_y/c_y > h$ and $e_z/c_z(1 - f) > h$. The host plant population is sufficiently sparse that neither luteovirus alone, nor the luteovirus–umbravirus pair are able to invade the population.

2 $e_y/c_y < h$ and $e_z/c_z(1 - f) > h$. The first condition means that the luteovirus can invade and reach a density of y^+, but the second condition means that the umbravirus is not able to spread. For the reasons given at the beginning of the chapter we expect the colonization ability of viruses from a joint luteovirus and umbravirus infection to be superior to viruses from a pure luteovirus infection ($c_z > c_y$). However, the spread of the umbravirus depends not on c_z but on $c_z(1 - f)$ and a low efficiency of co-transmission can obviate the benefits of better colonization.

3 $e_y/c_y > h$ and $e_z/c_z(1 - f) < h$. Here the luteovirus is unable to invade the system by itself, but can do so in the presence of the umbravirus. We still expect to observe some luteovirus-only infected plants, but these purely occur because of the failure of co-transmission; in the absence of the umbravirus they would be absent.

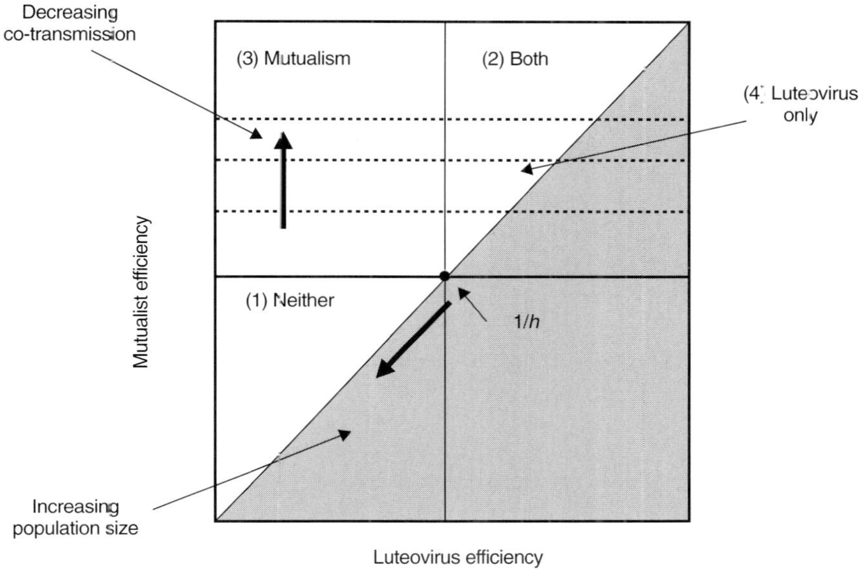

Figure 12.2 Summary of the results of the carrot mutualism model. The axes of the box represent luteovirus efficiency in single infections (c_y/e_y) and the efficiency of joint luteovirus/umbravirus infections (c_z/e_z). Efficiency here is the product of the average length of the infection ($1/e$) and transmission rate (c). We assume that the mutualism is more efficient than the single luteovirus infection and hence ignore the shaded lower triangle. When transmission is perfect $f = 0$ the model predicts three outcomes: (1) neither the luteovirus nor the mutualist pair can persist; (2) both can persist; or (3) only the mutualist pair can persist. The three regions of parameter space are delimited by a pair of lines that cross at the point $1/h$, the reciprocal of population size. The effect of increasing population size is to move the intersection of the two lines towards the bottom left, as indicated by the heavy arrow. When transmission is imperfect ($f > 0$) the horizontal line separating regions (1) and (3) moves upwards (illustrated by the succession of dotted lines), and a new region (4) in which only the luteovirus alone can persist, appears.

4 $e_y/c_y < h$ and $e_z/c_z(1 - f) < h$. In these circumstances the luteovirus could persist by itself, but the umbravirus is also able to invade and as in the last example we expect to see both viruses present.

We can make some other observations from the model. First, the density of plants containing the umbravirus is not influenced by the properties of the luteovirus in single infections (z^* is not a function of e_y or c_y). The intuitive reason for this is that umbraviruses successfully infect plants in category x and y at an identical rate determined by $c_z(1 - f)$. Were x or y plants to differ in the ease with which they could be infected then the dynamics of luteovirus-only infections would impinge on the umbravirus.

Second, the densities of luteovirus-only and jointly infected plants vary smoothly with total plant density as in the single-virus model (i) but unlike Nee's TRV-motivated mutualism model (ii). The mathematical reason for this is that the equivalent expression for z^* in Nee's model depends quadratically on h while in this model the dependence is linear. Biologically, the reason is that the luteovirus, the equivalent of RNA1, is not trapped in the plant in the absence of the umbravirus (the RNA2), and this prevents the catastrophic collapse of the system.

The model can be used to predict the density of luteovirus-only infected plants at equilibrium (y^*). Unfortunately the expression is less transparent than that for z^*:

$$y^* = \frac{A - \sqrt{A^2 - 4c_y c_z e_z f(e_z - c_z(1-f)h)}}{2c_y c_z(1-f)} \tag{12.6}$$

where $A = c_y e_z + c_z(e_z - e_y(1-f) - c_z(1-f)h)$.

A limiting case of this expression occurs when $e_y/c_y = e_z/c_z$; i.e. when both the colonization and clearance rates of the two viruses are identical, or when the improvement in the colonization rate due to the presence of the umbravirus is precisely compensated for by a deterioration (from the viruses point of view) in the clearance rate. In these circumstances, Equation (12.6) becomes

$$y^* = \frac{e_y}{c_y}\frac{f}{(1-f)} \tag{12.7}$$

and $y^* + z^*$, the density of plants infected by luteovirus (irrespective of whether they also contain an umbravirus) is $h - e_y/c_y$; in other words exactly what it would have been in the absence of the umbravirus ($y^+ = y^* + z^*$). In the more general case when $e_y/c_y \neq e_z/c_z$, numerical exploration of Equation (12.4) suggests that the fraction of plants with luteovirus-only infections increases with f and when e_y/c_y increases relative to e_z/c_z.

To conclude, to understand the persistence of the mutualistic interaction, we need to know relative colonization and clearance rates of single and joint infections, and the probability of co-transmission. Whether one or both viruses will be found in the system depends on these parameters on the density of hostplants.

Stability to invasion by a virus with both phenotypes

As discussed in the first section, we believe that the basis of the mutualistic interaction between luteoviruses and umbraviruses in general, and between CtRLV and CtMoV in particular, is that luteoviruses possess a coat protein gene and umbraviruses possess a transport protein gene that facilitates escape from histological constraint. The obvious question is under what circumstances could a virus which carried genes for both proteins invade the system (Nee (2000) explores a similar question within the context of the model described under 'Nee's mutualism model' above). Here, we do not attempt a full analysis of this question, which requires extensive numerical solutions of a variety of assumptions about the precise natural history of the system. Instead we sketch the approach, analyse a simple limiting case, and provide conjectures about the more complex situation.

Consider a virus (we shall label it the w-virus) that possesses the joint luteovirus and umbravirus phenotypes. Call the density of plants infected with this virus w and assume that its clearance and colonization rates are e_w and c_w respectively. We now have to state what happens when a luteovirus or luteovirus/umbravirus pair attempts to colonize a plant containing a w-virus, and reciprocally what happens when a w-virus attempts to infect a plant in category y or z. We make the simplest assumption, that of pre-emption; a plant in category w cannot move into the y or z classes or vice versa. With this simplifying assumption we can write down a series of equations to describe the system (illustrated in Figure 12.3):

$$\begin{aligned} \frac{dx}{dt} &= -c_y xy - c_z xz - c_w xw + e_y y + e_z z + e_w w \\ \frac{dy}{dt} &= c_y xy + f c_z xz - (1-f) c_z yz - e_y y \\ \frac{dz}{dt} &= (1-f) c_z z(x+y) - e_z z \\ \frac{dw}{dt} &= c_w xw - e_w w \end{aligned} \tag{12.8}$$

Now consider whether a mutant w-virus will spread through a population of luteoviruses and umbraviruses at the equilibrium described above. By definition this requires

$$\frac{dw}{dt} > 0 \Rightarrow x^* > \frac{e_w}{c_w} \tag{12.9}$$

In words, the equilibrium density of uninfected hosts, x^*, must exceed e_w/c_w. x^* is simply $h - y^* - z^*$ which we derived in the last section, but unfortunately y^* is give by Equation (12.6) and further analytical progress looks bleak.

But we can go forward in the special case where $e_y/c_y = e_z/c_z$ which requires for its persistence that $e_z/c_z(1-f) < h$. In this situation the umbravirus is neither a parasite nor a mutualist but a commensal, and as described in the last section $y^* + z^* = h - e_y/c_y$ and hence $x^* = e_y/c_y$. The w-virus thus spreads if

$$\frac{e_y}{c_y} = \frac{e_z}{c_z} > \frac{e_w}{c_w} \tag{12.10}$$

Now e/c is an inverse measure of fitness. Slower clearance rates or higher colonization rates both lead to smaller values of this parameter combination. Equation (12.10) is telling us that for the w-virus to spread it must be more efficient and have higher fitness. Any reduction in efficiency consequent on the increased size of the virus required to code for the two phenotypes will prevent the spread of the new variant.

We can also ask the reverse question, will the umbravirus/luteovirus pair invade a

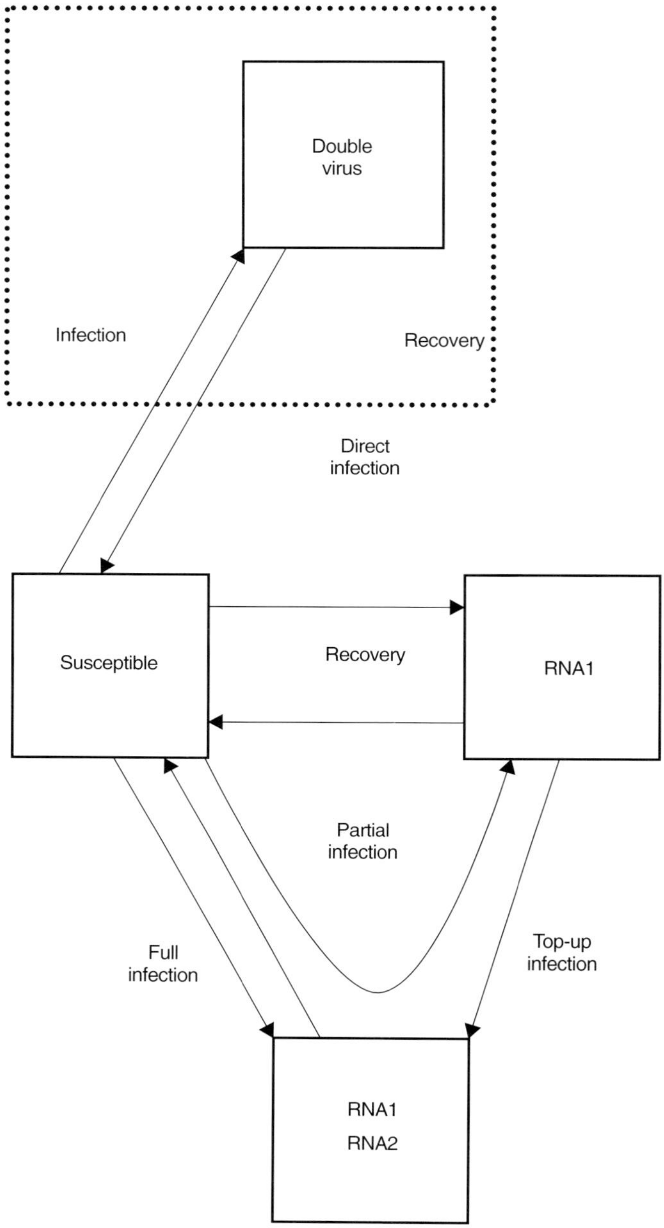

Figure 12.3 Representation of the model that includes a virus which expresses both phenotypes. In addition to the three categories of plants in the model represented in Figure 12.1, there is a new category of plants, *w*, infected by the double-phenotype virus.

population composed solely of w-viruses. The equilibrium density of w-viruses when alone is $h - e_w/c_w$ and hence x^* is e_w/c_w. From Equations (12.8)

$$\frac{\mathrm{d}(y+z)}{\mathrm{d}t} > 0 \Rightarrow c_y(y+z)x^* - e_y(y+z) > 0 \Rightarrow x^* > \frac{e_y}{c_y} \tag{12.11}$$

The condition for the spread of the luteovirus/ umbravirus combination is thus

$$\frac{e_y}{c_y} = \frac{e_z}{c_z} < \frac{e_w}{c_w} \tag{12.12}$$

which is the reverse inequality to that given by Equation (12.10).

We have thus shown that the luteovirus/ umbravirus combination and the w-virus cannot coexist, and that the winner is the virus with higher fitness (lower clearance rate divided by colonization rate). Nee (2000) derived conditions when a virus with both phenotypes could invade the RNA1 and RNA2 of a TRV-like virus, and he too found that coexistence was impossible. Numerical study of more complex instances of Equation (12.8) where $e_y/c_y \neq e_z/c_z$ suggests that this may be a general feature of this model, though we have not been able to prove this formally. We conjecture that in these cases, similar fitness criteria to Equations (12.10) and (12.12) determine the outcome of the competition between the viral types, but that other factors such as the frequency of co-transmission influence the outcome, with low values of f, and relatively high values of e_y/c_y favouring the two phenotype w-virus.

To conclude, the behaviour of this idealized system is determined by the relative values of e/c, h and f. We derived the conditions under which we expect the different viruses and virus combinations to be present, and speculated that a virus combining both luteovirus and umbravirus phenotypes could not coexist with the individual viruses. But we stress that the models developed here are a first step, still a long way from real plant disease systems. One of the major assumptions we made was that viruses are cleared from the plant at a rate e. This assumption was made in Nee's mutualism model, and we followed it here for reasons of tractability and comparison. But do plants ever completely rid themselves of infection? This is a question that requires experiment. Even if 'cleansing' was not observed, our model would still apply if new growth or modular plant units were subject to infection. We would reinterpret the state variables to represent the number of plant parts in different infection classes. Alternatively, we may assume the e parameters measure death rates in a population with perfect compensating density dependence (and with new individuals entering only the uninfected class).

Experimental studies of the carrot-virus system

Before assigning ecological and evolutionary significance to data from field populations, we needed to develop appropriate systems for umbravirus or luteovirus detection.

Virus detection by reverse transcriptase PCR (RT-PCR)

The possession of an RNA-directed RNA polymerase enzyme is one of the few features that all viruses with positive strand RNA genomes have in common. Koonin (1991) showed that the polymerase coding sequences of our target viruses lie in two distinct sets and we exploited the differences between polymerases in CtMoV and CtRLV when studying the dynamics of genes in the environment. For the molecular detection of CtRLV, RT-PCR primers were designed by aligning the predicted amino acid sequences in a number of different luteovirus polymerase regions. Sequence obtained at CEH Oxford from a CtRLV cDNA library was included in the alignment. Three areas of amino acid conservation found in this way were used to design highly degenerate RT-PCR primers. One primer was used to initiate cDNA synthesis and a second pair of oligomers was used for PCR amplification of the cDNA.

The molecular detection method for CtMoV was assembled in two stages. Initially, the predicted amino acid sequence of CtMoMV was aligned with the polymerase regions of other carmo-like viruses and regions of amino acid conservation revealed in this way were the basis for two highly degenerate RT-PCR primers. In the second stage, non-degenerate RT-PCR primers were designed by utilizing nucleic acid sequence obtained from isolates of CtMoV we found in wild carrots sampled from Berkshire.

Co-occurrence of CtRLV and CtMoV

At the outset, we suspected that CtRLV might gain through an increased ability to permeate plant tissues and that any such increased concentration might be reflected in transmission frequency. However, we were aware of reports of enhanced disease when umbraviruses and luteoviruses occur together in crops (Murant *et al.* 1969). Conceivably this could result in transmission disadvantage and/or premature death of infected wild relatives. Thus, we expected natural populations of luteoviruses and umbraviruses to be structured but did not know whether they consisted of stable pairs of coevolved nucleic acids or were characterized by frequent interchange of the putative partners. Furthermore, we did not know how often CtRLV and CtMoV occur alone. To test the predictions of the model, we therefore used a series of RT-PCR tests to establish the baseline frequencies of infection and the numbers of individual 'wild' carrots (*Daucus carota carota*) and aphids (*Cavariella aegopodii*) from wild carrots that contained CtRLV and CtMoV, separately or together.

Elnagar and Murant (1978) reported a few instances of CtMoV transmission by *C. aegopodii* alone. Furthermore, although groundnut rosette umbravirus was experimentally transmitted by aphids alone and in Africa was detected without the luteovirus transmission assistor in field-grown groundnuts (Naidu *et al.* 1999), we never found CtMoV in wild carrot without CtRLV. Over the period 1999–2000, wild carrots from three widely geographically separate sites (in Berkshire, Dorset and Oxfordshire) of roughly similar area (0.5 ha) were tested by RT-PCR for the presence of CtRLV and CtMoV (Figure 12.4). During these two years, the incidence of CtRLV in carrots from the three sites was not significantly different ($\chi^2_{[2]} = 4.78$) although incidence was greater in carrots from Oxfordshire (67% (49/73)), than in Berkshire

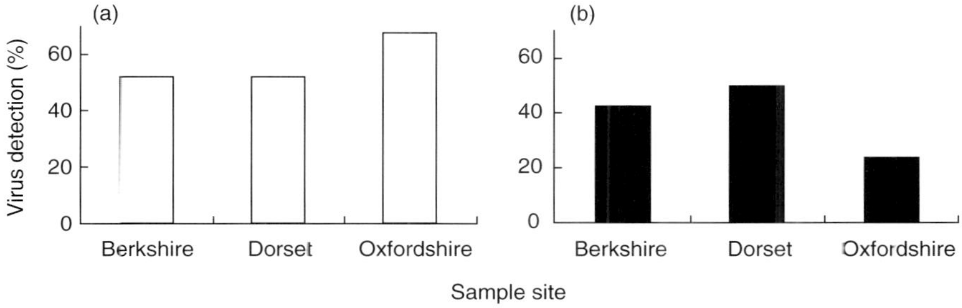

Figure 12.4 Incidence of wild carrots, harvested from three UK sites during 1999–2000, with carrot red leaf virus only (a) or both carrot red leaf virus and carrot mottle virus (b).

(50% (34/67)) and Dorset (52% (42/80)). However, there was a significant difference ($\chi^2_{[2]} = 6.26$, $P < 0.05$) in the incidence of CtMoV in CtRLV-infected plants at each site (Oxfordshire, 24% (12/49), Berkshire 50% (17/34) and Dorset 43% (18/42)). Presumably the variations in infection frequency reflect impacts of local weather conditions on vector abundance/activity.

Despite the name given half a century ago to one of the viruses we studied (CtRLV), we observed no strong correlation of either CtRLV ($\chi^2_{[1]} = 1.85$) or CtMoV ($\chi^2_{[1]} = 0.11$) with leaf colour. Of 99 wild carrots tested, 56% (15/27) of the carrots with red/purple foliage tested positive for CtRLV and 40% (6/14) positive for CtMoV. Of the carrots with green foliage, 40% (29/72) tested positive for CtRLV and 37% (11/29) for CtMoV.

Complementary tests were done to determine how many individual *C. aegopodii* aphids collected from wild umbellifers contained CtRLV and/or CtMoV (as determined by RT-PCR). Again CtMoV was not found except when CtRLV was also detected. Broadly, CtRLV was detected at similar frequencies in aphids and in wild carrot foliage samples; 6/8 (75%) contained CtRLV and, of these, 5/8 (62%) also contained CtMoV.

Do luteoviruses other than CtRLV occur in wild carrots?

It was known that CtMoV occurs in commercial carrot crops in the UK and it was expected that the umbravirus would occur with a luteovirus because of transmission dependence. However, at the outset of this study it was not known whether any luteovirus other than CtRLV naturally invaded carrots, although it was known that some umbraviruses are helped by luteoviruses other than their 'natural' helper (Adams & Hull 1972). To obtain a reference point for our coevolution studies, we screened wild carrots at the three previously mentioned sites using serological methods that, in our hands, unambiguously distinguished between potato leaf roll, beet western yellows and carrot luteoviruses. Carrot foliage was either tested directly by Enzyme-Linked Immunosorbent Assay (ELISA) or by aphid transmission. Leaves were used as virus sources on which aphids were given acquisition access periods (AAP) of 72 hours before being transferred to healthy test plants and given 72 hours inoculation access periods (IAP). Groups of ten aphids, either *Myzus persicae* or *C. aegopodii*, were given access to wild carrot foliage and then transferred to seedling indicators; carrots, coriander or chervil. Neither aphid transmission nor direct serological testing revealed the presence of any luteovirus in wild carrot other than the virus we called CtRLV.

Are the phylogenies of CtRLV and CtMoV correlated?

CtRLV was not uncommonly found in the absence of CtMoV and, to investigate partner exchange, we contrasted the sequences in umbraviruses and luteoviruses sampled in wild carrots. Evidence of partner exchange was sought by constructing phylogenetic trees from base sequences obtained using an Applied Biosystems Automated Sequencer model 373A. Standard enzymatic chain termination methods and PRISM kits (Applied Biosystems) were used after RT-PCR amplifying polymerase coding regions of the genomes of luteovirus and umbravirus that occurred together. Sequences were aligned using the CLUSTAL W program contained within the GCG Wisconsin software package. For tree construction, classifications were inferred using PAUP (Phylogenetic Analysis Using Parsimony) v. 4.0 and tree bisection-reconnection to search for the most parsimonious trees from 100 bootstrap replications.

The phylogenetic trees (Figure 12.5) constructed for CtRLV and CtMoV from Berkshire or Dorset were incongruent (indicating partner exchange). Tree branch lengths were directly proportional to sequence change and we observed different branch lengths for CtRLV and CtMoV pairs. Therefore, within the region of polymerase genes illustrated here (363 bp for CtRLV and 322 bp for CtMoV), rates of molecular evolution of CtRLV and CtMoV appeared to differ substantially. We found not only evidence of CtRLV/CtMoV partner exchange but also evidence implying different rates of molecular evolution for the two viruses. However, alternative explanations are hard to exclude. Thus, it is conceivable that differing branch lengths reflect differences in the histories of the two viruses such that, for example, CtMoV could have invaded wild carrot more recently than CtRLV. Nevertheless, it seems that even these key (polymerase) genes in CtRLV and CtMoV exist in a state of dynamic flux.

The luteovirus/umbravirus coevolution trial

To test experimentally the hypothesis that a synthetic umbravirus/luteovirus pair will evolve to enhance their joint fitness during a series of cotransmissions, we initially addressed the question in *Nicotiana clevelandii* using beet western yellows (BWYV) as the luteovirus partner with CtMoV. Laboratory experiments had shown that CtMoV is encapsidated in BWYV capsid proteins and is consequently transmitted by aphids (Waterhouse & Murant 1983). Polyphagous aphids (*M. persicae*) were used as vectors and cotransmissions were successful but too erratic to form the basis for a long-term experiment. Therefore to investigate at the genetic and popula-

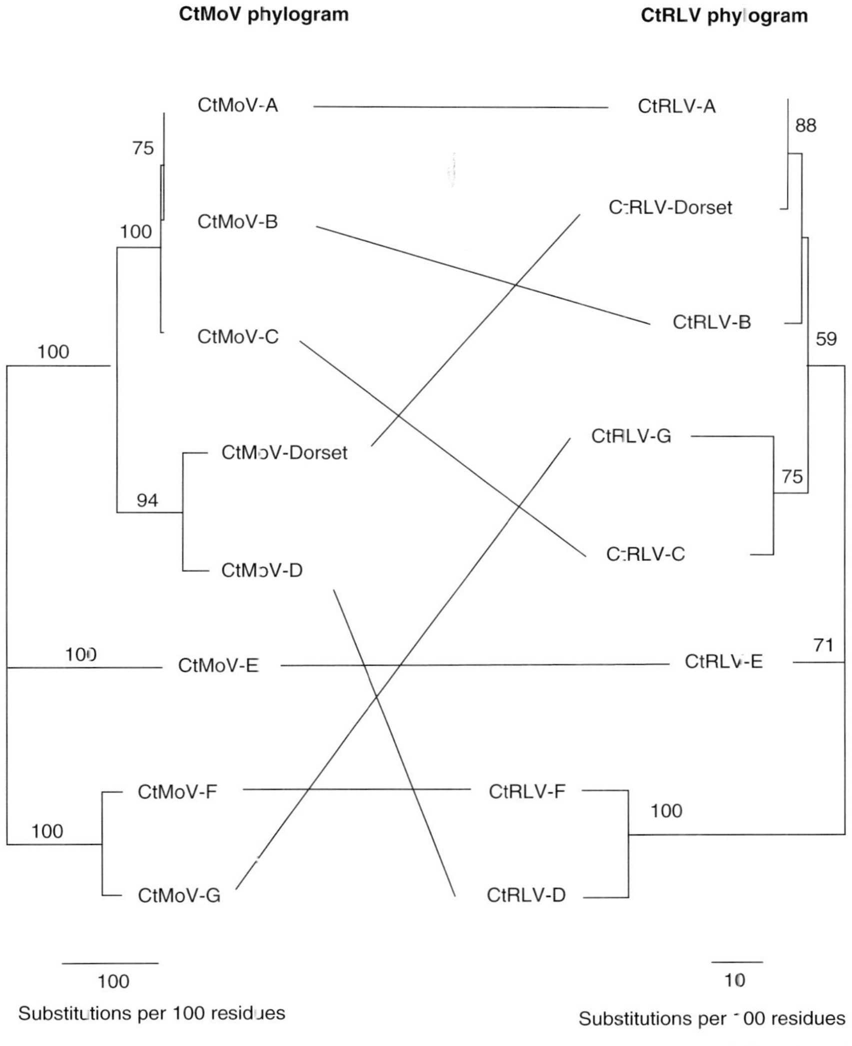

Figure 12.5 Phylogenetic trees constructed for CtRLV and CtMoV from Berkshire (A–G) or Dorset. For tree construction, classifications were inferred using PAUP (Phylogenetic Analysis Using Parsimony) v. 4.0 and tree bisection-reconnection to search for the most parsimonious trees from 100 bootstrap replications. Comparison of these trees shows that there has been partner exchange between the two viruses.

tion levels, the evolution and the dynamics of cotransmission via aphids (*C. aegopodii*), we initiated a further series of laboratory and glasshouse experiments.

This coevolution experiment was initiated in carrot seedlings using CtRLV obtained from wild carrot growing in Berkshire and a CtMoV isolate from Northumberland, a site 500 km distant. The CtMoV had been maintained by serial manual passage in *N. clevelandii* before it was transmitted manually into one carrot seedling. This 'culture' was maintained in carrot by manual passage. Systemic invasion was revealed by RT-PCR 4 weeks post inoculation. That individual carrot seedling was subsequently inoculated with CtRLV, using *C. aegopodii* under the conditions previously described. The experimental colony had been initiated from one individual obtained from horticultural carrot growing at a site in Oxford where both CtRLV and CtMoV routinely occur. On the same day, as the secondary challenge, a second carrot seedling was inoculated with the same CtRLV isolate (alone). Not less than 4 weeks later, the carrot seedlings inoculated in this way were tested for CtRLV and CtMoV by RT-PCR. When it had been established which plants were infected, the cycle of aphid transmissions resumed. Five lines (five carrot seedlings acted as sources for CtRLV alone and five for both viruses together) were maintained in parallel. *C. aegopodii* were again used for the inoculation. This is an on-going investigation, but the results so far are as follows:

1 When CtRLV was serially transmitted from the singly infected carrot, the transmission rate decreased (not significantly, $\chi^2_{[2]} = 4.81$) over three generations. Transmission from the parent plant to the first generation of carrots was 100% (11/11), at the second generation this had decreased to 86% (19/22) and to 71% (17/24) after the third transmission cycle.

2 When CtRLV was transmitted from carrots infected with both viruses, the transmission rate decreased significantly ($\chi^2_{[2]} = 7.9$, $P < 0.05$) from 100% (7/7) in the first generation, to 86% (18/21) in the second and to 54% (14/26) by the third.

3 Over three transmission cycles, the transmission of CtMoV, from doubly infected carrots also diminished significantly ($\chi^2_{[2]} = 6.98$, $P < 0.05$) and, at a faster rate than that of CtRLV. Transmission of CtMoV from the parent plant to the first generation was 79% (11/14), from first to second generation was 64% (14/22) and was 31% (8/26) from the second to the third generation.

Conclusions

Plant viruses show a bewildering array of natural histories that present exciting challenges for evolutionary biologists and ecologists. More and more viruses are being discovered in plants, and many seem to modulate damaging effects.

Viruses not only interact with their hosts but also with other viruses, and simpler nucleic acids. In this chapter we have concentrated on two viruses for which the initial evidence was consistent with a mutualistic relationship. One of these viruses possesses a coat protein that can encapsidate both itself and the second virus, and which is essential for aphid-vectored transmission between plants. The second virus lacks a coat protein and is incapable of being transmitted between plants (other than mechanically which is not important in the field). Importantly, the latter has a movement protein that allows both viruses to move systemically through the plant and, although this is not yet proved, presumably that increases their probability of acquisition by aphids. We have shown that in the carrot system these two viruses have only a loose partnership. We provide the first data from the field that show the degree to which they are associated, but using sequence data we also showed that there is exchange of partners. In experiments that are still running we explore whether the two viruses can coevolve together to enhance each other's fitness. In addition to our experimental work, we explore theoretically the conditions under which a mutualistic interaction between two viruses such as these may evolve into a more permanent union with both phenotypes being coded by a single nucleic acid. Mutualistic interactions are just one of a broad spectrum of ways in which viruses and other nucleic acids may interact.

Our data show that CtMoV lessens transmission of CtRLV from carrot. On that basis it seems that the luteovirus and the umbravirus are not in a mutualistic relationship as was first suspected on the basis of Barker's (1989) observations.

The experience that led to the view that mutualism was operating in the CtRLV/CtMoV system was short-term observations in manually inoculated tobacco plants. Our tests span several months, are continuing and are different in that they use aphid-inoculated carrots. We speculate that the amount of CtMoV introduced into a plant will vary with the means of inoculation and it is arguable that, when fingers are used, the primary inoculum is relatively great and 4 weeks later the benefit of that primary inoculum has been retained. By contrast, when aphids introduce inoculum they deposit only a little CtMoV (although they deliver that inoculum with precision). In any event it seems, 4 weeks later, that amount of CtMoV replication is (just) adequate to impact on the transmission of CtRLV. Our current data suggest a one way parasitism rather than a mutualistic interaction, i.e. CtMoV did not strongly and sustainably enhance the transmission of CtRLV in carrot. We would not be surprised if other hosts differed from these systems but, in any event, our observations seem the more pertinent to the natural environment.

To conclude, we believe plant viruses will continue to be a source of opportunity in ecology and evolutionary biology; a dynamic perspective is key to the understanding of viral genome structure and disease impacts in wild and domesticated plants.

Acknowledgements

This work was funded by the Natural Environment Research Council's thematic programme EDGE (Ecological Dynamics and Genes).

References

Adams, A.N. & Hull, R. (1972) Tobacco yellow vein, a virus dependent on assistor viruses for its transmission by aphids. *Annals of Applied Biology* **71**, 135–140.

Anderson, R.M. & May, R.M. (1979) Population biology of infectious diseases. Part I. *Nature* **280**, 361–367.

Barker, H. (1989) Specificity of the effect of sap-transmissible viruses in increasing accumulation of luteoviruses in co-infected plants. *Annals of Applied Biology* **115**, 71–78.

Baulcombe, D.C. (1994) Novel strategies for engineering virus resistance in plants. *Current Opinion in Biotechnology* **5**, 117–124.

Chao, L. (1992) Evolution of sex in RNA viruses. *Trends in Ecology and Evolution* **7**, 147–151.

Chela-Flores, J. (1994) Are viroids molecular fossils of the RNA world? *Journal of Theoretical Biology* **166**, 163–166.

Cooper, J.I. & Harrison, B.D. (1973) The role of weed hosts and the distribution and activity of vector nematodes in the ecology of tobacco rattle virus. *Annals of Applied Biology* **73**, 53–66.

Cooper, J.I., Edwards, M.L., Rosenwasser, O. & Scott, N.W. (1994) Transgenic resistance genes from nepoviruses: efficacy and other properties. *New Zealand Journal of Crop and Horticultural Science* **22**, 129–137.

Demler, S.A. & de Zoeten, G.A. (1991) The nucleotide sequence and luteovirus-like nature of RNA1 of an aphid non-transmissible strain of pea enation mosaic virus. *Journal of General Virology* **72**, 1819–1834.

Demler, S.A., Borkhsenious, O.N., Rucker, D.G. & de Zoeten, G.A. (1994a) Assessment of the autonomy of replicative and structural functions encoded by the luteophase of pea enation mosaic virus. *Journal of General Virology* **75**, 997–1007.

Demler, S.A., Rucker, D.G., Nooruddin, L. & de Zoeten, G.A. (1994b) Replication of the satellite RNA of pea enation mosaic virus is controlled by RNA2 functions. *Journal of General Virology* **75**, 1399–1406.

Domingo, E. & Holland, J.J. (1994) Mutation rates and rapid evolution of RNA viruses. In: The *Evolutionary Biology of Viruses* (ed. S. Morse), pp. 161–184. Raven Press, New York.

Elena, S.F., Dopazo, J., Flores, R., Diener, T.O. & Moya, A. (1991) Phylogeny of viroids, viroid-like satellite RNAs and viroid-like domain of hepatitis delta virus RNA. *Proceedings of the National Academy of Sciences of the United States of America* **88**, 5631–5634.

Elnagar, S. & Murant, A.F. (1978) Relations of carrot red leaf and carrot mottle viruses with their aphid vector, *Cavariella aegopodii*. *Annals of Applied Biology* **89**, 237–244.

Fitchen, J.H. & Beachy, R.N. (1993) Genetically engineered protection against viruses in transgenic plants. *Annual Review of Microbiology* **47**, 739–763.

German, T.L. & de Zoeten, G.A. (1975) Purification and properties of the replicative forms and replicative intermediates of pea enation mosaic virus. *Virology* **66**, 172–184.

Gibbs, M.J. & Cooper, J.I. (1995) A recombinational event in the history of luteoviruses probably induced by base-pairing between the genomes of two distinct viruses. *Virology* **206**, 1129–1132.

Gibbs, M.J., Cooper, J.I. & Waterhouse, P.M. (1996a) The genome organization and affinities of an Australian isolate of carrot mottle umbravirus. *Virology* **224**, 310–313.

Gibbs, M.J., Ziegler, A., Robinson, D.J., Waterhouse, P.M. & Cooper, J.I. (1996b) Carrot mottle mimic virus (CMoMV): a second umbravirus associated with carrot motley dwarf disease recognized by nucleic acid hybridization. Molecular Plant Pathology On-Line. Available on: http://www.bspp.org.uk/mppol/1996/1111gibbs

Hanski, I. (1999) *Metapopulation Ecology*. Oxford University Press, Oxford

Harper, G., Osuji, J.O., Heslop-Harrison, J.S. & Hull, R. (1999) Integration of banana streak badnavirus into the *Musa* genome: molecular and cytogenetic evidence. *Virology* **255**, 207–213.

Koonin, E.V. (1991) The phylogeny of RNA-dependent RNA polymerases of positive-strand RNA viruses. *Journal of General Virology* **72**, 2197–2206.

Liu, Y.Y. & Cooper, J.I. (1994) Satellites of plant viruses. *Review of Plant Pathology* **73**, 371–387.

Mayo, M. A. & Jolly, C.A. (1991) The 5′-terminal sequence of potato leaf roll virus RNA: evidence of recombination between virus and host RNA. *Journal of General Virology* **72**, 2591–2595.

Murant, A.F., Goold, R.A., Roberts, I.M. & Cathro, J. (1969) Carrot mottle—a persistent aphid borne virus with unusual properties and particles. *Journal of General Virology* **4**, 329–341.

Naidu, R.A., Kimmins, F.M., Holt, J. *et al.* (1999) Spatiotemporal separation of groundnut rosette disease agents. *Phytopathology* **89**, 934–941.

Nee, S. (2000) Mutualism, parasitism and competition in the evolution of coviruses. *Philosophical Transactions of the Royal Society of London Series B* **355**, 1607–1613.

Nee, S., May, R.M. & Hassell, M.P. (1997) Two-species metapopulation models. In: *Metapopulation Biology* (eds I. Hanski & M. E. Gilpin), pp. 123–147. Academic Press, San Diego, CA.

Palukaitis, P. & Zaitlin, M. (1997) Replicase-mediated resistance to plant disease. *Advances in Virus Research* **48**, 349–377.

Pirone, T.P. & Blanc, S. (1996) Helper-dependent vector transmission of plant viruses. *Annual Review of Phytopathology* **34**, 227–247.

Querci, M., Owens, R.A., Bartolini, I., Lazarte, V. & Salazar, L.F. (1997) Evidence for heterologous encapsidation of potato spindle tuber viroid in particles of potato leaf roll virus. *Journal of General Virology* **78**, 1207–1211.

Scholthoff, K-B. G., Shaw, J G. & Zaitlin, M. (1999) *Tobacco Mosaic Virus. One Hundred Years of Contributions to Virology*. APS Press, The American Phytopathological Society, St Paul, Minnesota.

Taliansky, M.E., Robinson, D.J. & Murant, A.F. (2000) Groundnut rosette disease complex: biology and molecular biology. *Advances in Virus Research* **55**, 357–400.

Van Regenmortel, M.H.V., Fauquet, C.M., Bishop, D.H.L. *et al.* (2000) *Virus Taxonomy Classification and Nomenclature of Viruses.* Seventh Report of the International Committee on Taxonomy of Viruses. Academic Press, San Diego.

Vance, V. & Vaucheret, H. (2001) RNA silencing in plants—defense and counterdefense. *Science* **292**, 2277–2280.

Waterhouse, P.M. & Murant, A.F. (1983) Further evidence on the nature of the dependence of carrot mottle virus and carrot redleaf virus for transmission by aphids. *Annals of Applied Biology* **103**, 455–464.

Waterhouse, P.M., Wang, M.-B. & Lough, T. (2001) Gene silencing as an adaptive defence against viruses. *Nature* **411**, 834–842.

Watson, M.T. & Falk, B.W. (1994) Ecological and epidemiological factors affecting carrot motley dwarf development in carrots grown in the Salinas Valley of California. *Plant Disease* **78**, 477–481.

Watson, M.A., Serjeant, E.P. & Lennon, E.A. (1964) Carrot motley dwarf and parsnip mottle viruses. *Annals of Applied Biology* **54**, 153–166.

Watson, M.T., Tian, T., Estabrook, E. & Falk, B.W. (1998) A small RNA resembling the beet western yellows luteovirus ST9-associated RNA is a component of California carrot motley dwarf complex. *Phytopathology* **88**, 164–170.

de Zoeten, G.A. & Skaf, J.S. (2001) Pea enation mosaic and the vagaries of a plant virus. *Advances in Virus Research* **57**, 323–350.

Chapter 13

The ecology of turnip mosaic virus in wild populations of *Brassica* species

A. F. Raybould, M. J. Alexander, E. Mitchell, M. I. Thurston, D. W. Pallett, P. Hunter, J. A. Walsh, M-L. Edwards, A. M. E. Jones, C. L. Moyes, A. J. Gray and J. I. Cooper*

Introduction

Virus diseases of plants have serious economic impact worldwide, but losses are difficult to quantify and are probably underestimated. Infected plants may grow slower, be smaller, die younger and produce a lower yield (perhaps of lower quality) than uninfected plants. However, such effects may be apparent only when infected and uninfected plants are compared under controlled conditions. In addition, virus infections may be mistaken for effects of pesticides, genetic abnormalities or nutrient imbalances (Waterworth & Hadidi 1998), and so relatively sophisticated techniques such as enzyme-linked immunosorbent assay (ELISA), reverse transcriptase polymerase chain reaction (RT-PCR) or electron microscopy are required to confirm diagnoses.

For the same reasons, the effects of virus infections in natural plant populations may be underestimated. While obvious epidemics of virus disease are rare in wild plants (e.g. Burdon 1987), we have little information about whether viruses have chronic sublethal effects. One result of this lack of information is that we are unable to predict the effects of virus resistance genes on populations of feral crops or wild crop relatives (e.g. Cooper & Raybould 1997).

In this chapter, we summarize work on the ecology of turnip mosaic virus (TuMV) in wild populations of *Brassica* species in the south of England. After cucumber mosaic virus, TuMV is the most serious virus disease of field-grown vegetables (Tomlinson 1987). TuMV infects many crops, including cabbage (and other *Brassica* vegetables), oilseed rape, chicory, horseradish, lettuce, peas and rhubarb. In the UK, the most important effect of TuMV is in cabbage production; infection causes internal ('cigar burns') and external necrotic lesions on cabbage heads. Necrotic ringspots can develop in heads after 2–5 months in cold storage (Shattuck 1992).

The stimulus to begin work on TuMV in wild brassicas was the wish to predict the effects of introgression of TuMV resistance genes on the population dynamics

* Correspondence author: alan.raybould@syngenta.com

('weediness') of wild relatives of *Brassica* crops. Although we were concerned with genetically modified resistance, the conclusions would apply to conventional major gene resistance. At the beginning of the work, there was no clear risk assessment framework for predicting the effects of new resistance genes in natural populations. Therefore, we studied a large number of factors of potential relevance, while attempting to think through what an efficient risk assessment procedure should comprise. This chapter therefore has two primary objectives. The first objective is to report data to ecologists and virologists who are interested in how an important disease of crop plants interacts with the wild relatives of those crops. The second objective is to discuss which of these data have most relevance for predicting changes in population dynamics following resistance gene introgression, and therefore to make recommendations about how other plant–virus combinations should be approached.

The chapter is divided into three main parts. In the first, we report data on the incidence of TuMV in wild *Brassicas* and factors that might control the distribution of the virus. These factors include genetic variation in the host and virus and the distribution of vectors of TuMV. In the second part, we report observations on the effects of TuMV infection on wild *Brassicas*. These comprise laboratory and field studies and comparisons with the effects of other viruses that often infect commercial *Brassica* production. Finally, we assess whether our data really do assist in predicting the effects of TuMV resistance genes on the population dynamics of wild *Brassicas* and make recommendations about how future studies might be streamlined. The paper consists of a mixture of published data collected before, or in the early days of the EDGE programme, and data collected later in the EDGE programme that appear for the first time. The aim is to provide a comprehensive survey of the state of knowledge of the impacts of TuMV in wild plants thus forming the basis for discussion about how we should predict the impacts of resistance genes.

Occurrence of TuMV in wild *Brassica* species

We studied TuMV in three wild *Brassica* hosts: *B. oleracea*, a long-lived perennial that grows on sea cliffs; *B. nigra*, an annual that grows in ruderal habitats; and *B. rapa* an annual or biennial species that grows on riverbanks. At various times between 1995 and 2000, we estimated the frequency of TuMV in populations of *B. oleracea* and *B. nigra* in Dorset (Figure 13.1) (Raybould *et al.* 1999a; Thurston *et al.* 2001), and in populations of *B. rapa* on the banks of the River Thames in Oxfordshire, using the ELISA method of Converse and Martin (1990). The method of Edwards and Cooper (1985) was used to test for viruses other than TuMV.

Over this period, TuMV occurred in about 22% of the *B. oleracea* plants tested (158/723). There were statistically significant differences in the proportion of plants infected among sites in 1995 ($\chi^2_{[4]} = 16.66$, $P = 0.002$) and 1998 ($\chi^2_{[2]} = 22.88$, $P < 0.0005$), but not in 1999 ($\chi^2_{[3]} = 6.21$, $P = 0.102$) (Figure 13.2). At the two sites where we have 3 years' data, there were significant differences among years (Chapman's Pool $\chi^2_{[2]} = 25.45$, $P < 0.0005$; Kimmeridge $\chi^2_{[2]} = 21.18$, $P < 0.0005$).

Figure 13.1 Location of *Brassica* populations studied.

The number of TuMV infections of *B. nigra* was extremely low. We sampled 597 plants from four populations between Ringstead and Holes Bay (see Figure 13.1) in 1998, 1999 and 2000, and only five gave positive ELISA readings for TuMV (0.8%) (Thurston *et al.* 2001). With so few positive samples, it was not possible to test for differences among sites or years. In 2000, we also collected a total of 115 plants of *B. rapa* from Culham and Abingdon in Oxfordshire (UK); none tested positive for TuMV.

Factors associated with variation in TuMV incidence in *B. oleracea*

Field observations

In 1995, we classified each plant tested into one of four life-history stages. There were significant differences in the age structure of the populations among sites. However, we found that within each site, and over all sites, there was no tendency for a greater

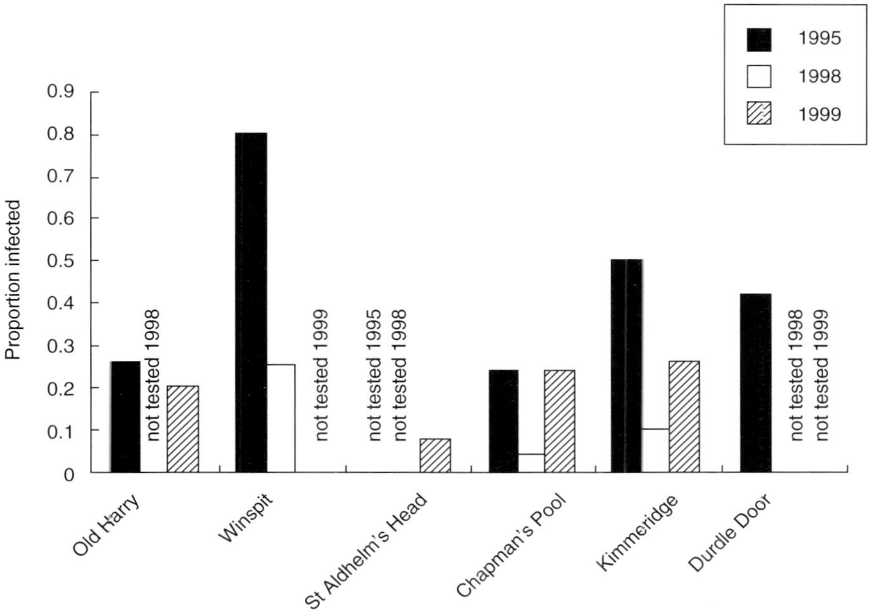

Figure 13.2 The proportion of plants of *Brassica oleracea* that tested positive for TuMV in populations on the Dorset coast between 1995 and 1999.

proportion of older plants to be infected with TuMV (all sites $\chi^2_{[3]} = 5.94$, $P = 0.115$). Controlling for age differences among sites, the effect of site was highly significant ($\chi^2_{[3]} = 31.12$, $P < 0.001$), whereas there was no effect of age when the effect of site was controlled ($\chi^2_{[3]} = 6.67$, $P = 0.083$). This suggests that the differences among sites in the proportion of plants infected with TuMV were due to factors unrelated to the age of the plants.

We detected significant associations between TuMV and three other viruses found in *B. oleracea* populations in 1995. Across all sites there was a positive association between TuMV and cauliflower mosaic virus (CaMV) ($\chi^2_{[1]} = 9.48$, $P = 0.002$) and beet western yellows virus (BWYV) ($\chi^2_{[1]} = 15.89$, $P < 0.001$), whereas there was a negative association between TuMV and turnip yellow mosaic virus (TYMV). CaMV, BWYV and TuMV are aphid-transmitted viruses and TYMV is transmitted by biting insects such as beetles (e.g. Paul & Rawlinson 1992). The positive association between TuMV and both CaMV and BWYV could be due to spread by a common vector and therefore vector behaviour might be an important determinant of the distribution of TuMV among wild cabbage plants. The negative association between TuMV and TYMV could be due to high mortality of plants infected with both viruses.

The main vectors of TuMV in commercial *Brassica* production are thought to be the peach potato aphid (*Myzus persicae*) and the cabbage aphid (*Brevicoryne brassicae*) (e.g. Broadbent 1957; Paul & Rawlinson 1992), although 89 species of aphid have been recorded as vectors (Shattuck 1992). Over a 7-year period, we have made continual observations throughout Dorset of insects feeding on *B. oleracea* and have seen one peach potato aphid on *B. oleracea* (Moyes *et al.* 2000). *Brevicoryne* is quite common on wild *B. oleracea* in Dorset. In June 1995, over 70% of plants were infested with *Brevicoryne* at Durdle Door, Kimmeridge and Old Harry, but fewer than 25% were infested at St Aldhelm's Head. The peak infestation time in 1996 was in July, and far fewer plants were infested than in 1995. *Brevicoryne* was found on about 30% of plants at Durdle Door, about 10% at Kimmeridge and fewer than 5% at Old Harry. No *Brevicoryne* were found at St Aldhelm's Head (Moyes 1997). In both years, St Aldhelm's Head had the fewest aphids, and had the lowest frequency of TuMV infection detected in *B. oleracea* in Dorset (8% of plants in 1999).

Plant genotype may influence the efficiency of *Brevicoryne* as a vector of TuMV. Špak (1992) found that *Brevicoryne* acquired TuMV more efficiently from *B. nigra* than *Sinapis alba* and attributed the difference to the much higher concentration of sinigrin (2-propenyl glucosinolate) in *B. nigra*. The possibility of a link between sinigrin and the incidence of TuMV is interesting and warrants further study. There are significant differences in the average concentrations of sinigrin among *B. oleracea* populations in Dorset (Mithen *et al.* 1995; Moyes *et al.* 2000; Moyes & Raybould 2001) and these populations would be an excellent system to test whether there is a relationship between intraspecific differences in sinigrin concentration and transmission efficiency of TuMV in the field.

Laboratory and field experiments

To test for differences among *B. oleracea* populations in reactions to TuMV, we grew seeds from Old Harry, Winspit and Kimmeridge in the glasshouse and mechanically inoculated them with a TuMV isolate from Kimmeridge when they were 4 weeks old. We used 12 individuals from each of 60 open-pollinated families (see Raybould *et al.* 2000). One month after inoculation, we removed a fully expanded leaf and estimated the concentration of TuMV using the methods of Converse and Martin (1990).

An important observation was that after 4 weeks no seedlings had died; this is in marked contrast to *B. nigra* and *B. rapa*. There was a large amount of variation in TuMV concentration among plants; the highest concentration was about 200 000 times that of the lowest. About 40–80% of the variation was due to differences among families, indicating that *B. oleracea* populations contain a large amount of genetic variation for response to challenge with TuMV. Differences in average concentration among populations accounted for only 6.2% of the variation (compared with about 20% at microsatellite loci, Raybould *et al.* 1999b). These results suggest that genetic differences among populations play only a small role (if any) in the difference in the proportion of plants infected.

The experiment described above estimated the extent to which plant genotype determines the ability of a TuMV isolate to replicate in *B. oleracea*. However in terms of determining the ecological significance of TuMV infections in the field, the experiment was quite limited. For example, we have no information about whether

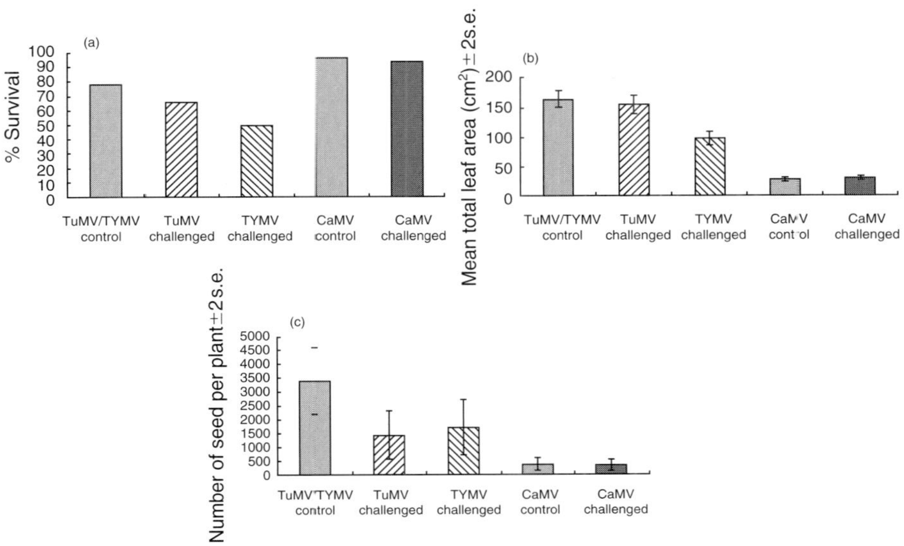

Figure 13.3 Survival (a), growth (b) and reproduction (c) of *Brassica oleracea* plants challenged with TuMV, TYMV or CaMV and control plants mock-inoculated with water.

virus concentration 4 weeks after inoculation is a good predictor of effects on long-term survival or fecundity. We therefore carried out a second experiment to examine the effects of TuMV on *B. oleracea* in conditions similar to those in the wild.

We took seed from several Dorset populations of *B. oleracea*, germinated them in a glasshouse and mechanically inoculated 185 seedlings with TuMV (same isolate as above), 187 with TYMV and 203 with water, when they had between three and five fully expanded leaves. The plants were grown in the glasshouse for 4 weeks, and then transplanted to a fallow field about 300 m away from a natural population of *B. oleracea* at the beginning of June 1996. The plants were arranged randomly in both the glasshouse and the field. Various aspects of the performance of the plants were recorded over the subsequent 2 years (see Maskell *et al.* 1999).

After 18 months, a significantly higher percentage of TuMV-challenged plants had died compared with controls (Figure 13.3). In the year following transplanting, the surviving TuMV-challenged plants produced fewer seed than the surviving controls, even though there was no difference in plant size (estimated as dry weight or leaf area) (Figure 13.3). We also found a significant effect of TYMV on survival and fecundity, and TYMV significantly reduced plant size (Figure 13.3). We used the same experimental procedure to test for effects of a CaMV isolate from Winspit, except that in this experiment the plants were grown in a plunge bed rather than in a field. For each treatment we used 25 plants from each of four Dorset populations. We found no effects of CaMV on survival, growth or fecundity (Figure 13.3). There was no population by virus interaction for any of the viruses in these experiments; in other words, plants in each population reacted in the same way to virus challenge compared with the controls from the same population.

Genetic variation in TuMV

To test whether the incidence of TuMV among *B. oleracea* populations could be due to genetic variation among TuMV populations we isolated the virus from 15 wild cabbages, five each from Kimmeridge, Chapman's Pool and Winspit. Pathotypes of the isolates were determined by the responses they induced in four lines of *B. napus* (Jenner & Walsh 1996). Plant responses were confirmed by testing the differential plants for TuMV by ELISA. All the isolates were found to be pathotype 1, which is the commonest pathotype in Europe (Jenner & Walsh 1996).

The coat protein genes from the isolates were amplified using RT-PCR and the amplified regions sequenced. PCR primers specific to the 3′ untranslated region and the *NIb* gene of TuMV were designed at Horticulture Research International from published sequences (Jenner *et al.* 2000). The RT-PCR products were sequenced directly by PCR cycle sequencing using the Big Dye terminator cycle FS Ready Reaction Kit (PE-Applied Biosystems), following the manufacturer's protocol. The coat protein gene nucleotide sequences were aligned using the CLUSTAL W program (Thompson *et al.* 1994) with default parameters. The dendrogram was produced and analysed using PHYLIP (Phylogeny Inference Package) v. 3.57c (Felsenstein 1993). The homologous regions of plum pox potyvirus were used as an outlier for the analysis. Trees were constructed using the parsimony method (program DNAPARS) and the consensus tree (Figure 13.4) was produced using CONSENSE in PHYLIP. A bootstrap value for each internal node was calculated from 100 random resamplings using SEQBOOT in PHYLIP (Felsenstein 1985). Comparisons were also made with the coat protein sequence published for TuMV isolate UK 1, a pathotype 1 isolate (GenBank accession number X65978) isolated from swede (*B. napus* ssp. *napobrassica*) pre-1975 in Warwickshire, UK. This isolate was not included in the tree.

The isolates all had high degrees of nucleotide sequence homology (the lowest homology recorded was 97.0%) with even higher levels of homology at the amino acid level. Several of the isolates from all sites, including all those from Winspit, and UK 1 had identical amino acid sequences.

An alignment of the sequence data indicated that the majority of the variation occurred in the 5′ end of the coat protein gene, corresponding to the N-terminal region of the protein. This region has been previously noted as the most variable region of the coat protein of TuMV (Lehmann *et al.* 1997) and other potyviruses (Shukla *et al.* 1994).

A dendrogram constructed from nucleotide sequences (see Figure 13.4) showed no statistically significant branching (i.e. not significant at 95% likelihood) of the isolates. There was an indication of possible internal structure within the group. Two of the five isolates from the Kimmeridge population (GBR 20 and GBR 24) grouped with each other, at a likelihood of 82%. Three of the five isolates from the Winspit population (GBR 37, GBR 43 and GBR 44) grouped together with 67% likelihood. The other two Winspit isolates (along with four of the Chapman's Pool and one of

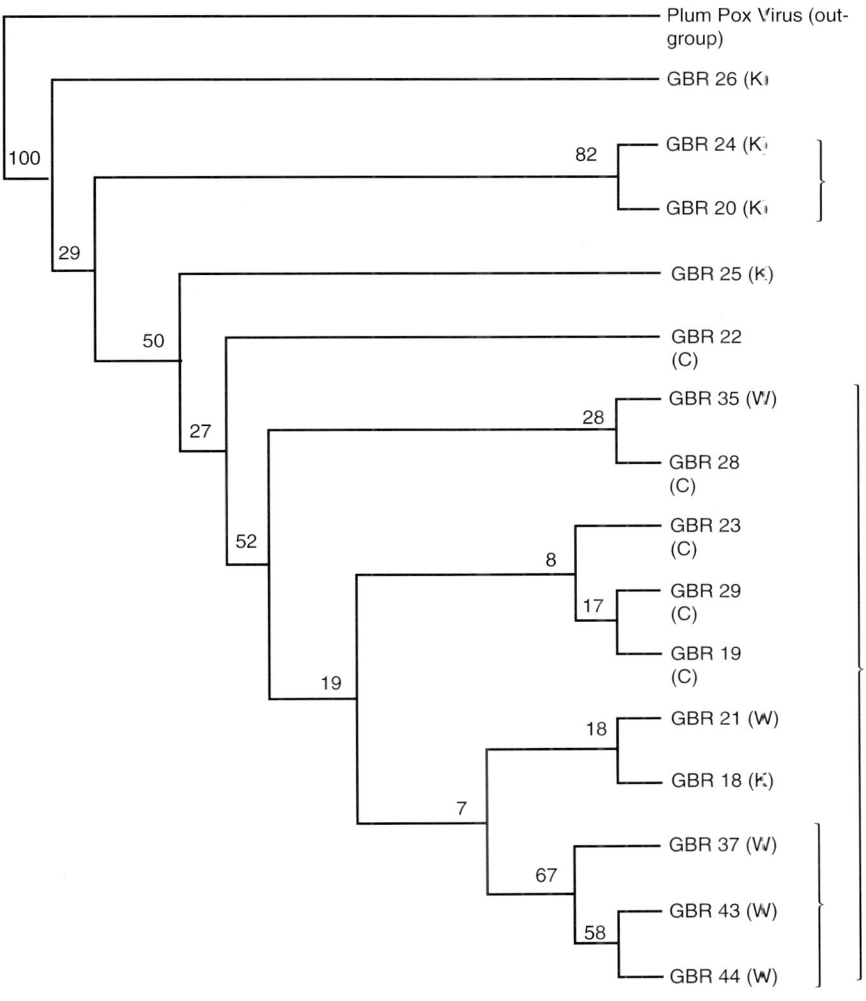

Figure 13.4 Dendrogram of the nucleotide sequences of the isolates of TuMV from three sites on the Dorset coast. The figures are bootstrap values and the brackets indicate groupings at >50% likelihood. The three sites are represented by the letters C (Chapman's Pool), K (Kimmeridge) and W (Winspit).

the Kimmeridge isolates) were associated with them at a likelihood of 52%. The overwhelming picture is one of a high degree of similarity between isolates from Dorset.

A further point of interest is the high homology between these Dorset isolates and UK 1, which was isolated over 25 years earlier from over 250 km away. UK 1 was isolated from a commercial swede crop rather than a wild *Brassica*. As part of the coat protein is reported to be one of the most variable regions of the TuMV genome

(Lehmann *et al.* 1997), the lack of variation observed in isolates from wild *B. oleracea* might indicate selection pressure to maintain certain features of the coat protein in this host.

The only virulence determinant of TuMV identified so far is the cylindrical inclusion protein, not the coat protein (Jenner *et al.* 2000). Reported functions of potyvirus coat proteins include aphid transmission, movement within the plant, assembly of the virus particle and regulation of viral RNA amplification (Urcuqui-Inchima *et al.* 2001). It is likely that there are selection pressures applied to maintain features associated with some or all of these functions. The predominance of the UK 1 coat protein genotype in many of the TuMV isolates, along with the fact that the same genotype was in existence more than 25 years ago, suggests that this genotype is particularly well adapted to *B. oleracea*.

In summary, there is very little genetic differentiation among *B. oleracea* populations in response to challenge with a single isolate of TuMV; there is also little genetic variation in the coat protein of the sequenced isolates. In other words, the variation in the abundance of TuMV in populations of wild cabbage in Dorset appears to be unrelated to genetic variation in either the pathogen or host.

Variation in vector abundance and/or ability to transmit the virus might be a better predictor of spatial and temporal variation in the incidence of TuMV infections in wild cabbage. Vector behaviour could be controlled by host plant genotype, abundance of other food plants (e.g. crops such as oilseed rape or weedy plants like charlock) and/or climate. We have no direct evidence about which aphid species is the vector of TuMV in natural populations of *B. oleracea*. One candidate, *B. brassicae*, is abundant on wild cabbage and does show spatial and temporal variation among populations (see above). However, establishing a link between TuMV and *Brevicoryne* abundance is difficult, as wild cabbage is perennial, and present infection frequency may be the result of vector behaviour over many years. Another candidate vector, *M. persicae*, is very rarely seen in wild cabbage populations, but this might reflect high mobility. If so, it may be a more important vector of TuMV than the rather sedentary *Brevicoryne*. If we are to predict the impacts of TuMV on wild cabbage, for example after the introduction of virus-resistant crops or under different climates, we need to establish the vector(s) and how its behaviour is correlated with environmental and genetic variables.

TuMV in wild *B. nigra* and *B. rapa*

Pathology of TuMV

The near total absence of TuMV from *B. nigra* could be due to several factors. First, the species may be almost immune or highly resistant to TuMV (immunity means that the plant cannot be infected, whereas resistance means that the plant can be infected but the replication and spread of the virus is restricted) (Cooper & Jones 1983). Second, *B. nigra* may be infectible, but vectors of TuMV avoid the species or are unable to transmit the virus. Third, *B. nigra* plants may be susceptible and highly sensitive to TuMV, such that plants die very quickly following infection.

To test whether *B. nigra* is immune to TuMV, we grew seedlings from several different populations in a glasshouse and challenged them with a TuMV isolate from Chapman's Pool when they were 4 weeks old. Over 90% of seedlings became infected, whether inoculated mechanically by rubbing sap from infected plants into the leaves (17/18) or by exposing plants to aphids that had fed recently on infected *Brassica juncea* plants (20/22) (see Thurston *et al.* 2001). All infected seedlings developed systemic necrosis within 10 days and were dead 2–3 weeks after inoculation. Eight-week-old plants were also challenged; necrosis took longer to develop, but 8 weeks after inoculation 15/19 plants had died and the remaining four were stunted and had severe necrosis of the apical meristem (Thurston *et al.* 2001).

The data show that *B. nigra* genotypes from the UK are not immune to TuMV and are highly susceptible to the virus when infection occurs. Observations made in California revealed a potyvirus-like agent (Takahashi 1949) and TuMV (Arnold & Bald 1960) in *B. nigra*. These studies did not use immunological methods to characterize the viruses, but taken at face value the data indicate that *B. nigra* in the USA is not killed by TuMV. The differences between our findings and those from the USA could be due to variation in TuMV between the two regions, differences in the *B. nigra* genotypes and/or the influence of climate (e.g. temperature).

It is not possible to be sure of the reason, but TuMV variation is a strong possibility. Jenner and Walsh (1996) found that the relative frequency of TuMV pathotypes varied with location. Two out of three USA isolates tested were pathotype 5, which has never been recorded in the UK. Preliminary studies of the effects of other TuMV isolates held by Horticultural Research International, Wellesbourne indicate variation in the lethality of TuMV isolates to *B. nigra* from Dorset. A survey of microsatellite variation in *B. nigra* suggests that the species was introduced to the USA from Europe and that European and North American populations are closely related (Westmann & Kresovich 1999). However, there is significant genetic variation among *B. nigra* populations growing only a few miles apart (Lowe *et al.* 2002), so genetic differences in response to TuMV cannot be ruled out completely.

Studies using the same methods as those used for the *B. nigra* study show that about 80% of *B. rapa* seedlings die following mechanical inoculation with TuMV. The surviving 20% show severe symptoms and do not flower.

The infection process in the field

The data on incidence of TuMV in *B. nigra* and *B. rapa* are 'snapshots'; it is not possible to infer from these data whether plants become infected rarely, or whether they are infected frequently and die quickly. To distinguish between these possibilities for *B. nigra*, we regularly tested natural seedling cohorts at two sites for the presence of virus infection to determine when plants became infected with viruses and whether infected plants were less likely to survive.

At both sites we counted the number of seedlings in a 20 cm × 20 cm 'demography quadrat' in early September (these seedlings comprise 'cohort 1'). (*B. nigra* sheds

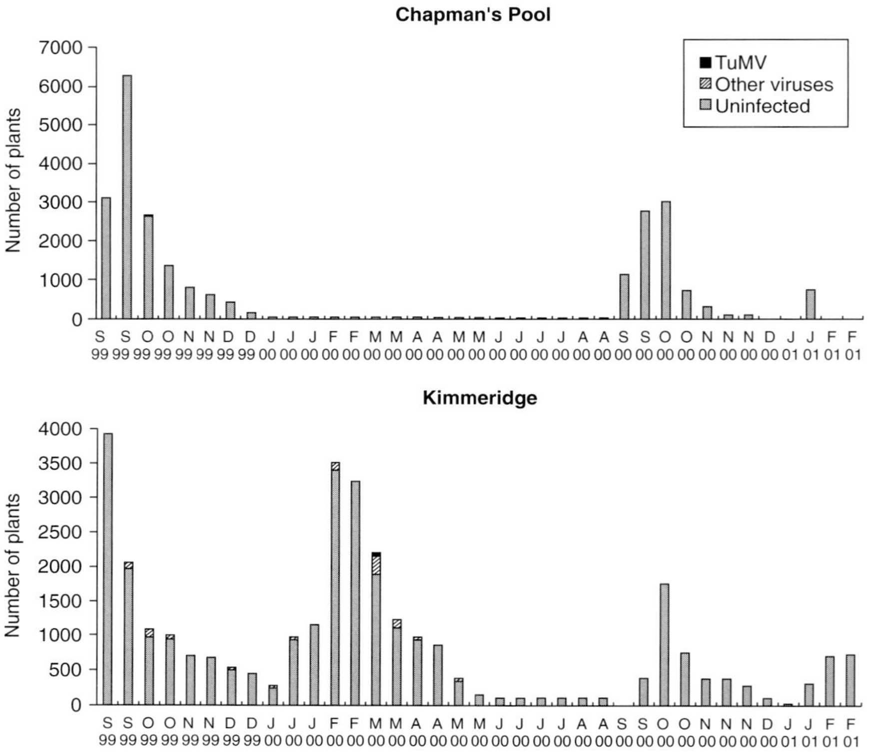

Figure 13.5 Pattern of infection of *Brassica nigra* with TuMV and other viruses in populations on the Dorset coast. Bars represent all plants present at a given time.

seed in August and germination begins in early September.) At 2-week intervals we counted the number of seedlings of the cohort that were still alive, and recorded the presence of new seedlings (cohort 2, etc.). In a neighbouring 'test quadrat', we sampled seedlings of the same age (inferred from size) to test for the presence of viruses and assumed that equal proportions of seedlings in particular cohorts were infected in the test and demography quadrats (this assumption is being examined by testing the frequency of virus infection in several neighbouring quadrats). We could not test the plants in the demography quadrat because sampling is destructive when seedlings are very small. By May, we judged that we could remove tissue from plants in the demography quadrat without affecting their chances of survival and were therefore able to test for viruses in plants in the demography quadrat directly.

At Chapman's Pool, we detected very few seedlings with viruses. We found a higher proportion of infected seedlings at Kimmeridge, but very few of these infections were TuMV (Figure 13.5). All the data suggest that TuMV rarely infects *B. nigra* in Dorset (or that it kills plants virtually instantaneously). The data also suggest that other viruses are a minor cause of mortality in *B. nigra* populations.

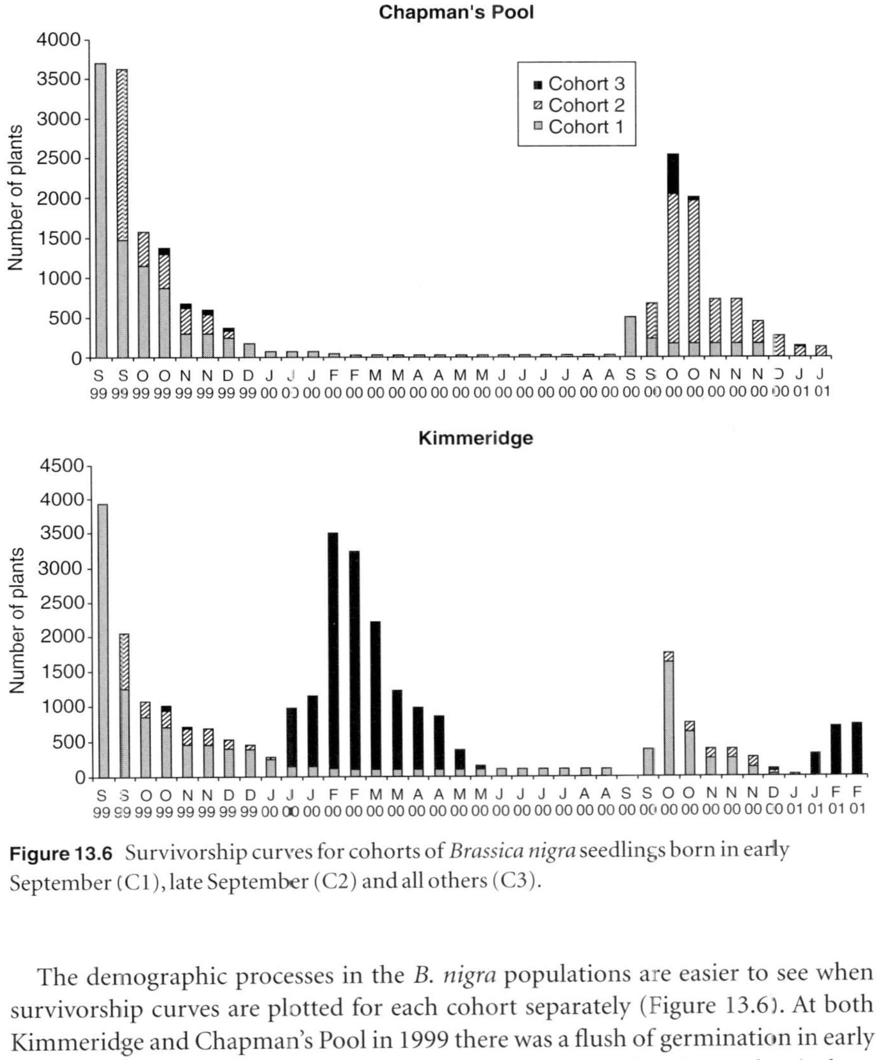

Figure 13.6 Survivorship curves for cohorts of *Brassica nigra* seedlings born in early September (C1), late September (C2) and all others (C3).

The demographic processes in the *B. nigra* populations are easier to see when survivorship curves are plotted for each cohort separately (Figure 13.6). At both Kimmeridge and Chapman's Pool in 1999 there was a flush of germination in early September (cohort 1) and seeds continued to germinate into late September (cohort 2). All seedlings that germinated after September are referred to as cohort 3. At Chapman's Pool, cohort 3 was limited to a few seedlings that germinate in early October. At Kimmeridge, on the other hand, there was some germination in early October, but a huge flush of germination in January and February 2000. At both sites, all flowering plants in 2000 were recruited from cohort 1.

In 2000–2001, the Chapman's Pool population followed a similar pattern of germination to 1999–2000, but the number of seedlings was about half that in 1999. In February 2001, all surviving plants were from cohort 2 and there was little or no germination in January. At Kimmeridge, seedling numbers were also lower than in 1999. By February 2001, all plants of cohorts 1 and 2 were dead and a flush of seedlings germinated in January and February. We could not monitor the progress of the populations after February 2001 because of movement restrictions imposed following the outbreak of foot and mouth disease. However, there were seed-producing *B. nigra* plants at both sites in August 2001, and we assume that these are cohort 2 at Chapman's Pool and early spring germinating cohort 3 at Kimmeridge.

The different patterns in 1999–2000 and 2000–2001 are probably due to the extremely wet autumn in 2000. Early germinating seeds were washed away by heavy rain in early September, which left space for later germinating seedlings to be recruited. At Kimmeridge conditions were so wet that all seedlings that germinated before the end of 2000 had disappeared by early 2001, allowing seedlings to be recruited from seed germinating in January and February. Across years there was a consistent difference between Kimmeridge and Chapman's Pool—Kimmeridge produces a second strong flush of germination in early spring. We have no information on whether the difference is genetically and/or environmentally determined.

At both sites in 1999, cohort 1 survived and subsequent cohorts were unable to establish. In 2000, cohort 1 was destroyed by heavy rain, leaving room for recruitment from later cohorts. Taken together with the data in Figure 13.5, we can suggest that climatic and density dependent factors are far more important than viruses in controlling recruitment. It is interesting to note that estimates of total population size were little different between years: Kimmeridge 2000 = 1300, 2001 = 20 700; Chapman's Pool 2000 = 2400, 2001 = 1900). Even very high seedling mortality in a wet autumn did not reduce the adult population size. We have no data on *B. rapa*, but we are undertaking a similar study to estimate the effects of TuMV infection on population dynamics of this species.

Predicting the effects of TuMV resistance genes in wild *Brassica* species

Brassica nigra

Although TuMV is lethal to *B. nigra* from Dorset, the demographic studies show that it rarely infects plants in our study populations. The reason for this is not clear, but is perhaps due to lack of suitable vectors. The alternative of infection followed by instant death seems unlikely from laboratory observations on time to death after challenge. The low frequency of infection means that there is likely to be little selective advantage to any TuMV resistance genes that move into the population. The high mortality due to other factors also means that even if resistance did spread through the population, it might not have a significant effect on population dynamics.

Brassica oleracea

TuMV does infect *B. oleracea* and does decrease survival and fecundity. Therefore one might expect selection for uniformly high resistance or tolerance. However, there is a very large amount of heritable variation in the reaction of *B. oleracea* seedlings to challenge with TuMV. There are several possible reasons for this. First, we estimated the heritability of control of virus replication in young seedlings, which could be a poor predictor of whether the plant will survive to reproduction, or of how many seeds it will produce. We need to look at the heritability of response to TuMV through the whole life-cycle of the plant, however this would be a very large and long-term experiment.

Another possibility is that the variation in response of *B. oleracea* to a single isolate of TuMV is due to 'trench warfare' between plant resistance genes and pathogen avirulence genes (e.g. Stahl *et al.* 1999). The coat protein gene sequences of different Dorset isolates are very similar, however the coat protein gene is not known to determine TuMV virulence. Sequences of potential virulence genes are needed to test whether TuMV variation is maintaining genetic variation in *B. oleracea.*

Another possible explanation for the variation in *B. oleracea* is that there is a 'cost' to resistance and that TuMV is distributed patchily in space and/or time so that susceptible plants gain a fitness advantage over resistant plants in the absence of the pathogen. We know that TuMV is patchily distributed, however we have no data on whether, for instance, unchallenged susceptible plants produce more seeds than unchallenged resistant plants. Part of the problem is that we can only identify resistant and susceptible genotypes by challenging them with TuMV. One solution would be to propagate plants clonally, or challenge plants at the end of the experiment and hope that we had happened to choose a suitable array of genotypes.

The potential cost of resistance

We have made a test of the general hypothesis that there is a cost to virus resistance. *TuRB01* is a *Brassica* A-genome dominant gene that confers resistance to certain isolates of TuMV, including pathotype 1 (Walsh *et al.* 1999). During the winter and spring 2000–2001 we carried out an experimental test into whether there is a 'cost' of this gene (lower fitness than the allele for susceptibility in the absence of TuMV). We used a pair of near isogenic oilseed rape lines that differ only at and around the *TuRB01* locus. One line ('R') is homozygous for resistance at *TuRB01* and other line ('S') is susceptible to TuMV (homozygous recessive). We inoculated 50 seedlings of the R line and 50 seedlings of the S line with a pathotype 1 isolate of TuMV from Warwickshire (RI = inoculated resistant, SI = inoculated susceptible). We also mock-inoculated 50 R and 50 S seedlings with inoculation buffer (RC = resistant control, SC = susceptible control). The plants were planted in a randomized design in an outdoor plunge bed at CEH Dorset. Plants were scored for mortality, height, seed production and seed germination. Using ELISA, we detected TuMV in all inoculated plants, but we were unable to detect it in any of the control plants.

After 26 weeks there was no significant difference in the proportion of plants surviving among treatments ($\chi^2_{[3]} = 2.85$, $P = 0.415$) (Figure 13.7). The height of plants at 26 weeks is also shown in Figure 13.7. The data were log-transformed for homogeneity of variance and analysed by one-way ANOVA. There were significant differences among treatments ($F_{[3,122]} = 108.9$, $P < 0.0005$). We obtained confidence intervals for all pairwise differences between treatment means using Tukey's method, with a family error rate of 0.05. The SI treatment was significantly different

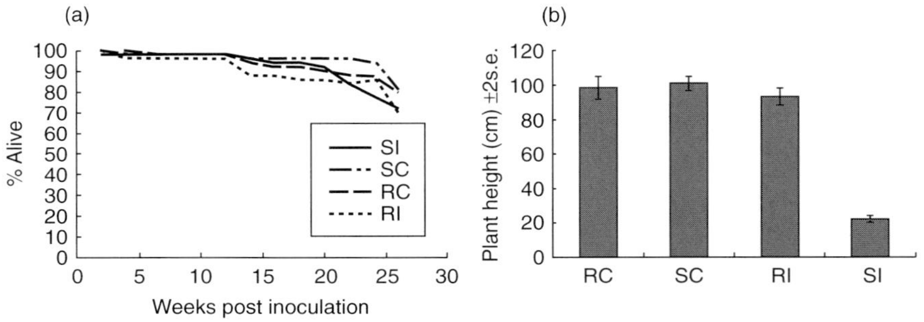

Figure 13.7 Survival (a) and growth (b) of a *Brassica napus* line with the *TuRBO1* gene for resistance to TuMV (R) and a near isogenic line without the resistance gene (S). I, inoculated with TuMV; C, control (mock-inoculated with water).

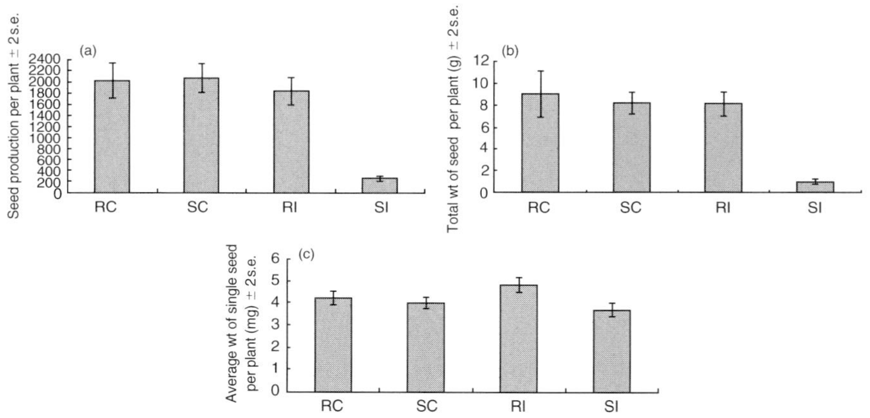

Figure 13.8 (a–c) Fecundity of a *Brassica napus* line with the *TuRBO1* gene for resistance to TuMV (R) and a near isogenic line without the resistance gene (S). I, inoculated with TuMV; C, control (mock-inoculated with water).

from all other treatments, but there were no significant differences among the other treatments.

ANOVA was also carried out on total seed number and total seed weight. Both variables were log-transformed. There were significant differences among treatments for both variables (seed number: $F_{[3,115]} = 18.54$, $P < 0.0005$; seed weight: $F_{[3,115]} = 17.62$, $P < 0.0005$), with SI significantly different from the others (Figure 13.8). We also calculated the average weight of single seeds on each plant. One-way ANOVA on untransformed data revealed no significant difference among treatments ($F_{[3,115]} = 2.30$, $P < 0.081$) (Figure 13.8).

Finally, we tested for differences in seed quality by placing ten seeds from each of ten plants of each treatment on moistened filter paper in Petri dishes. The dishes

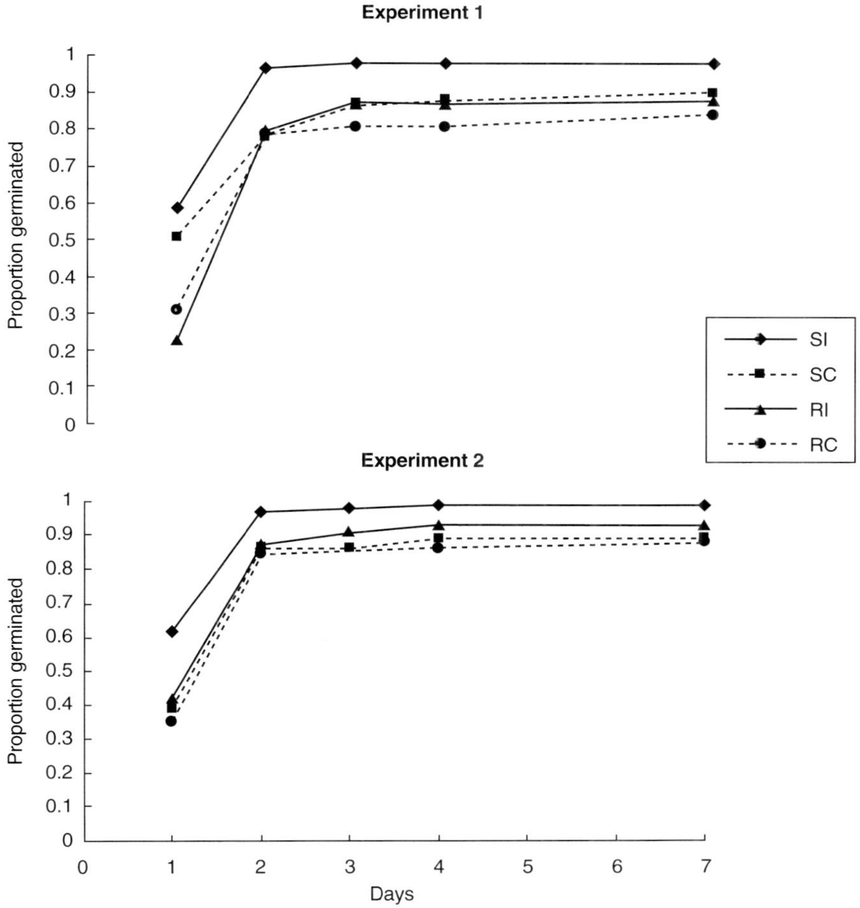

Figure 13.9 Germination of seed produced by a *Brassica napus* line with the *TuRBO1* gene for resistance to TuMV (R) and a near isogenic line without the resistance gene (S). I, inoculated with TuMV; C, control (mock-inoculated with water).

were randomized on a bench in the laboratory and the number of seed that germinated was recorded after 1, 2, 3, 4 and 7 days. The experiment was done twice and the results are shown in Figure 13.9. The largest difference between the treatments was seen after 1 day; the seeds from the SI treatment seemed to be germinating more rapidly. Using arcsine square root transformed data, we combined the results of the two experiments and tested for pairwise differences between the treatments using Tukey's test for multiple comparisons. The test was performed using a general linear model (GLM) procedure in SAS that controlled for differences among experiments. The SI treatment was significantly different from the RC and RI treatments ($P < 0.001$ in both cases).

We detected no cost of resistance (performance of SC > RC) for any variable. The main differences were the higher mortality, smaller size and lower seed production of the SI plants. Perhaps surprisingly, seeds from the SI treatment germinate faster than seeds from the RI and RC treatments. Given the importance of rapid germination in *Brassica* populations (see *B. nigra* demography study above), the faster germination of seeds from infected susceptible plants might help maintain polymorphism for virus resistance in *Brassica* species. Further work is needed to establish the generality of this phenomenon.

TuMV infection can reduce survival and fecundity in *B. oleracea*. There appears to be a high amount of variation in accumulation of TuMV in *B. oleracea* and this variation has a strong genetic component. We have discussed above the reasons why genetic variation for TuMV might be maintained in *B. oleracea* populations. A single resistance gene with no cost might behave very differently from the quantitative resistance we detected, and might spread through a population. The long life-cycle of *B. oleracea* makes it difficult to predict whether a TuMV resistance gene will change the population dynamics of the plant.

Brassica rapa

We have only just begun studies on the effect of TuMV on the population dynamics of *B. rapa* and TuMV. Preliminary observations suggest a pattern more similar to *B. nigra* than *B. oleracea*.

Conclusions

We began this study knowing nothing about the interaction between TuMV and wild *Brassicas*. We soon established that many *B. oleracea* plants are infected with TuMV. Subsequent work suggested that vector behaviour is a stronger determinant of virus frequency in a population than genetic variation in either the host or the pathogen. Other experiments showed that TuMV could reduce the seed output of infected plants. There is potential, therefore, for a TuMV resistance gene to increase the population growth rate of *B. oleracea* if the growth rate is seed limited. Regardless of any effect on population dynamics, a TuMV resistance gene may spread through *B. oleracea* populations provided there is little or no cost of resistance.

By contrast, TuMV was almost completely absent from *B. nigra* and *B. rapa* populations. However, from a risk assessment point of view, this information was ambiguous because extreme susceptibility and rapid death, immunity, resistance or lack of a suitable vector could have caused absence of the virus. These alternatives needed to be tested in laboratory experiments and both species were shown to be highly susceptible. This information necessitated a further set of (very labour intensive) field experiments to determine whether susceptibility and death or lack of vectors was the main factor in the field. We established, albeit over quite a limited area, that *B. nigra* almost never became infected with TuMV. These data suggested that even though *B. nigra* is highly susceptible to TuMV infection, a TuMV resistance gene would have little or no effect on population dynamics. Work to determine infection rates in the field for *B. rapa* is in progress.

These specific conclusions suggest that while generic risk assessments for virus resistance genes are not possible, a common experimental approach could be devised. First, the host should be challenged with the virus in the laboratory. If there is pathotype variation, then a range of virus isolates should be tested. If the host proves to be immune to all isolates, then the experimental work can stop, as the resistance gene will have no effect. If the virus does infect the host, the next step will depend on how easy it is to study the effect of the virus on the population dynamics of the host. For an annual such as *B. nigra*, it may be possible to study the effect directly by marking plants, recording whether they become infected, and whether infected plants have lower survival or reproductive output. For a perennial, such as *B. oleracea*, it may be more efficient to use a combination of surveys of the frequency of virus infection and experimental inoculations to determine effects on life-history parameters. Under no circumstances is a 'snapshot' survey of virus incidence sufficient for risk assessment purposes.

Acknowledgements

This work was funded by the Natural Environment Research Council's thematic programme EDGE (Ecological Dynamics and Genes).

References

Arnold, W.N. & Bald, J.G. (1960) Turnip mosaic virus from two weed hosts. *Phytopathology* **50**, 578–581.

Broadbent, L. (1957) *Investigation of Virus Diseases of* Brassica *Crops*. Cambridge University Press, Cambridge.

Burdon, J.J. (1987) *Diseases and Plant Population Biology*. Cambridge University Press, Cambridge.

Converse, R.H & Martin, R.R. (1990) ELISA methods for plant viruses. In: *Serological Methods for Detection and Identification of Viral and Bacterial Plant Pathogens*, pp. 179–196, American Phytopathological Society, St Paul, MN.

Cooper, J.I. & Jones, A.T. (1983) Responses of plants to viruses: proposals for the use of terms. *Phytopathology* **73**, 127–128.

Cooper, J.I. & Raybould, A.F. (1997) Transgenes for stress tolerance: consequences for weed evolution. In: *Proceedings of the Brighton Crop Protection Conference—Weeds*, pp. 265–272.

Edwards, M-L. & Cooper, J.I. (1985) Virus detection using a new form of indirect ELISA. *Journal of Virological Methods* **11**, 309–319.

Felsenstein, J. (1985) Confidence limits on phylogenies: an approach using the bootstrap. *Evolution* **39**, 783–791.

Felsenstein, J. (1993) Phylogeny inference package, version 3.5. Department of Genetics, University of Seattle, WA.

Jenner, C.E. & Walsh, J.A. (1996) Pathotype variation in turnip mosaic virus with special reference to European isolates. *Plant Pathology* **45**, 848–856.

Jenner, C.E., Sanchez, F., Nettleship, S.B. *et al.* (2000) The cylindrical inclusion gene of turnip mosaic polyvirus encodes a pathogenic determinant to the brassica resistance gene *TuRB01*. *Molecular Plant–Microbe Interactions* **13**, 1102–1108.

Lehmann, P., Petrzik, K., Jenner, C.E. *et al.* (1997) Nucleotide and amino acid variation in the coat protein coding region of turnip mosaic virus isolates and possible involvement in the interaction with the brassica resistance gene *TuRB01*. *Physiological and Molecular Plant Pathology* **51**, 195–208.

Lowe, A.J., Jones, A.M.E., Raybould, A.F. *et al.*

(2002) Transferability and genome specificity of a new set of microsatellite primers among *Brassica* species of the U triangle. *Molecular Ecology Notes* **2**, 7–11.

Maskell, L.C., Raybould, A.F., Cooper, J.I., Edwards, M-L. & Gray, A.J. (1999) Effects of turnip mosaic virus and turnip yellow mosaic virus on the survival, growth and reproduction of wild cabbage (*Brassica oleracea*). *Annals of Applied Biology* **135**, 401–407.

Mithen, R., Raybould, A.F. & Giamoustaris, A. (1995) Divergent selection for secondary metabolites between wild populations of *Brassica oleracea* and its implications for plant herbivore interactions. *Heredity* **75**, 472–484.

Moyes, C.L. (1997) *The role of glucosinolate production in populations of wild cabbage.* PhD Thesis, University of Liverpool.

Moyes, C.L. & Raybould, A.F. (2001) The role of spatial scale and intraspecific variation in secondary chemistry in host plant location by *Ceutorhynchus assimilis* (Coleoptera: Curculionidae). *Proceedings of the Royal Society of London Series B* **268**, 1567–1573.

Moyes, C.L., Collin, H.A., Britton, G. & Raybould, A.F. (2000) Glucosinolates and differential herbivory in wild populations of *Brassica oleracea. Journal of Chemical Ecology* **26**, 2625–2641.

Paul, V.H. & Rawlinson, C.J. (1992) *Disease and Pests of Oilseed Rape.* Verlag Th. Mann, Gelsenkirchen-Buer.

Raybould, A.F., Maskell, L.C., Edwards, M-L., Cooper, J.I. & Gray, A.J. (1999a) The prevalence and spatial distribution of virus infections in natural populations of *Brassica oleracea. New Phytologist* **141,** 265–275.

Raybould, A.F., Mogg, R.J., Clarke, R.T., Gliddon, C.J. & Gray, A.J. (1999b) Variation and population structure at microsatellite and isozyme loci in wild cabbage (*Brassica oleracea* L.) in Dorset (UK). *Genetic Resources and Crop Evolution* **46**, 351–360.

Raybould, A.F., Edwards, M-L., Clarke, R.T., Pallett, D.W. & Cooper, J.I. (2000) Heritable variation for the control of turnip mosaic virus and cauliflower mosaic virus replication in wild cabbage. *Beiträge zur Züchtungsforschung—Bundesanstalt für Züchtungsforschung an Kulturpflanzen. Proceedings of the 7th Aschersleben Symposium—New Aspects of Resistance Research on Cultivated Plants* 4–8.

Shattuck, V.I., (1992) The biology, epidemiology and control of turnip mosaic virus. *Offprints from Plant Breeding Reviews* **14**, 199–238.

Shukla, D.D., Ward, C.W. & Brunt, A.A. (1994) *The Potyviridae.* CAB International/Cambridge University Press, Cambridge.

Špak, J. (1992) Effect of sinigrin on the efficiency of acquisition of turnip mosaic virus by *Myzus persicae* and *Brevicoryne brassicae. Biologia Plantarum* **34**, 451–455.

Stahl, E.A., Dwyer, G., Mauricio, R., Kreitman, M. & Bergelson, J. (1999) Dynamics of disease resistance polymorphism at the *Rpm1* locus of *Arabidopsis. Nature* **400**, 667–671.

Takahashi, W.N. (1949) The morphology and host range of *Brassica nigra* virus. *American Journal of Botany* **36**, 533–535.

Thompson, J.D., Higgins, D.G. & Gibson, T.J. (1994) CLUSTAL W. Improving the sensitivity of progressive multiple sequence alignment through sequence weighting, position-specific gap penalties and weight matrix choice. *Nucleic Acids Research* **22**, 4673–4680.

Thurston, M.I., Pallett, D.W., Cortina-Borja, M. *et al.* (2001) The incidence of viruses in wild *Brassica nigra* in Dorset (UK). *Annals of Applied Biology* **139**, 277–284.

Tomlinson, J.A. (1987) Epidemiology and control of virus diseases of vegetables. *Annals of Applied Biology* **101**, 43–50.

Urcuqui-Inchima, S., Haenni, A-L. & Bernardi, F. (2001) Potyvirus proteins: a wealth of functions. *Virus Research* **74**, 157–175.

Walsh, J.A., Sharpe, A.G., Jenner, C.E. & Lydiate, D.J. (1999) Characterisation of resistance to turnip mosiac virus in oilseed rape (*Brassica napus*) and genetic mapping of *TuRB01. Theoretical and Applied Genetics* **99**, 1149–1154.

Waterworth, H.E. & Hadidi, A. (1998) Economic losses due to plant viruses. In: *Plant Virus Disease Control* (eds A Hadidi, R. K. Khetarpal & H. Koganezawa), pp. 1–13. APS Press, St Paul, MN.

Westmann, A.L. & Kresovich, S. (1999) Simple sequence repeat (SSR)-based marker variation in *Brassica nigra* genebank accessions and weed populations. *Euphytica* **109**, 85–92.

Chapter 14

The molecular evolution of host specificity in the rhizobium–legume symbiosis

J. P. W. Young, L. A. Mutch, D. A. Ashford, A. Zézé and K. E. Mutch*

Molecules are important determinants of ecological specificity

Much of ecology is concerned with the interactions between organisms, whether of the same or different species, and direct interactions are always specific to a greater or lesser degree. The recognition between organisms may be mediated through light, sound, touch and so on, but recognition at the molecular level is undoubtedly the most universal and important mode. All organisms respond to chemicals that are produced by other organisms. In some cases the result may be mutually beneficial: a female moth attracts a mate with her pheromone, a pollinator is enticed by a floral scent. In other cases, the creator of the tell-tale signal does not benefit, though the recipient does: a mosquito homes in on its victim, the immune system responds to the foreign antigens of a pathogen, the style of a flower rejects incompatible pollen. A knowledge of the capabilities and limitations of such recognition systems is important for our understanding of adaptation. Is the degree of specificity limited by the intrinsic properties of the signalling system, or by the relative advantage of 'generalist' versus 'specialist' strategies?

While all organisms make use of molecular recognition, bacteria offer some particularly accessible model systems in which genetics, biochemistry and ecology can be brought together. Our knowledge of the genetic basis of some important ecological traits, including signalling, means that explicit causal links can be made between genes and ecology. Rhizobia, the nitrogen-fixing symbionts of legumes, are good candidates for such studies because they are of ecological and biotechnological importance and there is a wealth of agronomic, genetic, biochemical and taxonomic information. The emphasis to date has been on agricultural rather than natural systems, but this solid foundation of knowledge can now be exploited to tackle fundamental ecological and evolutionary questions. There may also be an applied spin-off, because host range is an aspect of niche breadth of widespread relevance to pests, parasites, pathogens, and so on.

* Correspondence author: jpy1@york.ac.uk

The rhizobium–legume symbiosis depends on specific molecular signals

We have chosen to investigate the rhizobium–legume symbiosis because of the solid foundation of genetic and biochemical information available about the specificity determinants. Rhizobia are bacteria that live as a normal component of the soil and rhizosphere microbial population but have the unique ability to stimulate the formation of special root organs (nodules), which they colonize through an intracellular infection, and in which they fix atmospheric nitrogen into organic forms that are passed on to the plant. There are currently around 30 recognized species of rhizobia, and the number is growing as new isolates are characterized. All belong to a particular clade in the alpha subdivision of the *Proteobacteria*, with the exception of a few isolates recently reported to be in the beta subdivision (Chen *et al.* 2001; Moulin *et al.* 2001; Tripathi 2002). Most rhizobia are in the genera *Rhizobium*, *Sinorhizobium*, *Mesorhizobium*, *Bradyrhizobium* and *Azorhizobium*. With the single exception of *Parasponia* (Ulmaceae), all known plant hosts that are nodulated by rhizobia are legumes (family Fabaceae), and the majority of the 18 000 or so legume species are believed to form the symbiosis (Sprent 2001).

Nodulation has a major effect on rhizobial population dynamics: the abundance of rhizobia in the soil is greatly increased after a compatible host plant has been grown. Host range is therefore important to the bacterium, but it varies enormously among rhizobia, with strains of some species able to nodulate hundreds of diverse legume species while others are confined to a single species (Young & Johnston 1989). At the functional level, host range is determined by a 'molecular conversation' that is increasingly well understood both genetically and biochemically: signals from legume roots activate bacterial *nod* genes that synthesize 'Nod factors'; these stimulate nodule development in the plant (Spaink *et al.* 1991; Dénarié *et al.* 1996).

Legume roots and seeds exude a number of organic compounds, especially flavonoids and related compounds, in mixtures that differ between species. Rhizobia respond to these because they have one or more *nodD* genes, which encode regulator proteins that activate the other *nod* genes when they interact with appropriate flavonoid or other signal compounds from the plant. Once activated, the *nod* genes direct the synthesis of 'Nod factors', a family of lipochitin oligomers (Figure 14.1). Structurally, all Nod factors are based on short chains of β-1,4-linked *N*-acetylglucosamine residues (usually 4 or 5). The distal or non-reducing glucosamine residue is *N*-acylated. This common lipo-chitooligosaccharide core (produced by the enzymes encoded by the *nodABC* genes) may be modified by various specific substituents on the distal or reducing sugars. These modifications are governed by additional *nod* genes. Rhizobial species and genotypes differ in their complement of *nod* genes and in allelic forms of shared *nod* genes, and these differences lead to predictable differences in the structure of the Nod factors.

At the evolutionary level, the pressures and constraints on host range are less well explained. The benefit of a generalist strategy may be a higher chance of finding a host, but the costs may include increased biochemical investment or reduced yield from nodules. Alternatively, the variation in host specificity may result from con-

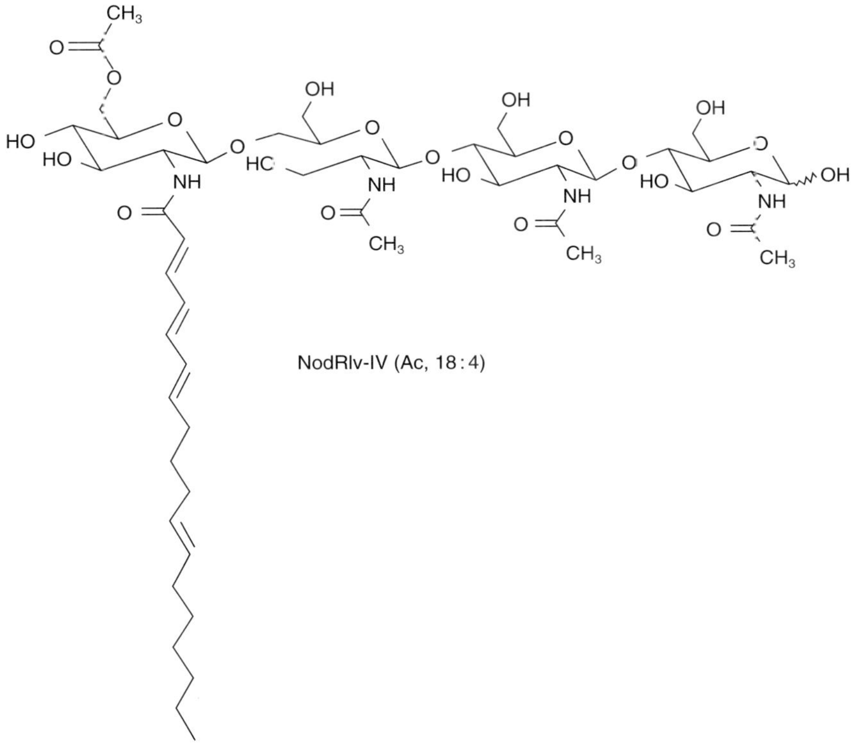

Figure 14.1 One of the main lipochitin oligosaccharides (LCOs, Nod factors) made by *Rhizobium leguminosarum* biovar *viciae* (*Rlv*). Other major components have five sugar residues rather than four, or a monounsaturated acyl chain rather than polyunsaturated.

straints in the evolutionary process imposed by the genetic structure, with limited opportunities to assemble the necessary sets of genes.

One of the merits of rhizobia as a model for the evolution of host range is that detailed studies have been carried out on a number of distinct species, including the following: *Rhizobium leguminosarum* biovar *viciae* (*Rlv*; hosts—*Pisum, Vicia, Lathyrus, Lens*), *R. leguminosarum* biovar *trifolii* (*Rlt*; hosts—*Trifolium* species), *R. leguminosarum* biovar *phaseoli* (*Rlp*; host—*Phaseolus*), *R. etli* (host—*Phaseolus*), *R. tropici* (Hosts: *Phaseolus* and many other legumes), *Sinorhizobium meliloti* (hosts—*Medicago* and *Melilotus*) and *S. fredii* (hosts—*Glycine* and, for some strains, a very large number of other genera) (Young 1996). *Rlv*, *Rlt* and *Rlp* all share a common pool of background genotypes, which is why they are regarded as a single species despite having different host ranges. The *nod* genes of *Rlv* and *Rlt* are fairly similar, while those of *Rlp* are very different from these but almost identical to those of *R. etli*, a related but distinct species. *R. tropici* has a related background but quite different *nod* genes. The genetic backgrounds of *S. meliloti* and *S. fredii* are very similar to each other but very different from the *Rhizobium* species, while the *nod* genes of *S. meliloti* are relatively similar to those of *Rlv* and *Rlt*, and those of *S. fredii* resemble *R. tropici* and *R. etli*. This complex pattern of relationships strongly suggests that nodulation genes have at some time been transferred between different genetic backgrounds, but they are certainly not distributed randomly. Genome organization is consistent with this conclusion, in that the genes for nodulation are carried on plasmids in all the species mentioned so far, and are therefore potentially transmissible.

Although, to a first approximation, all isolates of a given species and biovar nodulate all the nominal hosts, nuances of specificity have been reported whenever this has been investigated. A particularly well-studied example involves symbionts of the cultivated pea (*Pisum sativum*). Certain peas from Western Asia (notably 'cultivar Afghanistan') are unable to form a symbiosis with the widespread forms of *Rlv* that nodulate most peas (Young & Matthews 1982; Young *et al.* 1982; Lie *et al.* 1987). Instead, they require a specific rhizobial variant (e.g. strain TOM) that is also uncommon except in Western Asia and the Middle East. These broader-range rhizobia (which are also able to nodulate the normal peas) have an extra nodulation gene, *nodX*, which confers the extended host range and encodes an enzyme that acetylates the reducing terminal of the Nod factor (Firmin *et al.* 1993).

Other studies of *Rlv* have identified differences in host preference, if not absolute specificity. Hynes and O'Connell (1990), who studied nodule isolates from peas, lentils (*Lens esculenta*) and faba beans (broad beans, *Vicia faba*), found a particular rhizobial genotype that was common in pea nodules but never identified among isolates from *V. faba*, and vice versa. Evans *et al.* (1996) showed that *V. faba* and *P. sativum*, although belonging to the same cross-inoculation group, select significantly for different *Rlv* genotypes. Handley *et al.* (1998), who studied *Rlv* isolates from *V. hirsuta*, *V. sativa*, *Lathyrus odoratus* and *L. sylvestris*, as well as peas, lentils and faba beans, confirmed that our choice of host plant can alter our perspective on the population diversity of *Rlv* in soil. Different trap plants were nodulated by isolates with different genetic profiles, suggesting host specificity of the *Rlv* in the populations studied.

A case study: *Rhizobium leguminosarum* biovars *viciae* and *trifolii*

The aim of our project was to gain an insight into the evolution of host specificity by exploring whether lack of suitable genetic variation might be a constraint that limited fine-tuning of specificity. The project integrated studies of genetic variation (using DNA methods), of molecular signal diversity (using analytical biochemistry) and of ecological niche (using host-range tests). We decided to look in more detail at the genetic basis for small differences in host range, because these differences provide the raw material for natural selection. We chose to study two biovars of *R. leguminosarum*. In the case of biovar *trifolii* (*Rlt*), we hoped to see specialization to genotypes within one host species, while for biovar *viciae* (*Rlv*) we investigated interactions with a number of local and non-local host species in the legume tribe Vicieae.

Strains differ in competitiveness for nodulation

Initial work looked at the variation of *Rlt* and a single host species, *Trifolium repens* (white clover), coexisting in a permanent grassland (Zézé *et al.* 2001). This led to an interesting comparison of the bacterial genotype frequencies in nodules with those in the soil population, as assayed by direct PCR from the community DNA, using primers designed to match the *nodD* genes of *Rlt* and *Rlv*. From this, the selectivity of the plant can be determined, and this is the first time such a comparison has been made at the genetic level. Genotype frequencies in the soil DNA were different from those in nodule isolates, implying that there were differences among the rhizobia in 'competitiveness' for nodulation. Differences in competitiveness were also demonstrated in pairwise competition assays between isolates (unpublished observations). However, there was no clear evidence for comparable genetic variation among the clover plants, so coevolution could not readily be investigated and this *Rlt* system was not pursued further.

Host range is a genetically polymorphic character

The study of *Rlv* was based on a sample of 625 isolates obtained from nodules on eight host species grown in the same soil. The soil was from rough grassland on the University of York campus. Four of the legume species grew wild on the site: *Vicia cracca*, *V. hirsuta*, *V. sativa* and *Lathyrus pratensis*. Two of the others, *L. aphaca* and *L. nissolia*, are British natives but not found locally, while the others were crop plants originally domesticated in the Middle East: garden pea (*Pisum sativum*) and faba bean (*Vicia faba*). All these species belong to the legume tribe Vicieae, which is regarded as the normal host range of *Rlv*. The aim was to explore the microheterogeneity in host specificity within *Rlv*, and its genetic and biochemical basis.

The 625 isolates were typed for chromosomal and nodulation gene restriction fragment polymorphism, based on PCR of the intergenic spacer between the 16S and 23S ribosomal RNA genes, and of *nodD* and the spacer between *nodD* and *nodF*, respectively. Eighty of the isolates were tested for host range on seven of the host plants (L. A. Mutch and J. P. W. Young, unpublished observations). While the majority of strains were able to nodulate all the native hosts, a few showed clear-cut limitations in host range, being unable to nodulate certain species despite repeated tests (Table 14.1). Many more strains showed specificity when non-native hosts were tested. In fact, only a small minority of the strains isolated from other species were able to nodulate faba bean (*V. faba*). The specificity of *L. nissolia* was more subtle: all isolates could nodulate it except for seven of the 10 strains isolated from *V. cracca*. A number of strains were able to nodulate a particular host, but not to fix nitrogen in the nodules. This clearly shows that there are species-specific events well after nodule initiation. Although these are beyond the scope of our study, which focused on the initial recognition, a failure to fix nitrogen has important consequences for the fitness of both partners. Not only does the symbiosis become parasitic, since the main benefit to the plant is lost, but the plant's feedback regulation of nodule development is responsive to nitrogen supply from the nodule, so that a non-fixing

Table 14.1 Host range of *Rhizobium leguminosarum* biovar *viciae* isolates.

	Test host						
Original host	*V.c.*	*V.h.*	*V.s.*	*L.n.*	*L.p.*	*V.f.*	*P.s.*
Vicia cracca	10	9(1)	8(2)	1(2)	8	2	9(1)
V. hirsuta	10	10	9(1)	10	10	1	6(1)
V. sativa	8	10	10	10	10	3	10
Lathyrus aphaca	2(5)	10	10	10	10	4	10
L. nissolia	5(5)	7(3)	10	10	9	2	3(4)
L. pratensis	9(1)	10	9(1)	8(2)	10	2(1)	9(1)
V. faba	10	9	10	10	10	10	6(3)
Pisum sativum	9(1)	10	8(2)	10	10	2	10

Figures given are number of isolates, out of 10 tested, that formed a nitrogen-fixing symbiosis. Figures in parentheses are the number of additional isolates that formed nodules but did not fix nitrogen.

Table 14.2 Rhizobia isolated from the same soil via nodules on each of eight legume species.

nodD	Host plant species								
type*	*V. cracca*	*V. hirsuta*	*V. sativa*	*L. aphaca*	*L. nissolia*	*L. pratensis*	*V. faba*	*P. sativum*	Total
1	46	13	19	13	4	7	4	29	135
2	4	4	4	4	0	0	83	8	107
3	25	37	26	33	2	14	0	61	198
4	5	5	10	13	0	2	2	12	49
5	0	0	0	0	0	3	0	1	4
6	0	0	0	1	0	0	0	0	1
7	1	39	29	28	4	21	1	8	131
Total	81	98	88	92	10	47	90	119	625

Figures shown are number of isolates.
* RFLP type of the *nodD* region.

nodule will be aborted by the plant and the bacterial population within it will be much smaller than normal.

Genetic variation within the sample tells a similar story of specificity. Chromosomal RFLP variant 1 was the most abundant overall (25%), but not a single one of the 90 isolates from *V. faba* was of this type (data not shown). Conversely, 83 of these 90 isolates were of a single RFLP type for the *nodD* region (type 2), though this was relatively rarely isolated from the other species (Table 14.2). Type 2 *nod* patterns were found in association with chromosomal types 2, 3, 4, 5, 6 and 8, but never in the type 1 chromosomal background. This illustrates the non-random association between plasmid and chromosomal markers that has been observed previously in this species (Young & Wexler 1988). Strains of *nod* type 2 nodulated *V. faba* regardless of their chromosomal type, suggesting that it is the genotype of the *nod* region rather than the chromosomal background that determines species specificity. The next phase of the study focused on the products created by the *nod* genes, aiming to establish the relationship between *nod* gene variation and variation in Nod factor molecules.

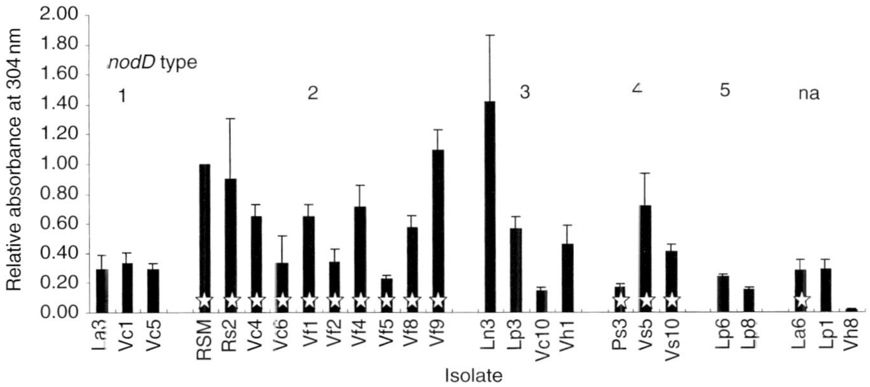

Figure 14.2 Production of polyunsaturated Nod factors by isolates of different *nodD* RFLP types, assayed by absorbance at 304 nm after HPLC (scaled relative to strain RSM2004). Starred isolates are capable of nodulating faba bean (*Vicia faba*).

Rhizobium strains differ in the amount of Nodulation Factor signal they produce

We developed new protocols for the analysis of Nod factors (lipochitin oligomers) with greatly enhanced sensitivity and without the need for radioactive labelling (D. A. Ashford *et al.*, unpublished observations). After fractionation and HPLC, the components were detected either by intrinsic UV absorbance at 304 nm (sensitive primarily to factors with polyunsaturated acyl moieties) or by fluorescence after labelling with 2-aminobenzimide (2-AB, which labels Nod factors at the reducing terminal). The 2-AB labelling approach (Bigge *et al.* 1995) has not been routinely used for Nod factors, but gives picomolar sensitivity for detection of oligosaccharides (Costa *et al.* 1997). Most Nod factors, including all those known from *Rlv*, have a free reducing terminal and should be labelled equally by this reagent. Both methods yielded useful results using small amounts of material, and avoiding the need for metabolic labelling of Nod factors with [^{14}C]acetate (Price *et al.* 1992; Bec-Ferté *et al.* 1994; Lortet *et al.* 1996). The identity of the major peaks was confirmed by mass spectrometry. When induced with naringenin, all strains produced a mixture of Nod factors, including those with four or five sugars, and with C18 : 1 or C18 : 4 acyl groups. The composition was relatively constant, but there were large and repeatable differences in the amount of factor produced by different strains (Figure 14.2). On average, strains of *nodD* type 2 produced more Nod factor than other genotypes (this does not mean, of course, that the *nodD* gene itself is responsible for the difference, it is merely serving as a marker for genetically distinct nodulation gene regions) The quantity of Nod factor has been shown to be important in the interaction of *Rlv* with *Pisum sativum* cv. Afghanistan (Hogg *et al.* 2002).

As indicated in the previous section, strains of *nodD* type 2, which produced more Nod factor than average, show a strong association with *Vicia faba* (faba bean). This is clear evidence for genetic differences in host range associated with the *nod* genes, but it seems that the quantity of Nod factor may not be the only, or even the main, relevant variable, because some strains with quite modest Nod factor levels are nevertheless able to nodulate *V. faba*, while others fail to nodulate despite having higher levels (see Figure 14.2). An important caveat, though, is that so far we have only examined the Nod factors produced in response to pure naringenin. It is possible that the response to natural inducers is different, a question that is addressed in the next section.

Legume species produce different signals and rhizobium strains differ in their response

The activation of the *nod* genes by the NodD protein in response to plant exudates provides another information-rich opportunity for the development of specificity. There could be differences in the flavonoids or other inducers exuded by the plants, and differences in the relative sensitivity of the NodD proteins to these. To assess this we introduced a reporter plasmid pIJ1477 (Firmin *et al.* 1986) into representative strains of each *nodD* RFLP type (L. A. Mutch *et al.*, unpublished observations). The reporter had a *nodC-lacZ* fusion downstream of a nod box regulatory sequence, so that activation of the *nod* genes could be assayed by a colorimetric assay for beta-galactosidase. As a standard we used the laboratory strain A34, also carrying pIJ1477. This strain has the symbiosis plasmid pRL1JI (*nodD* RFLP type 2), which was originally identified in an *Rlv* strain isolated from a *V. faba* nodule.

Root exudates from each of the legume species were used as inducers, as well as the pure flavonoid naringenin. Strikingly, the two *nodD* type 2 strains were induced by faba bean as well or better than by pea exudate, whereas for all other genotypes pea was a far more active inducer (Figure 14.3). This is a direct biochemical correlate (and possible explanation) of the observed strong bias of faba bean towards nodulation by type 2 strains. In order to investigate the differences in root exudates, we fractionated each exudate by HPLC and tested each fraction for induction of A34 (Figure 14.4). A unique major inducer eluted from faba bean exudate in fraction 9. All the *Vicia* species, including faba bean, had a major peak in fraction 30. We ran all the readily available flavonoids as standards, but none corresponded to either of these peaks. The peak of activity in fraction 11 from pea and *Lathyrus* exudates may correspond to eriodictyol and/or an apigenin-*O*-glycoside, as previously reported (Firmin *et al.* 1986). When pure flavonoids were tested (data not shown), hesperitin was the most potent inducer of all genotypes, followed by homoeriodictyol, but it seems as though neither of these compounds is a major natural inducer in our hosts (Figure 14.4). Clearly more work is needed to determine the true natural inducers, and then to examine Nod factor production using these inducers.

Are host specificity differences caused by differences in NodD sequence?

We found large differences in the response to root exudates (and large differences in the composition of root exudates), and some of these differences were in the direction expected from the observed differences in host specificity. It seems clear that

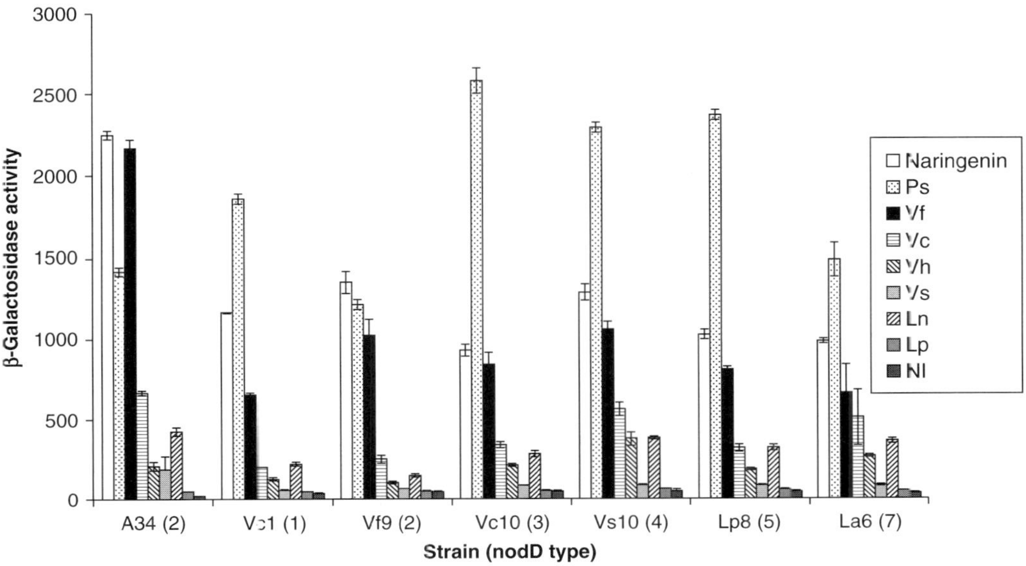

Figure 14.3 Induction of *nod* genes of representative isolates by the flavonoid naringenin or by root exudates from each of seven legume species: Lp, *Lathyrus pratensis*; Vf, *Vicia faba*; Ln, *L. nissolia*; Ps, *Pisum sativum*; Vc, *V. cracca*; Vh, *V. hirsuta*; Vs, *V. sativa*; NI, not induced.

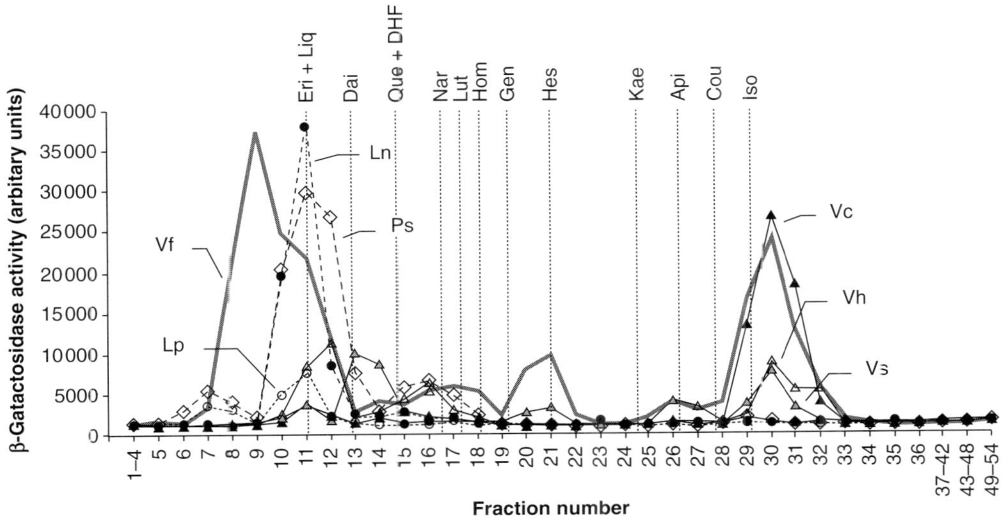

Figure 14.4 Induction of *nod* genes of A34 (*nodD* type 2) by HPLC fractions of root exudates from each of seven legume species (as in Figure 14.3). The elution positions of various pure flavonoids and related compounds are also indicated: Eri, eriodictyol; Liq, liquiritigenin; Dai, daidzein; Que, quercetin; DHF, 4′,7-dihydroxyflavone; Nar, naringenin; Lut, luteolin; Hom, homoeriodictyol; Gen, genistein; Hes, hesperetin; Kae, kaempferol; Api, apigenin; Cou, coumestrol; Iso, isoliquiritigenin.

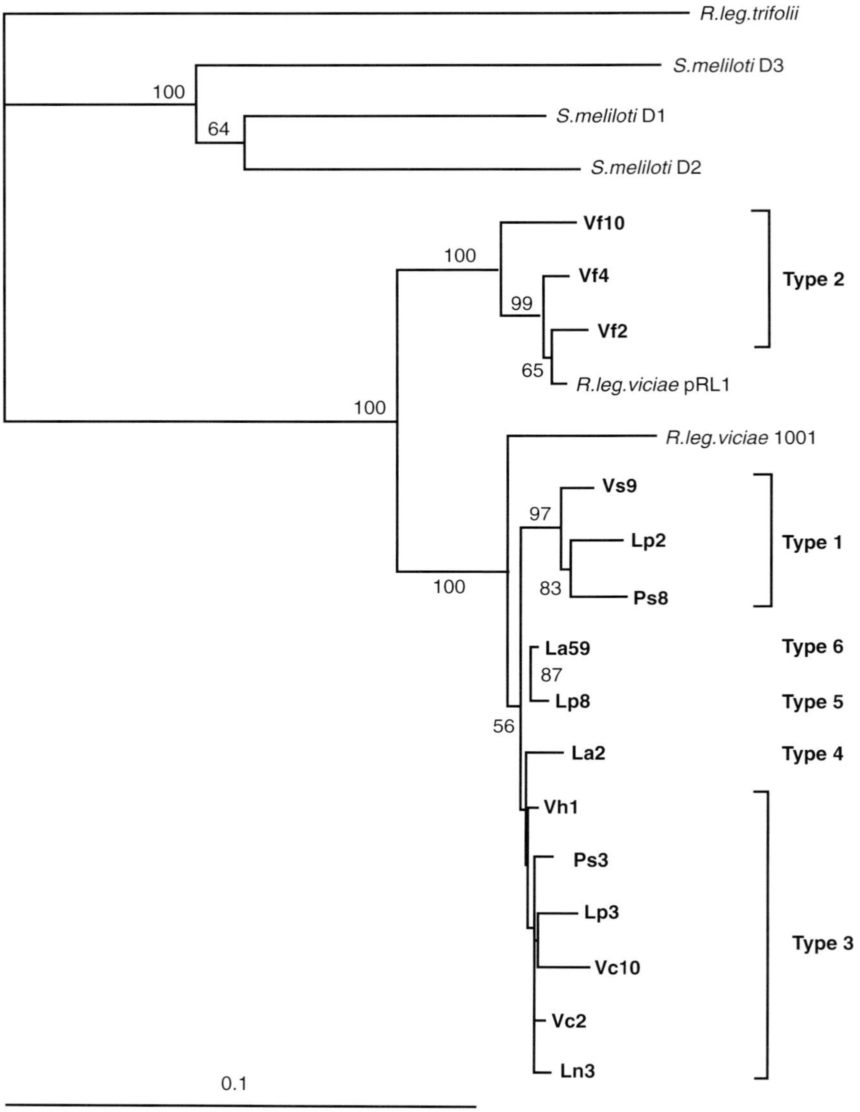

Figure 14.5 Phylogeny of *nodD* sequences of *Rlv* isolates representing different *nodD* RFLP types (neighbour-joining method, scale indicates substitutions per nucleotide, per cent bootstrap support indicated).

some of the significant ecological differences, in particular in the interaction with faba bean, stem from this phase of the interaction. In a way, this is surprising, because the major component on the bacterial side is a single protein, the product of the *nodD* gene and a member of the large LysR family of bacterial regulatory proteins.

Strain	Type	18	22	24	29	61	113	117	174	176	184	197	214	218	222	234	239	247	249	251	253	254	255	256	259	260	261	263	264	266	269	283	299
Lps	1	M	K	T	S	I	E	E	R	S	K	A	L	L	T	P	T	L	V	R	Y	A	Q	D	L	R	I	K	H	L	L	S	S
Vc3-1	1	M	K	T	N	I	E	E	G	S	K	A	L	L	I	P	T	L	V	R	Y	A	Q	N	R	R	I	E	H	L	L	S	S
Vs9	1	V	K	T	S	I	E	E	G	S	K	A	L	L	I	P	T	M	V	R	Y	A	Q	N	R	R	I	E	H	W	L	S	S
pRL1JI	2	M	K	T	S	V	D	L	G	P	Q	A	M	L	I	P	T	L	L	K	Y	E	R	T	L	R	I	E	H	L	V	T	L
Vf10	2	M	K	T	S	V	D	L	G	P	Q	A	M	F	I	R	T	V	C	K	Y	G	Q	S	V	R	I	E	Y	L	V	T	L
Vf2	2	M	R	T	S	V	D	L	G	P	Q	A	M	L	I	P	T	V	L	K	Y	G	R	T	L	R	I	E	H	L	V	T	L
Vf4	2	M	R	T	S	V	D	L	R	P	Q	A	M	L	I	P	T	V	L	K	Y	E	R	T	L	K	I	E	H	L	V	T	L
Ln3	3	M	K	T	S	I	E	V	G	S	K	A	L	L	I	P	T	L	L	K	Y	E	Q	S	V	R	S	E	H	L	L	S	S
Lp3	3	M	K	T	S	I	E	V	G	S	K	R	L	L	I	P	S	L	L	K	Y	E	Q	T	M	R	I	E	H	L	L	S	S
Ps3	3	M	K	T	S	I	E	V	G	S	K	A	L	L	I	P	T	V	L	K	F	E	Q	T	L	R	I	E	H	L	L	S	S
Vc10	3	V	K	T	S	I	E	V	G	S	K	A	L	L	I	P	S	M	L	K	Y	E	Q	R	M	R	I	E	H	F	L	S	S
Vc2	3	M	K	T	S	I	E	V	G	S	K	A	L	L	T	P	T	L	L	K	Y	E	Q	T	M	R	I	E	H	L	L	S	S
Vc3-3	3	M	K	T	N	I	E	V	G	S	K	A	L	L	I	P	S	L	L	K	Y	E	Q	T	M	R	I	E	H	L	L	S	S
Vh1	3	M	K	T	S	I	E	V	G	S	K	A	L	L	I	P	T	L	L	K	Y	E	R	T	L	R	I	E	H	L	L	S	S
La2	4	V	K	A	S	I	E	V	C	S	K	A	L	L	I	P	T	L	L	K	Y	E	Q	T	L	R	I	E	H	L	L	S	S
Lp8	5	M	K	T	S	I	E	V	G	S	K	A	L	L	I	P	T	L	L	K	Y	E	Q	T	L	R	I	E	H	L	L	S	S
La59	6	M	K	T	S	I	E	V	G	S	K	A	L	L	I	P	T	L	L	K	Y	E	Q	T	L	R	I	E	H	L	L	S	S

Figure 14.6 NodD sequence variation among isolates of *Rlv* representing different *nodD* RFLP types. Only the variable amino acid positions are shown; their positions in the peptide sequence are indicated by the vertical numbers at the top of the columns. Variants from the consensus are boxed.

We have sequenced *nodD* from examples of all the RFLP types, and find that type 2 sequences are substantially diverged from the others (Figure 14.5). There is also polymorphism in the predicted amino acid sequence of NodD, and the type 2 sequences are distinguished by a number of amino acid substitutions (Figure 14.6), particularly towards the C-terminal, in a region that is putatively involved in substrate binding. We can postulate that some of these changes are responsible for different flavonoid binding properties and hence for the difference in host range. Clearly this is a testable hypothesis, and we hope to test it in future.

Molecular ecology: using molecular biology to understand adaptation

In this chapter we have dissected the microvariation in host specificity in a collection of strains within one rhizobial biovar with a broadly similar host range confined to the legume tribe Vicieae. At the ecological level, we found the majority of strains were able to nodulate all the legume hosts tested, but we also found strains with some specificity in their choice of hosts. This demonstrates that the genetic potential exists for greater specialization, but presumably there is not strong ecological pressure in favour of specialists. At the genetic level, we found polymorphic markers of the nodulation gene region that correlated strongly with host range and host preference. At the biochemical level we found relatively little variation in the Nod factors produced by the rhizobia, despite the large number of *nod* genes that are involved in their production and might be targets for functional variation. Instead, we found indications that differences in specificity may lie in the production of inducers by the hosts and their reception by the NodD protein in the rhizobium. If validated, this will mean that we have traced natural variation in a significant ecological property (host range) back through the biochemical mechanism to the underlying DNA sequence differences. In the future it will be increasingly possible, and necessary, to link genes with ecology in this way.

Acknowledgements

We thank NERC for a research studentship (L.A.M) and for funding under the EDGE thematic programme.

References

Bec-Ferté, M.P., Krishnan, H.B., Prome, D. *et al.* (1994) Structures of nodulation factors from the nitrogen-fixing soybean symbiont *Rhizobium fredii* USDA257. *Biochemistry* **33**, 11782–11788.

Bigge, J.C., Patel, T.P., Bruce, J.A. *et al.* (1995) Nonselective and efficient fluorescent labelling of glycans using 2-amino benzamide and anthranilic acid. *Analytical Biochemistry* **230**, 229–238.

Chen, W.-M., Laevens, S., Lee, T-M. *et al.* (2001) *Ralstonia taiwanensis* sp. nov., isolated from root nodules of *Mimosa* species and sputum of a cystic fibrosis patient. *International Journal of Systematic and Evolutionary Microbiology* **51**, 1729–1735.

Costa, J., Ashford, D.A., Nimtz, M. *et al.* (1997) The glycosylation of barley (*Hordeum vulgare* L.) and cardoon (*Cynara cardunculus* L.) aspartic proteinases. *European Journal of Biochemistry* **243**, 695–700.

Dénarié, J., Debellé, F. & Promé, J.C. (1996) Rhizobium lipo-chitooligosaccharide nodulation factors: signaling molecules mediating recognition and morphogenesis. *Annual Review of Biochemistry* **65**, 503–535.

Evans, J., Gregory, A., Dobrowolski, N. *et al.* (1996) Nodulation of field-grown *Pisum sativum* and *Vicia faba*: competitiveness of inoculant strains of *Rhizobium leguminosarum* bv. *viciae* determined by an indirect, competitive ELISA method. *Soil Biology and Biochemistry* **28**, 247–255.

Firmin, J.L., Wilson, K.E., Rossen, L. & Johnston, A.W.B. (1986) Flavonoid activation of nodulation genes in *Rhizobium* reversed by other compounds present in plants. *Nature* **324**, 90–92.

Firmin, J.L., Wilson, K.E., Carlson, R.W., Davies, A.E. & Downie, J.A. (1993) Resistance to nodulation of cv. Afghanistan peas is overcome by *nodX*, which mediates an *O*-acetylation of the *Rhizobium leguminosarum* lipo-oligosaccharide nodulation factor. *Molecular Microbiology* **10**, 351–360.

Handley, B.A., Hedges, A.J. & Beringer, J.E. (1998) Importance of host plants for detecting the population diversity of *Rhizobium leguminosarum* biovar *viciae* in soil. *Soil Biology and Biochemistry* **30**, 241–249.

Hogg, B., Davies, A.E., Wilson, K.E., Bisseling, T. & Downie, J.A. (2002) Competitive nodulation blocking of cv. Afghanistan pea is related to high levels of nodulation factors made by some strains of *Rhizobium leguminosarum* bv. *viciae. Molecular Plant–Microbe Interactions* **15**, 60–68.

Hynes, M.F. & O'Connell, M.P. (1990) Host plant effect on competition among strains of *Rhizobium leguminosarum. Canadian Journal of Microbiology* **36**, 864–869.

Lie, T.A., Göktan, D., Engin, M., Pijnenborg, J.

& Anlarsal, E. (1987) Co-evolution of the legume–rhizobium association. *Plants and Soil* **100**, 171–181.

Lortet, G., Mear, N., Lorquin, J. *et al.* (1996) Nod factor thin-layer chromatography profiling as a tool to characterize symbiotic specificity of rhizobial strains—application to *Sinorhizobium saheli, Sinorhizobium teranga*, and *Rhizobium* spp strains isolated from *Acacia* and *Sesbania*. *Molecular Plant–Microbe Interactions* **9**, 736–747.

Moulin, L., Munive, A., Dreyfus, B. & Boivin-Masson, C. (2001) Nodulation of legumes by members of the β-subclass of Proteobacteria. *Nature* **411**, 948–950.

Price, N.P.J., Relic, B., Talmont, E. *et al.* (1992) Broad-host-range *Rhizobium* species strain NGR234 secretes a family of carbamoylated, and fucosylated, nodulation signals that are *O*-acetylated or sulfated. *Molecular Microbiology* **6**, 3575–3584.

Spaink, H.P., Sheeley, D.M., Van Brussel, A.A.N. *et al.* (1991) A novel highly unsaturated fatty-acid moiety of lipo-oligosaccharide signals determines host specificity of *Rhizobium*. *Nature* **354**, 125–130.

Sprent, J. (2001) *Nodulation in Legumes*. Royal Botanic Gardens, Kew.

Tripathi, A.K., (2002) Rhizobia of the β-subclass of proteobacteria: a tale of losing the race. *Current Science* **82**, 8–9.

Young, J.P.W. (1996) Phylogeny and taxonomy of rhizobia. *Plants and Soil* **186**, 45–52.

Young, J.P.W. & Johnston, A.W.B. (1989) The evolution of specificity in the legume–rhizobium symbiosis. *Trends in Ecology and Evolution* **4**, 341–349.

Young, J.P.W. & Matthews, P. (1982) A distinct class of peas (*Pisum sativum* L.) from Afghanistan that show strain specificity for symbiotic rhizobium. *Heredity* **48**, 203–210.

Young, J.P.W. & Wexler, M. (1988) Sym plasmid and chromosomal genotypes are correlated in field populations of *Rhizobium leguminosarum*. *Journal of General Microbiology* **134**, 2731–2739.

Young, J.P.W., Johnston, A.W.B. & Brewin, N.J. (1982) A search for peas (*Pisum sativum* L.) showing strain specificity for symbiotic *Rhizobium leguminosarum*. *Heredity* **48**, 197–201.

Zézé, A., Mutch, L.A. & Young, J.P.W. (2001) Direct amplification of *nodD* from community DNA reveals the genetic diversity of *Rhizobium leguminosarum* in soil. *Environmental Microbiology* **3**, 363–370.

Chapter 15

The existence and persistence of genotypic variation in nucleopolyhedrovirus populations

D. J. Hodgson, R. B. Hitchman, A. J. Vanbergen, R. S. Hails, S. E. Hartley, R. D. Possee, A. D. Watt and J. S. Cory*

Introduction

The importance of genotypic diversity to the structuring and coevolution of host–parasite population interactions is becoming increasingly apparent. The advent of sensitive molecular techniques for the differentiation of parasite genotypes has revealed complex genotypic diversity at several ecological scales, including within individual hosts. Multiple-genotype infections have been described in diverse host–pathogen systems, including rodent malaria parasites (Taylor *et al.* 1997, 1998), rat nematodes (Fisher & Viney 1998) and trematodes (Barral *et al.* 1996), trypanosomes in bumblebees (Imhoof & Schmid-Hempel 1998) and tsetse flies (MacLeod *et al.* 1999), and baculoviruses in Lepidoptera (Laitinen *et al.* 1996; Muñoz *et al.* 1998). Mixed-genotype infections are also known in plant–pathogen systems (e.g. Ennos 1992; Meijer & Leuchtmann 1999), and in human diseases (Gupta & Anderson 1999) such as malaria (Paul *et al.* 1999), hepatitis B (Santos *et al.* 1998) and C (Tuveri *et al.* 1997), and human papillomavirus (Tucker *et al.* 1993).

Genotypic variability is often taken for granted, since it is the expected norm in populations of sexual organisms. Asexual microparasites such as nucleopolyhedroviruses, however, utilize distinct and limited patches of resource (the host individual), and the conditions for genetic exchange between parasite genotypes are often very restricted. If parasite genotypes have different phenotypes (antigenic diversity or differences in quantitative life-history traits), then competition for limited resources within hosts should select for the fittest genotype and therefore reduce genotypic diversity. Why do we see examples of complex mixtures of parasite genotypes within individual hosts?

The answer to this question, as well as being a central issue in evolutionary ecological research, has important implications for applied biology. In most organisms, genotypic variation is viewed as beneficial, since it increases the opportunity for adaptation, and increases the survival probability of threatened species and populations. Genotypic variation in parasite populations, however, is usually seen as a prob-

* Correspondence author: djhod@ceh.ac.uk

lem, for two main reasons. First, genotypically diverse infections are difficult to vaccinate against since this process relies on antigen-specific responses by the host. Second, genetic diversity promotes genetic exchange between parasites: rapidly evolving parasite populations may quickly gain resistance to specific treatments. There exists however an important class of microparasites, those used as biopesticides (Rodgers 1993), in which both predicted features of genotypically diverse infections can be advantageous because they may increase their effectiveness against pests. In this chapter we focus on nucleopolyhedroviruses (NPVs: members of the virus family Baculoviridae), in particular the NPV found to attack the pine beauty moth, *Panolis flammea* D&S (Lepidoptera: Noctuidae). This virus exists as a complex mixture of genotypes within individual host larvae (Paul 1997; Hodgson *et al.* 2001). We discuss several features of NPV biology that predispose them to genotypic variation. We present empirical data on epidemiological and ecological mechanisms that further promote genotypic variation in this host–parasite interaction, and review evidence for similar processes in other systems. Finally, we discuss the implications of genotypic variation and the processes that create and maintain it for the effectiveness of NPV biopesticides, and for the evolutionary ecology of parasites in general.

Nucleopolyhedrovirus biology

NPVs are arthropod-specific double-stranded DNA viruses with relatively large genomes (approx. 150 kbp) (Miller 1997). Their normal mode of transmission is horizontal, via the production of occlusion bodies (OBs) that protect enclosed virus genomes from degradation. OBs adhere to plant surfaces and are consumed by hosts. Virus particles are then released from OBs by dissolution in the alkaline conditions of the insect midgut (Cory *et al.* 1997). The virus particles fuse with epithelial midgut cells and liberate nucleocapsids into the cytoplasm. Virus DNA is then transported to the cell nucleus, where replication and virus gene expression is initiated. NPVs produce a high yield of progeny (up to 10^{10} OBs per host), but only upon host death, when they produce enzymes that liquefy the host cuticle and spread OBs onto surrounding vegetation. These transmission stages are degraded by ultraviolet radiation and weathering (Cory *et al.* 1997) but can persist between generations of univoltine hosts (Carruthers *et al.* 1988). The best known NPVs are those that infect pest species of Lepidoptera.

A further biological feature of NPVs that directly influences genotypic complexity is the co-occlusion of numerous virus genomes within each OB. OBs are formed within infected host cell nuclei, and we assume that if a nucleus is infected by multiple virus genotypes then each OB can contain a mixture of genotypes. Circumstantial evidence for this comes from the natural persistence in mixed genotype infections of defective NPV genotypes that produce no OBs and therefore have zero fitness when propagated in pure infections (e.g. Muñoz *et al.* 1998).

***Panolis flammea*–NPV interaction**

The pine beauty moth, *P. flammea* (Figure 15.1a), is a herbivore of two-needle pine species. In northern Europe it is a pest of *Pinus sylvestris* (Scots pine) plantations. In

(a) (b)

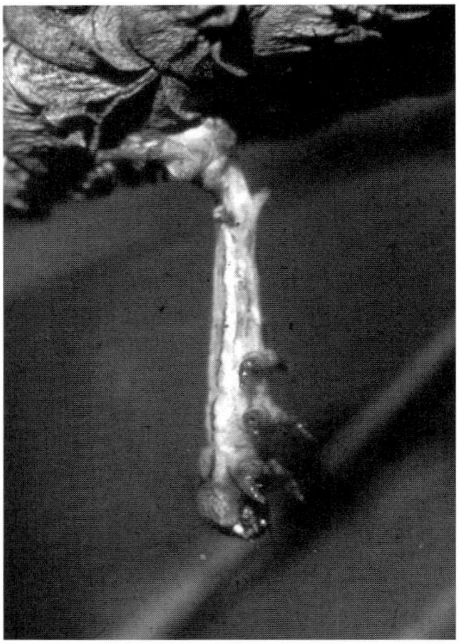

Figure 15.1 *Panolis flammea* larvae, the natural host of *Pf*NPV, feeding on pine foliage: (a) a healthy but susceptible host; (b) a virus-killed cadaver, starting to drip OBs onto the foliage below.

Scotland it is native on *P. sylvestris* but has become an economically important outbreak pest of the introduced *Pinus contorta* (lodgepole pine) (Watt & Hicks 2000). During an outbreak in 1976, NPV-killed larvae (see Figure 15.1b) were collected. The virus has since been used as a biopesticide to control *P. flammea* outbreaks (Evans *et al.* 1991). Molecular analysis of the virus isolate suggested a complex mixture of genotypes. To confirm this, individual genotypes originating from a single host larva were isolated by dilution cloning in *P. flammea* larvae. Restriction fragment length polymorphism (RFLP) of viral DNA from individual larvae was used to analyse virus variation. Only those isolates showing no genotypic variation (i.e. no obvious submolar bands after RFLP analysis and southern blotting), were compared by restriction mapping (J. S. Cory and B. M. Green, unpublished results). The analysis demonstrated that at least 24 NPV genotypes were present in the virus yield from the original host larva, and no single genotype dominated the virus yield.

Nature of the variation

RFLP analysis

Genotypic variants of *Pf*NPV were distinguished from each other by RFLP due to altered restriction enzyme sites or additions or deletions of DNA within the virus

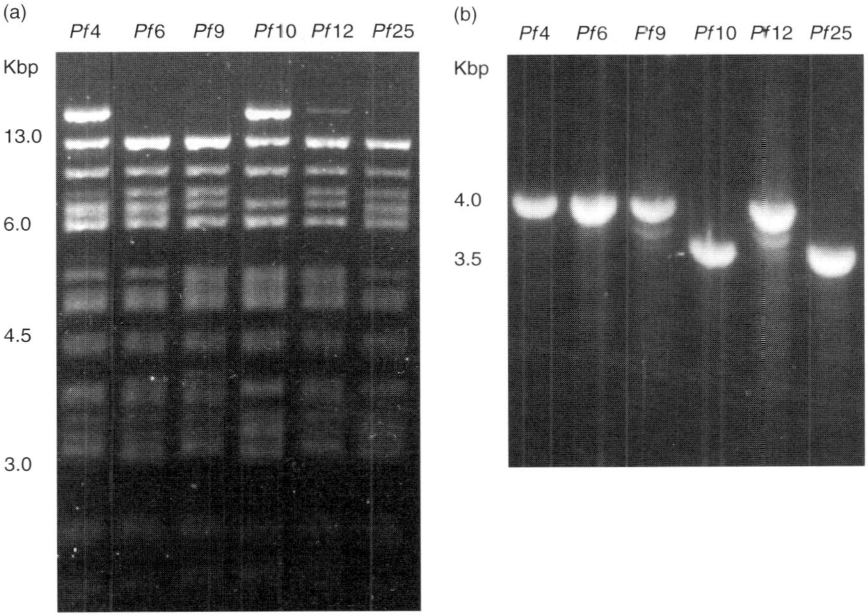

Figure 15.2 RFLP electrophoretic profiles of several *Pf*NPV genotypes, digested with *Hin*dIII restriction enzyme. (a) Profiles of entire NPV genomes. Different genetic profiles are indicated by the presence and absence of bands and the existence of different sized DNA fragments. (b) Comparison of the size of the R fragment (the 18th largest *Hin*d III fragment identified by Possee & Kelly (1988) between the same *Pf*NPV genotypes as in (a)).

genomes (Figure 15.2a). For example, note the larger size of one of the DNA fragments within variants 1 and 4 at the top of the gel in Figure 15.2a. Most of the genotypic variation appeared to be located on three regions of the virus genome (Figure 15.3a). Restriction of the *Pf*NPV genotypes with *Hin*d III revealed at least eight fragments that differed in size between the variants. Restriction mapping of these variable fragments showed that some of them contained large (approximately 300–1000 bp) inserts (Hitchman 2002; J. S. Cory and B. M. Green, unpublished results). For example, Figure 15.2b shows the amplification of the *Hin*d III R fragment using specific oligonucleotide primers (see Figure 15.3b for the position of these), and demonstrates that *Pf*NPV genotypes 4, 6, 9 and 12 contain an insertion of approximately 310 bp, compared to genotypes 10 and 25. However, RFLP analysis only identifies the presence or absence of specific restriction sites for each endonuclease, giving only a crude indication of the genotypic variation. Future sequence data from the different genotypes may reveal even greater levels of genomic variation.

Sequencing and gene identification

Several of the variable *Pf*NPV fragments (Figure 15.3a) were sequenced, revealing the presence of genes homologous to other baculovirus coding regions, and some with no similarity to other genes. For example, Figure 15.3b shows the major open

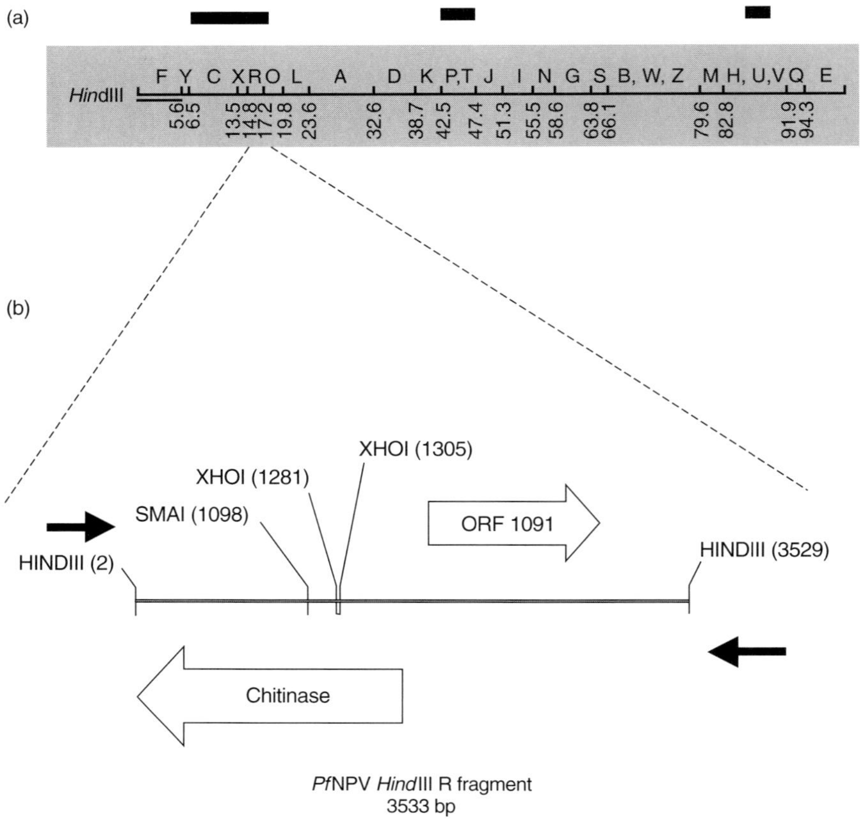

Figure 15.3 RFLP analysis allowed the mapping of the *Pf*NPV genome (adapted from Possee & Kelly, 1988), while sequencing of several variable regions of the genome has revealed the presence of important auxiliary genes and regions of repeated DNA sequence. (a) Fragments labelled according to their size (A largest) are arranged according to their position on the genome. Variable regions (fragments with different sizes in different genotypes) are highlighted with a black bar. (b) Open reading frames identified by sequencing of the X fragment.

reading frames (ORFs) identified on the *Hind* III R fragment. The putative *Pf* NPV chitinase (*chiA*) is 1.6 kbp long and has the potential to encode a 544 amino acid product. Pair-wise (GAP) and multiple pile-up analyses demonstrated a high degree of homology between the *Pf* NPV *chiA* amino acid sequence and other baculovirus *chiA* sequences. Hence, *Pf* NPV *chiA* fits well into the baculovirus chitinase family of proteins. Another potential ORF of 1091 base pairs was identified within the *Hind* III R fragment (Figure 15.3b), but similar sequences were not identified in genome databases.

ChiA is an important baculovirus auxiliary gene, which together with cathepsin (*v-cath*) facilitates host cuticle breakdown and tissue liquefaction in infected larvae, facilitating spread of OBs. Southern hybridization identified a *Pf*NPV cathepsin homologue on the *Pf*NPV *Hind* III K fragment that flanks the variable P fragment (Figure 15.3a). Within some of the *Pf*NPV variable regions direct repeat sequences were identified containing clusters of transcription binding sites (i.e. TATA) interspersed with multiple direct repeat regions and motifs (i.e. TTTGCTTT). These are structurally similar to described baculovirus origins of replication (*oris*) and may play a role in enhancing viral DNA replication (Kool *et al.* 1995).

To date auxiliary genes, potential ORFs with unknown functions and putative *oris* have all been identified within the *Pf*NPV variable regions. Structural genes have also been identified within fragments flanking these regions. All of these recognized and putative genes may be associated with increasing the replication efficiency of the *Pf*NPV variants. Future sequencing of the PCR-amplified variable regions of different variants may reveal differences between these regions and allow the characterization of the inserts maintained by some of the variants. These findings may translate to important qualitative phenotypic differences between *Pf*NPV genotypes.

Quantitative phenotypic traits

As well as qualitative phenotypic traits revealed by molecular analysis, bioassays of single genotype infections can reveal differences in quantitative infection traits that correlate with NPV fitness. Fitness is a difficult parameter to measure for any organism but especially parasites, since their growth and transmission strategies are intimately linked to host physiology and population dynamics. The generally accepted definition of microparasite fitness is R_0, the number of secondary infections resulting from infection of a single host in a natural population (Anderson & May 1981). For a parasite such as an NPV that releases infective transmission stages after killing the host, R_0 is a function of the rate of degradation of transmission stages, and the rate of uptake of these stages by susceptible hosts (Cory *et al.* 1997). Field experiments to accurately compare transmission parameters and degradation rates of purified NPV genotypes require large numbers of host larvae and have not yet been performed in the *Pf*NPV system. Simple laboratory bioassays, however, allow the comparison between purified NPV genotypes of phenotypic traits such as pathogenicity, speed of kill and yield. Here we define pathogenicity as the NPV's capacity to kill the host after ingestion.

Pathogenicity will correlate positively with NPV fitness because infection is required for replication. The rate of production of OBs will also correlate positively with fitness because NPV genotypes with faster replication rates will outcompete less 'virulent' genotypes. Rate of production is the ratio of yield to speed of kill, but these two traits are likely to be correlated since host larvae continue to grow during the infection cycle. NPVs tend to destroy the bulk of the host's internal tissues before lysing its cuticle, therefore in general the bigger the host, the higher the yield. Yields of *Pf*NPV correlate with the weight of the host when infected, and the duration of infection (Hodgson *et al.* 2001). We propose that short speed of kill will increase NPV fitness (due to the multiplicative fecundity of future generations), as long as it

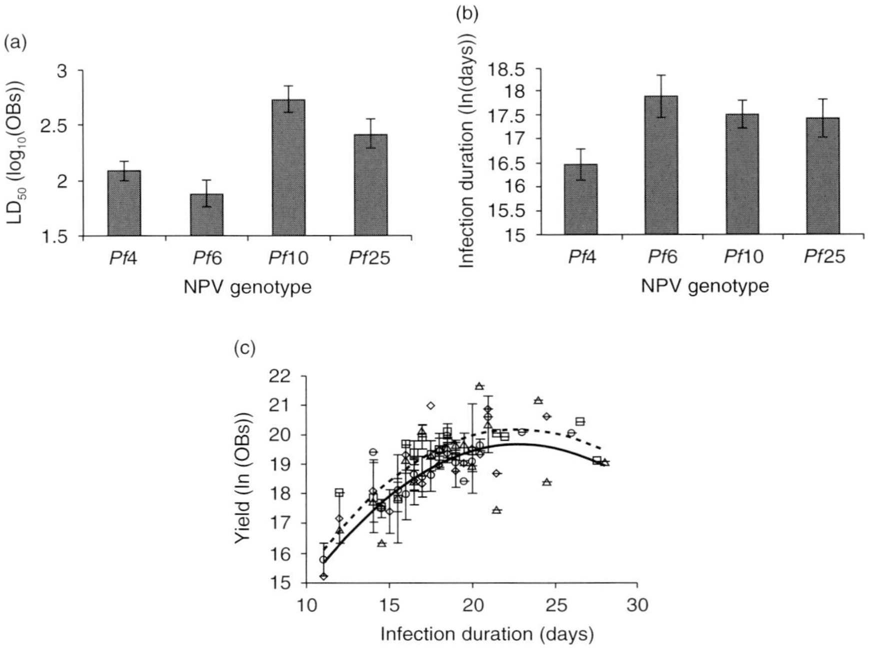

Figure 15.4 Differences in quantitative phenotypic traits between purified *Pf*NPV genotypes. (a) Pathogenicity is measured by LD_{50}, the dose required to kill 50% of the target host individuals. *Pf*4 and *Pf*6 are significantly more pathogenic than *Pf*10 and *Pf*25. (b) Infection duration is the time lapsed between dosing and host death. *Pf*4 kills hosts significantly faster than the other three genotypes. (c) Yield of occlusion bodies (OBs) from a dead host larva increases with increasing infection duration, but with diminishing returns. *Pf*6 and *Pf*10 (dotted fitted line; open diamonds and open triangles respectively) produce significantly higher yields, irrespective of infection duration, than *Pf*4 and *Pf*25 (solid fitted line; open circles and open squares respectively).

does not incur large costs in terms of yield. Similarly, higher yields will donate greater fitness if they can be gained without costs to speed of kill.

Comparisons of these three infection traits in pure infections have revealed significant phenotypic differences between four *Pf*NPV genotypes (Figure 15.4, Hodgson *et al.* 2001), and confirm a positive correlation between speed of kill and virus yield (but with diminishing returns) (Figure 15.4c). The quadratic curve fitted to the data in Figure 15.4c fits better than several saturating functions, indicating a real decline in yields with very slow killing infections, possibly due to a host's partial resistance to infection. The competitive superiority of a virus with a certain suite of phenotypic traits will depend on the relative contribution of each trait to fitness in both pure and mixed infections. The mere existence of differences in quantitative infection traits reinforces the focal question of this chapter: how are such complex mixtures of genotypes maintained in the face of selection for the fittest genotype?

Maintenance of variation

NPV population genetics

Cheating genotypes

During the propagation of NPVs in cell culture, relaxation of selection pressures often results in the appearance of mutant genotypes that have lost fragments of DNA, preventing the expression of auxiliary gene products such as the polyhedrin protein (King & Possee 1992). Such genomes may gain a reproductive rate advantage since smaller genomes take less time to replicate, and have no selective disadvantage in cell culture where OBs are not necessary for transmission (Godfray *et al.* 1997). 'Few-polyhedra' genotypes have been shown to exist (Muñoz *et al.* 1998) and persist (Muñoz & Caballero 2000) in natural NPV isolates. They cannot completely outcompete wildtype OB-forming genotypes because OBs are necessary for horizontal transmission, but a tradeoff between genome size and replication rate may allow them to be maintained in populations via frequency dependent selection. The model reported by Godfray *et al.* (1997) considered the few-polyhedra phenotype that has zero or very low fitness in pure infections, but similar arguments apply to mutant NPV genotypes that have any value of fitness lower than the wildtype genotype but gain a replication rate advantage within hosts. Such cheating genotypes have been observed in other host–parasite systems (e.g. defective interfering particles (Holland 1986)), and their maintenance in populations is comparable to classic negative frequency dependent selected behaviours in higher organisms, such as sneaky male strategies (e.g. Sinervo & Lively 1996; Widemo 1998) and Batesian mimicry (e.g. Lindstrom *et al.* 1997; Pfennig *et al.* 2001).

Mutation/selection equilibrium

The strategic model described above relies on the faster within-host replication of defective mutant genotypes to promote their persistence in NPV populations. The appearance and selective disappearance of defective mutant alleles, however, is a dynamic process: populations of any organism will contain low frequencies of individuals with deleterious mutations. We argue here that the packaging of multiple NPV genomes into each occlusion body makes NPV populations especially prone to the harbouring of defective mutants.

Haldane (cited in Maynard Smith 1998) produced a classic population genetics model that described the expected frequency of deleterious mutant alleles in a population of diploid organisms, assuming no linkage between alleles. He described the mutation/selection equilibrium, which defines the allele frequency at which new mutants are produced as quickly as they are selected out of a population. A logical assumption is that selection against deleterious alleles will be more rapid in haploid organisms (e.g. viruses) than in diploids. Haploids cannot be heterozygotes therefore there will be no masking of recessive alleles. The packaging of several NPV genomes into each OB, however, mimics a new genetic system that we have coined poly-haploidy. If we view the OB as a reproductive individual, then each virus genome within the OB represents a single chromosome in this poly-haploid individual. Virus genomes remain within the OB until it either degrades (and the virus dies), or is ingested by a new host (and the infection process begins again). Upon infection, each virus genome can segregate independently of the others, and new poly-haploid individuals are formed within the cell nuclei of an infected host. It is unknown whether each virus genome within an OB contributes to its production. Any level of sharing of this task may provide protection for rare mutant alleles, similar to the masking of recessive alleles in diploid genetics.

Haldane's analysis involved manipulation of the binomial distribution, assuming that chromosomes always occur in pairs. Thus the number of independently segregating alleles in a diploid individual is 2. If paired alleles are co-dominant, that is having an equal share in the expression of phenotype, the equilibrium frequency of the mutant allele, q, will be:

$$q = \frac{2\mu}{s} \tag{15.1}$$

where μ is the per-generation rate of mutation from wildtype allele p to defective mutant q, and s is the selective disadvantage of an individual expressing only the defective phenotype. If the mutant allele is recessive, it is masked in the heterozygote condition, thus gaining a higher equilibrium frequency in any population:

$$q = \sqrt{\frac{\mu}{s}} \tag{15.2}$$

A useful extension of Haldane's diploid model of the mutation/selection equilibrium is to allow n, the number of independently segregating alleles, to vary (D. J. Hodgson, unpublished observations). This model shows that poly-haploid organisms can sustain greater frequencies of deleterious mutations than diploids. If n virus genotypes segregate independently between successive rounds of OB-production, the extensions of equations (15.1) and (15.2), respectively, are:

$$q = \frac{n\mu}{s} \tag{15.3}$$

when each genome in the OB contributes equally to OB 'fitness', and

$$q = \left(\frac{\mu}{s}\right)^{\frac{1}{n}} \tag{15.4}$$

if defective mutant phenotypes are masked when at least one wildtype genome is present in the OB.

The generality of this result is emphasized by the equality of q for co-dominant and recessive alleles in haploid organisms ($n = 1$). Two main predictions follow. First, populations of NPV species that package large numbers of virus genomes within each OB may harbour higher frequencies of mutant genomes than haploids or diploids. Second, the more genomes are packaged into each OB, the greater the equilibrium frequency of defective mutants. This effect is exacerbated if genomes contribute non-additively to the production of the OB. These predictions so far lack empirical support.

Host immune system

Two popular hypotheses for the maintenance of genotypic variation in parasite populations focus on the coevolutionary interaction between antigens (physical or chemical features of the parasites that are recognized by host immune systems) and antibodies (chemical responses produced by hosts to prevent parasite infection). The Red Queen hypothesis was postulated as a selective force promoting sexual reproduction in hosts (see Ridley 1993; Lively 1999; Lively & Dybdahl 2000). More recently, antibody specificity has been suggested as a determinant of genotypic structuring in parasite populations (Gupta & Anderson 1999). In the absence of specific hypothesis-driven experiments, the relevance of the Red Queen and strain-structure hypotheses to the genotypic complexity of NPV–host interactions remains unknown. We propose that neither hypothesis alone is likely to result in the high observed levels of within-host diversity of *Pf*NPV, for two reasons. First, both hypotheses predict that parasite diversity is driven by frequency-dependent selection. This process should result in the existence of dominant parasite genotypes and the dynamic persistence of several low-frequency genotypes, as observed in other baculovirus populations (e.g. Brown *et al.* 1985; Laitinen *et al.* 1996; Muñoz *et al.* 1998; Stiles & Himmerich 1998), but not in *Pf*NPV which has no clear dominant genotype, at least on the basis of *in vivo* cloning results (J. S. Cory and B. M. Green, unpublished results). Second, there is little current evidence that insect immune systems are specific to parasite genotypes (Lafferty & Crichton 1973; Allander & Schmid-Hempel 2000). Resistance to NPV infection is usually via encapsulation of virions (Trudeau *et al.* 2001) or sloughing of the infected midgut cells (Hoover *et al.* 2000). Whether these responses can be specific to different genotypes of the same NPV species deserves further study.

Tradeoffs between phenotypic traits

Life-history theory predicts that genotypes with quantitative differences in phenotypic traits can coexist via correlations between traits that result in different phenotypes with equal fitness (Stearns 1992). Several phenotypic traits of pure NPV infections resemble fitness correlates in higher organisms that have well-established tradeoff relationships. Yield and speed of kill may correlate in the same way as fecundity and generation time in other organisms. Yield may correlate negatively with pathogenicity, degradation rate or transmission parameter, similar to tradeoffs between fecundity and offspring quality (e.g. Allan 1984; Smith *et al.* 1989; reviewed in Stearns 1992).

Most studies of tradeoffs between parasite infection traits have focused on the correlation between parasite transmission and infection virulence, defined as the impact of infection on host fitness (Read 1994). Mackinnon and Read (1999) found significant positive genetic correlations between malaria pathogenicity in rodent

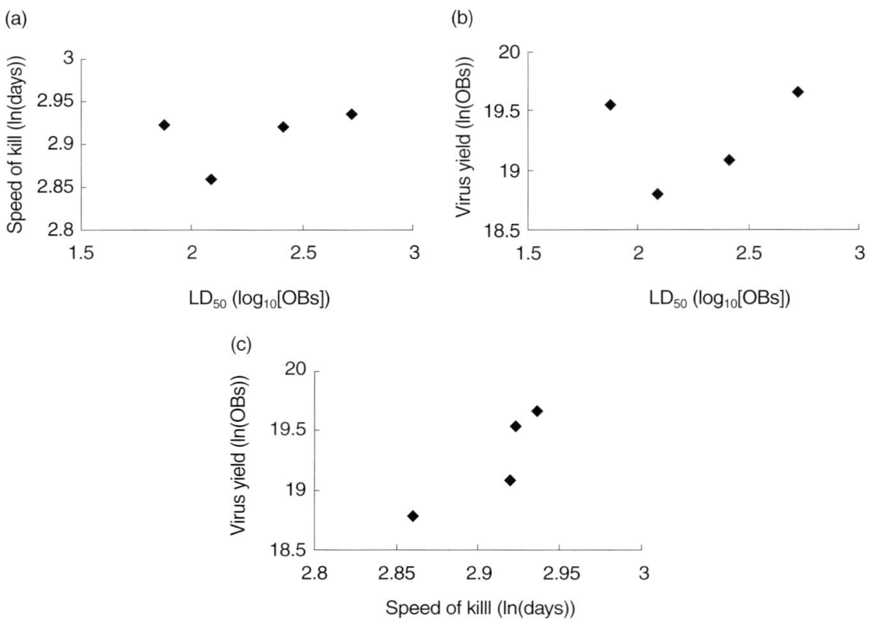

Figure 15.5 An examination of correlations between phenotypic traits of *Pf*NPV genotypes. (a) Pathogenicity vs infection duration. (b) Pathogenicity vs yield. (c) Infection duration vs yield. None of the correlations is statistically significant (sample size is very low), but the strongest correlation is between infection duration and yield. This relationship is a negative tradeoff because short infection duration is predicted to correlate positively with NPV fitness.

hosts and both within-host growth and transmission stage production, but no correlation between these latter variables. Positive correlations between fitness correlates will promote directional selection for fitter parasite genotypes, unless this results in increased virulence that adversely affects transmission success (Ewald 1995), or dramatically reduces host density resulting in the extinction of virulent parasite genotypes. Negative correlations between virulence and both pathogenicity and within-host growth were found by Mackinnon *et al.* (1999), but not between virulence and transmission stage production. These results highlight the complexity of relationships between phenotypic traits of parasites: general patterns are yet to emerge (Lipsitch & Moxon 1997).

The description of mean infection trait values of four purified *Pf*NPV genotypes (see above) provides an opportunity to examine tradeoffs between traits in this system (Figure 15.5). No significant correlations between infection traits were found, but sample sizes are small ($n = 4$) and we propose that the testing of further genotypes will reveal a significant correlation between speed of kill and yield. Similar correlations are apparent in other baculovirus systems (e.g. Muñoz *et al.* 2000; Hernández-Crespo *et al.* 2001), although none stands up to rigorous statistical analysis due to poor replication at the level of virus genotype. There is no indication of relationships between pathogenicity and other traits in *Pf*NPV. A negative trade-off between speed of kill and virus yield may promote the coexistence of NPV genotypes between hosts: fast-killing, low-yield genotypes may coexist alongside slow-killing, high-yield genotypes. The effect of these phenotypic differences on the outcome of within-host interactions is less clear and will depend on the relative contribution of each coinfecting genotype to the timing of host mortality.

Coinfection processes

Competition for host resources between coinfecting parasite genotypes has gained much theoretical, but very little experimental, attention (Bull 1994; Ewald 1995). Theoretical models predict various evolutionary responses by parasites to the presence of coinfecting genotypes. Although a prudent strategy of host resource use may maximize parasite fitness in single-genotype infections (Levin & Pimentel 1981), it is prone to cheating by other genotypes. This can lead to a race to exploit host resources fastest, resulting in high coinfection virulence (Bremermann & Pickering 1983; May & Nowak 1995) and may reduce parasite fitness (Levin & Bull 1994), or result in selection for interference mechanisms (Chao *et al.* 2000). High frequencies of mixed infections may also promote the evolution of cheating genotypes (see above; Holland 1986; Godfray *et al.* 1997; Turner & Chao 1999; Frank 2000).

Despite the diverse predictions made by theoretical studies, simple experimental manipulations of the genotypic diversity of parasite infections are rare, and have produced mixed results. Coinfection of parasite genotypes with different phenotypes leads to intermediate levels of parasite transmission or virulence in trypanosome infections of bumblebees (Imhoof & Schmid-Hempel 1998) and *Botrytis cinerea* on French beans (Weeds *et al.* 2000). Nakamura *et al.* (1992) showed that coinfecting coccidian infections in poultry had similar virulence to pure infections, but that the outcome of competition between genotypes could not be predicted on the basis of pure infection phenotypes. Taylor *et al.* (1997, 1998) demonstrated increased virulence and transmission stage production of malaria coinfections in mice, and ascribed their results to the difficulty of clearing an antigenically diverse infection by the mouse immune system.

To determine the nature of the interaction between *Pf*NPV genotypes within a host, we performed a simple bioassay experiment that exposed fourth instar host larvae to a fixed dose (1000 OBs) of either of two equally infective pure genotypes or a 1 : 1 ratio of both (500 OBs each) (D. J. Hodgson *et al.*, unpublished observations). Our predictions, based on conventional theory, were that pathogenicity would not change in mixed infections, and that competition within the host should result in increased rates of host resource consumption and therefore decreased yields (via the correlation between speed of kill and yield). In fact the opposite was true. Coinfections were more infectious to host larvae (Figure 15.6a), and coinfected larvae died in two waves, corresponding to the speed of kills of the two pure infection phenotypes (Figure 15.6b). Coinfections also gained high yields, both on average and for a given speed of kill (Figure 15.6c). RFLP analysis demonstrated that the lowest-

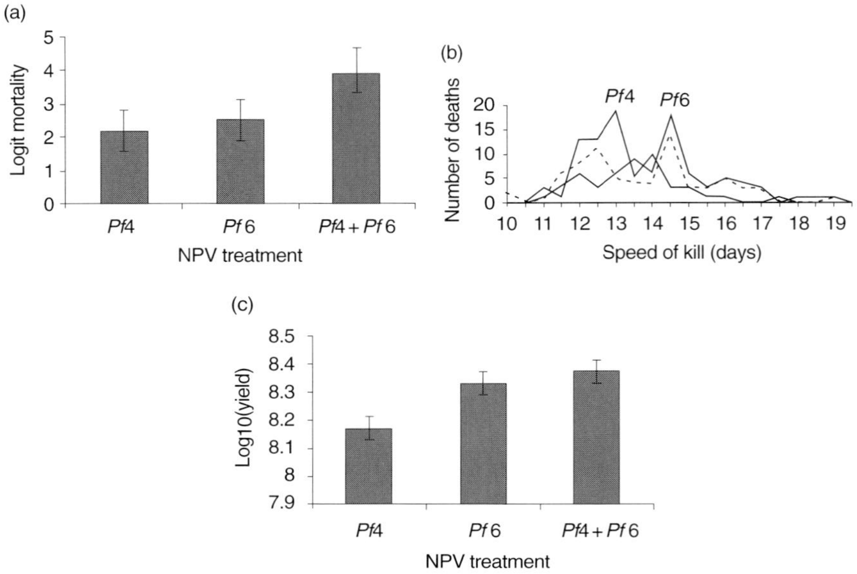

Figure 15.6 Graphical summaries of the infection traits of pure and mixed infections of *Pf*4 and *Pf*6 in fourth instar hosts. (a) Coinfections are significantly more pathogenic than either pure infection. (b) *Pf*4 killed larvae faster than *Pf*6 in pure infections, but mixed infection deaths were bimodal (dotted line). Therefore rapid kill in pure infections does not guarantee dominance of mixed infections. (c) Mixed infections also gained high yields (significantly higher than pure *Pf*4 infections).

yielding of the two pure genotypes was consistently represented in yields from coinfected larvae (mean representation 43%, absent from mixture in only 4% of cadavers; Figure 15.7). These results provide evidence that coinfecting *Pf*NPV genotypes can gain greater fitness per infecting OB than in pure infections, via high pathogenicity, lack of outcompetition of less fit genotypes and high yields.

The mechanisms that cause these results remain unclear. Any differences in timing or pathway of infection between *Pf*NPV genotypes will complicate the host's task of clearing a coinfection. Chemical factors in the polyhedrin protein of the OB are thought to promote pathogenicity by altering the biochemistry of host gut-cell membranes (Corsaro *et al.* 1993). These enhancing factors are known to differ between baculovirus species (Roelvink *et al.* 1995); they may also differ between conspecific genotypes, increasing the virulence and fitness of mixed infections. NPV genotypes may also differ in tissue tropism: if coinfecting genotypes employ different strategies of host tissue infection, the efficiency of this tradeoff may be maximized. Whatever the mechanism causing increased fitness in mixed-genotype infections, it will certainly promote genotypic diversity within hosts.

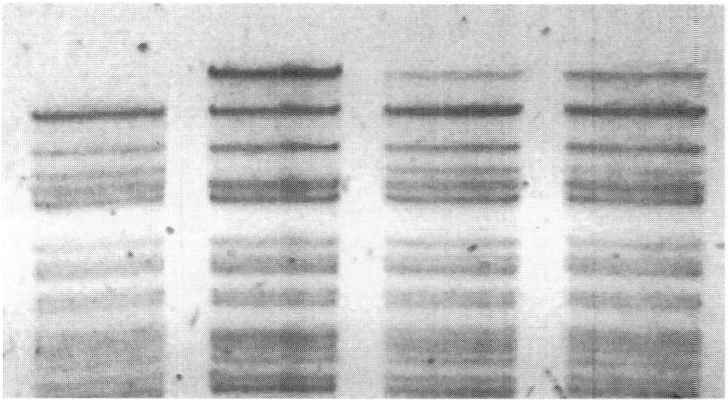

Figure 15.7 RFLP analysis of pure and mixed infections of *Pf*4 and *Pf*6, using *Hin*dIII restriction enzyme. *Pf*4 was identified by three equimolar DNA fragments of approximate sizes 22 kb (band A), 13.1 kb (band B) and 10.1 b (band C) (Possee & Kelly 1988). *Pf*6 also expressed band C but lacked band A and had two similar-sized long fragments, expressed as a dimolar band B (called bands A & B in Possee & Kelly (1988)). Electrophoresis analysis software was used to compare the intensities of bands A (*Pf*4 only) and band C (equimolar in both *Pf*4 and *Pf*6) to produce a ratio of yields from the two genotypes in each mixed infection cadaver. Lanes 3 and 4 in this gel represent mixed infections with low and high proportions of *Pf*4 DNA respectively.

Population dynamics

P. flammea is an outbreak pest, with huge fluctuations in population density between larval cohorts (Watt & Hicks 2000). Whether or not these cycles are driven by its interaction with *Pf*NPV, differences in host density will impact not only the virus' population dynamics but also its population genetic structure. High host densities promote high virus prevalence, and will therefore increase the probability of ingestion of multiple OBs. A similar increase in frequency of mixed infections with increasing disease prevalence has been noted in hyperendemic populations of malaria (Conway & McBride 1991). Under these conditions, selection for pathogenicity, yield and slow degradation outside the host will be greatly relaxed, allowing the coexistence of several NPV phenotypes. Host density may also provide a positive feedback to rates of recombination due to increased numbers of coinfections. Perhaps therefore we should not be surprised to find genotypic complexity during epizootics. When hosts are rare, there will be intense selection pressure either for phenotypic parameters that decrease the minimum host density required for virus persistence (e.g. slow degradation rates and high pathogenicity; Anderson & May 1981), or for low virulence infections and vertical transmission (Bull *et al.* 1991). Surveys of NPVs are logistically difficult when hosts are rare, but similar comparisons of high and low prevalence malaria populations in humans have revealed important differences in linkage disequilibrium and the average number of coinfecting genotypes per host (Anderson *et al.* 2000).

Ecological processes

The complexity of parasite ecology is often exacerbated by the ecology of the host. The *P. flammea-Pf*NPV interaction is no exception. *P. flammea* has two main food plant species, is univoltine but shows variation in the period of larval growth, and may itself adapt to local selection pressures caused by abiotic factors such as climate (Leather 1990), daylength, humidity and soil type (Leather 1993), and biotic interactions with parasites or other natural enemies (Watt 1989a). *Pf*NPV is not host specific and may find itself reproducing within other Lepidoptera species, sympatric with *P. flammea*. NPV fitness may be subject to very different selection pressures in each of these ecological categories. If different NPV genotypes have different fitness in different categories of host ecology, then host-ecology-mediated differential selection of NPV genotypes will determine the genotypic structure of the virus population. Any level of virus dispersal between host subpopulations will then promote within-host NPV genotypic diversity. This is the NPV version of Levene's (1953) classic selection/migration equilibrium. Here we focus on the relative fitness of *Pf*NPV genotypes infecting hosts feeding on *P. sylvestris* or *P. contorta*, and discuss the effect of differential selection pressures on within-host parasite diversity. We then discuss the potential importance of differential selection of NPV genotypes in other ecological categories.

Tritrophic interactions

To identify differential selection of *Pf*NPV genotypes according to host food plant species, we conducted two bioassays of pure infections in hosts feeding on either *P. sylvestris* or *P. contorta*. In the first, larvae reared on either tree species were exposed to a range of doses of either *Pf*4 or *Pf*25, then reared to death or pupation (Hodgson *et al.* 2002). In the second, larvae reared to fourth instar on *P. contorta* were given a fixed dose of either *Pf*4 or *Pf*6 (D. J. Hodgson *et al.*, unpublished observations). Differential selection is indicated by a genotype by environment interaction in statistical analyses of phenotypic traits. In the first bioassay, NPV genotypes were differentially infective in larvae reared on the two food plant species (Figure 15.8a). In the second, speed of kill was differentially affected (Figure 15.8b). Note that two pairs of genotypes were differentially selected for different phenotypic traits. This suggests that differential selection pressures between all possible pairs of genotypes in the *Pf*NPV population may result in a very complex set of competitive interactions.

What mechanisms drive differential selection mediated by the food plant species? *P. flammea* larvae grow faster and produce heavier pupae when feeding on *P. sylvestris* than on *P. contorta* (A. J. Vanbergen, unpublished results). Selection for pathogenicity against large, healthy hosts may incur costs associated with host physiology on other food plant species. Similarly, adaptation to host physiology may allow faster speed of kill on one food plant species but incur a cost on the other. Baculovirus fitness is affected by differences between food plants in acidity (Keating *et al.* 1988, 1989) and the production of secondary chemicals such as peroxidase (Hoover *et al.* 1998, 2000) and hydrolysable tannins (Cadman 1960, Keating *et al.* 1988). *P. sylvestris* foliage contains higher concentrations of phosphorus and nitro-

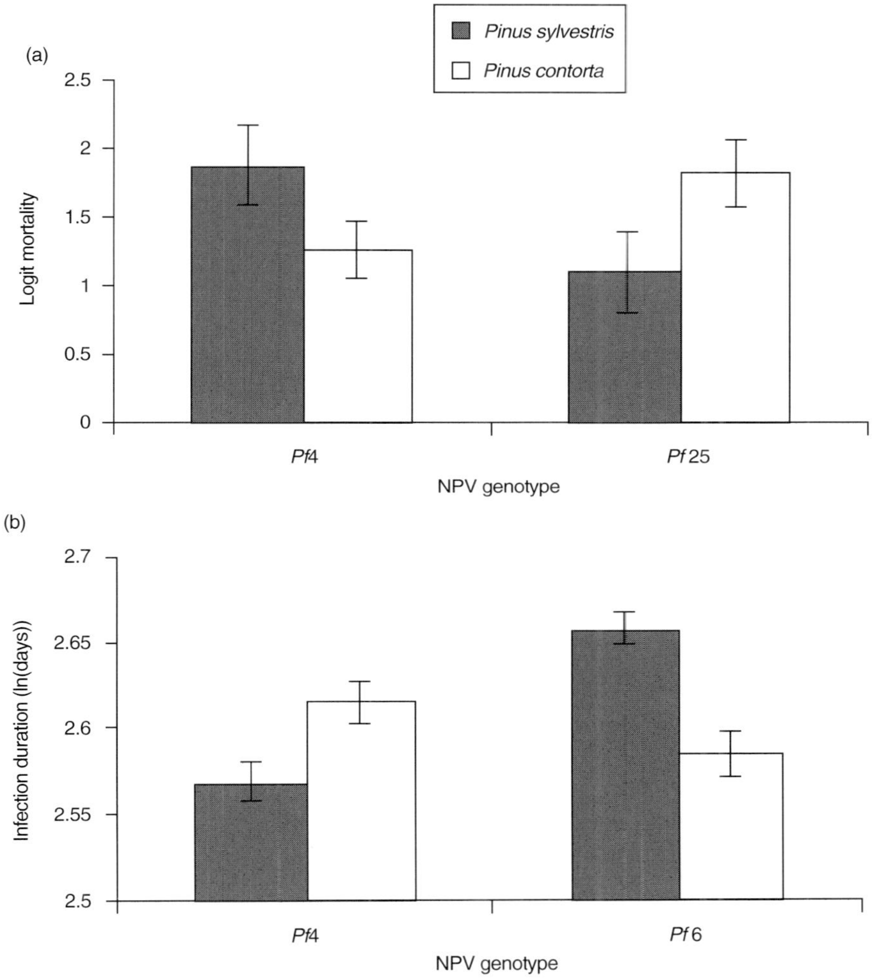

Figure 15.8 Differential selection of phenotypic traits of different *Pf*NPV genotypes, mediated by two species of host food plant. (a) Food plant species differentially affects the pathogenicity of *Pf*4 and *Pf*25. (b) Food plant species differentially affects the infection duration of *Pf*4 and *Pf*6 at a fixed dose of 1000 OBs.

gen, and lower tannin levels, than *P. contorta* (Watt 1989b). The chemical ecology of the *P. flammea–Pinus* sp. interaction may therefore provide sufficiently different gut and cell conditions in the host to influence selection pressures on genes for virus pathogenicity and within host replication.

The effect of food-plant-mediated differential selection on the genotypic diversity and structuring of the virus population will depend on the amount of dispersal of virus genotypes between host subpopulations, but will certainly promote diversity at a metapopulation scale. If host or virus dispersal is limited, then different virus genotypes will dominate different subpopulations. There is evidence for significant differentiation of *P. flammea* genotypes between plantation forests (Wainhouse & Jukes 1997) and between food plant species (Hicks 2001). Transfer of virus between subpopulations is therefore unlikely to occur via dispersal of infected larvae or sub-lethally infected adults, but can be achieved via predator dispersal (Entwistle *et al.* 1993; Vasconcelos *et al.* 1996) or abiotic processes (Olofsson 1988). Viruses can also persist outside the host for long periods, protected from UV degradation and weathering by the OB protein (Carruthers *et al.* 1988; Cory *et al.* 1997). Virus dispersal between subpopulations and the presence of a reservoir of OBs on tree foliage may promote the high levels of genotypic variation observed in the *Pf*NPV population at fine ecological scales. The level of mixing will also clearly depend on the spatial distribution of the host food plant species. This leads to the prediction that genotypic variation in natural *Pf*NPV populations will be higher in mixed species forests than in geographically isolated monocultures.

Other ecological categories

If differential selection of parasite genotypes is caused by ecological categories affecting the same host population or community, less dramatic dispersal events will be required to promote within-host genotypic diversity. Such categories include fluctuating host density, the infection of alternative but sympatric host species, and differences in the length of the larval growth season. We are currently analysing the effect of passaging the wildtype mixture of *Pf*NPV genotypes through several alternative hosts. The influence of season length is more speculative, but it is known that long seasons allow several rounds of NPV infection in natural *P. flammea* populations (J. S. Cory, unpublished results) and other NPV–host interactions (e.g. Woods & Elkinton 1987). If more than one round of infection is possible, rapid speed of kill may correlate positively with NPV fitness. In short seasons with no opportunity for secondary infections, genotypes with long speed of kill and therefore high yield may have higher fitness than fast-killing genotypes. Varying season length, selection for speed of kill within each season, and long-term persistence of OBs in the environment, may promote genotypic variation in NPV populations.

Discussion

Hypotheses for the maintenance of genotypic variation can be drawn from many fields of biological research. In this chapter we have presented theoretical population genetic models, empirical results of epidemiological and ecological experiments, and verbal models based on evidence from other parasite and non-parasite systems. Each hypothesis demonstrates the potential for high levels of genotypic variation in populations of *Pf*NPV in particular, and parasites in general (Figure 15.9). None of these hypotheses is exclusive of one another, and their relative importance will depend on the nature of interactions beween pairs and subgroups of the complex genotypic mixture observed in natural populations. Clearly the question, 'What processes promote genotypic variation in parasite populations?' is difficult to answer succinctly. Our attempts to answer the question have, however, revealed some

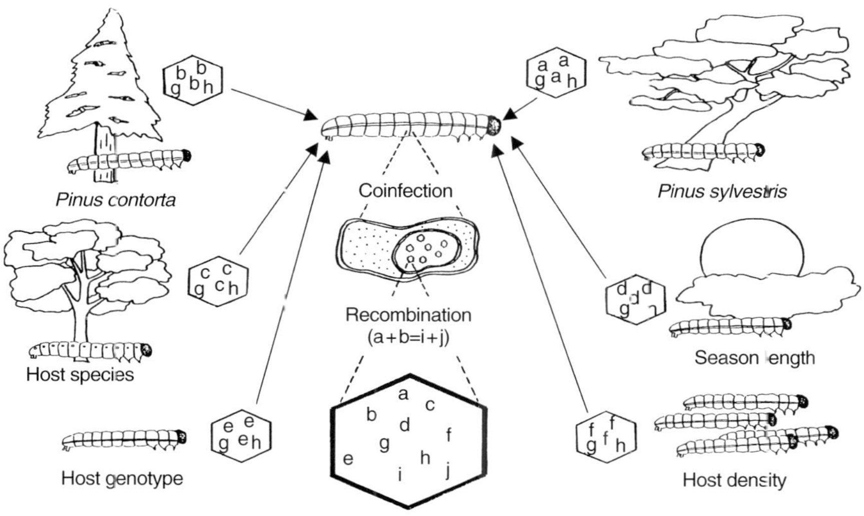

Figure 15.9 Diagram of the selection pressures promoting genotypic variation in *Pf*NPV populations. Different letters represent different genotypes. Genotypes are differentially selected in different ecological categories (a–f), or are cheating genotypes (g), defective mutants (h) or recombinants (i and j). Genotypes a–f dominate OBs due to outcompetition of less well adapted genotypes. Cheating genotypes (g) persist via frequency-dependent selection. Defective mutants (h) persist according to the mutation/selection equilibrium. Recombinant genotypes (i and j) appear during coinfection of host cell nuclei by two or more NPV genotypes. The resulting NPV yield from the focal host is a diverse array of NPV genotypes. (Diagram by Betty Bigson.)

fascinating aspects of NPV, especially *Pf*NPV, biology, which have direct relevance to the use and risk assessment of biopesticides. A great deal of research effort has been made to improve the pathogenicity and efficacy of biopesticides. Our results suggest that genotypic variation should be retained in biopesticides, for four reasons. First, the relative fitness of different NPV genotypes is closely linked to the local ecology of the pest species. Unless the ecology of the focal pest population is well defined, and the best NPV genotype for that ecology available, diverse biopesticide sprays will be most effective since they ensure the inclusion of locally 'adapted' NPV genotypes. Second, if mutualistic interactions are common in NPV coinfections, genotypic diversity in a biopesticide will increase pathogenicity and yield. Third, diverse biopesticides may persist longer due to differences in the component genotypes' fitness in different climates, season lengths and host densities. Fourth, on a longer timescale, genotypic diversity will increase a biopesticide population's ability to adapt to more extreme environmental changes. All these benefits of genotypic diversity should, however, be weighed against the risks posed to non-target and potentially endangered alternative host species.

As well as answering several questions concerning genotypic variation in *Pf*NPV populations, we have highlighted several future research questions for this specific system. The mechanisms driving mutualistic interactions between NPV genotypes in mixed infections remain unclear: the nature of interactions between other co-infecting genotypes should be studied and molecular analysis tools used to describe the growth rates of competing virus genotypes within hosts. Tradeoffs between quantitative traits can be more accurately analysed using bioassays of several more *Pf*NPV genotypes. Field experiments should be performed to describe the transmission parameters and degradation rates of different *Pf*NPV genotypes. Perhaps most important will be the comparison of further wildtype *Pf*NPV samples collected from *P. flammea* populations growing in different food plant populations, and from high and low density *P. flammea* populations. Such studies will be possible if *P. flammea* regains its pest status in Scottish and European pine forests.

Several of the hypotheses for the maintenance of genotypic variation presented here also apply to other microparasite species. The lack of empirical evidence for some of the hypotheses highlights important gaps in our understanding of parasite evolution and ecology. Throughout this chapter we have highlighted the fact that ecological and evolutionary mechanisms affecting parasites have analogies in higher, and often better studied, organisms. This is particularly the case for concepts of differential selection and dispersal, and for competition for limited resources in mixed genotype populations. We have demonstrated that differential selection may be an important determinant of NPV population genetics, and have provided a new empirical example of synergistic interactions between coinfecting parasites. These results, and the usefulness of NPVs and other microparasites for the testing of ecological and evolutionary theory, should promote further research into parasite genetic variation. Most importantly we should perform comparative studies of genotypically diverse and depauperate parasite populations, to determine the main causes of this difference (Ennos 1992). Such comparisons will be easiest in the best-studied host–parasite systems such as malaria, snail and bumblebee trypanosomes and the nucleopolyhedrovirus infecting *Panolis flammea*.

Acknowledgements

The authors greatly appreciated the practical assistance of Penny Luntz, Ben Woodcock, Nicola Goodship, Rachel King, Mo Docherty, Imogen Pearce, Maddy Thurlow and Richard Smith, all at CEH Banchory, and David Barbour of Aberlour. This chapter was greatly improved by the wise criticisms of Mike Boots, Greg Dwyer, Trevor Williams, Primitivo Caballero and Mike Begon. This work was funded by the Natural Environment Research Council's thematic programme EDGE (Ecological Dynamics and Genes).

References

Allan, J.D. (1984) Life history variation in a freshwater copepod: evidence from population crosses. *Evolution* **38**, 280–291.

Allander, K. & Schmid-Hempel, P. (2000) Immune defence reaction in bumble-bee workers after a previous challenge and parasitic coinfection. *Functional Ecology* **14**, 711–717.

Anderson, R.M. & May, R.M. (1981) The

population dynamics of microparasites and their invertebrate hosts. *Philosophical Transactions of the Royal Society Series B* **291**,451–524.

Anderson, T.J.C., Haubold, B., Williams, J.T. *et al.* (2000) Microsatellite markers reveal a spectrum of population structures in the malaria parasite *Plasmodium falciparum. Molecular Biology and Evolution* **17**, 1467–1482.

Barral, V., Morand, S., Pointier, J.P. & Théron, A. (1996) Distribution of schistosome genetic diversity within naturally infected *Rattus rattus* detected by RAPD markers. *Parasitology* **113**, 511–517.

Bremermann, H.J. & Pickering, J. (1983) A game-theoretical model of parasite virulence. *Journal of Theoretical Biology* **100**, 411–426.

Brown, S.E., Maruniak, J.E. & Knudson, D.L. (1985) Baculovirus (MNPV) genomic variants: characterization of *Spodoptera exempta* MNPV DNAs and comparison with other *Autographa californica* MNPV DNAs. *Journal of General Virology* **66**, 2431–2441.

Bull, J.J. (1994) Perspective—virulence. *Evolution* **48**, 1423–1437.

Bull, J.J., Molineux, I.J. & Rice, W.R. (1991) Selection of benevolence in a host–parasite system. *Evolution* **45**, 875–882.

Cadman, C.H. (1960) Inhibition of plant virus infection by tannins. In: *Phenolics in Plants in Health and Disease* (ed. J. B. Pridham), pp. 101–104. Pergamon, New York.

Carruthers, W.R., Cory, J.S. & Entwistle, P.F. (1988) Recovery of pine beauty moth (*Panolis flammea*) nuclear polyhedrosis-virus from pine foliage. *Journal of Invertebrate Pathology* **52**, 27–32.

Chao, L., Hanley, K.A., Burch, C.L., Dahlberg, C. & Turner, P.E. (2000) Kin selection and parasite evolution: higher and lower virulence with hard and soft selection. *Quarterly Review of Biology* **75**, 261–275.

Conway, D.J. & McBride, J.S. 1991 Population-genetics of *Plasmodium falciparum* within a malaria hyperendemic area. *Parasitology* **103**, 7–16.

Corsaro, B.G., Gijzen, M., Wang, P. & Granados, R.R. (1993) Baculovirus enhancing proteins as determinants of viral pathogenesis. In: *Parasites and Pathogens of Insects*, Vol. 2: *Pathogens* (eds N. E. Beckage, S. N. Thompson & B. A. Federici), pp. 127–145. Academic Press, San Diego, CA.

Cory, J.S., Hails, R.S. & Sait, S.M. (1997) Baculovirus ecology. In: *The Baculoviruses* (ed. L. K. Miller), pp. 301–339. Plenum Press, New York.

Ennos, R. (1992) Ecological genetics of parasitism. In: *Genes in Ecology* (eds R. J. Berry, T. J. Crawford & G. M. Hewitt), pp. 255–279. Blackwell Scientific Publications, Oxford.

Entwistle, P.F., Forkner, A.C., Green, B.M. & Cory, J.S. (1993) Avian dispersal of nuclear polyhedrosis viruses after induced epizootics in the pine beauty moth, *Panolis flammea* [Lepidoptera, Noctuidae]. *Biological Control* **3**, 61–69.

Evans, H.F., Stoakley, J.T., Leather, S.R. & Watt, A.D. (1991) Development of an integrated approach to control of pine beauty moth in Scotland. *Forest Ecology and Management* **39**,19–28.

Ewald, P.W. (1995) The evolution of virulence—a unifying link between parasitology and ecology. *Journal of Parasitology* **81**, 659–669.

Fisher, M.C. & Viney, M.E. (1998) The population genetic structure of the facultatively sexual parasitic nematode *Strongyloides ratti* in wild rats. *Proceedings of the Royal Society of London Series B* **265**, 703–709.

Frank, S.A. (2000) Within-host spatial dynamics of viruses and defective interfering particles. *Journal of Theoretical Biology* **206**, 279–290.

Godfray, H.C. J., O'Reilly, D.R. & Briggs, C.J. (1997) A model of nucleopolyhedrovirus (NPV) population genetics applied to co-occlusion and the spread of the few polyhedra (FP) phenotype. *Proceedings of the Royal Society of London Series B.* **264**, 315–322.

Gupta, S. & Anderson, R.M. (1999) Population structure of pathogens: the role of immune selection. *Parasitology Today* **15**,497–501.

Hernández-Crespo, P., Sait, S.M., Hails, R.S. & Cory, J.S. (2001) Behaviour of a recombinant baculovirus in Lepidopteran hosts of varying susceptibility. *Applied and Environmental Microbiology* **67**, 1140–1146.

Hicks, B.J. (2001) *The history, ecology and potential control of the pine beauty moth,* Panolis flammea *(Lepidoptera: Noctuidae) in Scotland.* PhD thesis, University of Aberdeen, UK.

Hitchman, R.B. (2002) *Pathogen variability and dynamics in insect populations.* PhD thesis, Brookes University, Oxford.

Hodgson, D.J., Vanbergen, A.J., Hartley, S.E., Hails, R.S. & Cory, J.S. (2002) Differential selection of baculovirus genotypes mediated by different species of host food plant. *Ecology Letters* **5**, (4), in press.

Hodgson, D.J., Vanbergen, A.J., Watt, A.D., Hails, R.S. & Cory, J.S. (2001) Phenotypic variation between naturally coexisting genotypes of a lepidopteran baculovirus. *Evolutionary Ecology Research* **3**, 687–701.

Holland, J. (1986) Defective viral genomes. In: *Fundamental Virology* (eds B. Fields & D. Knipe), pp. 77–99. Raven Press, New York.

Hoover, K., Stout, M.J., Alaniz, S.A., Hammock, B.D. & Duffey, S.S. (1998) Influence of induced plant defenses in cotton and tomato on the efficacy of baculoviruses on noctuid larvae. *Journal of Chemical Ecology* **24**, 253–271.

Hoover, K., Washburn, J.O. & Volkman, L.E. (2000) Midgut-based resistance of *Heliothis virescens* to baculovirus infection mediated by phytochemicals in cotton. *Journal of Insect Physiology* **46**, 999–1007.

Imhoof, B. & Schmid-Hempel, P. (1998) Single-clone and mixed-clone infections versus host environment in *Crithidia bombi* infecting bumblebees. *Parasitology* **117**, 331–336.

Keating, S.T., Yendol, W.G. & Schultz, J.C. (1988) Relationship between susceptibility of gypsy moth larvae (Lepidoptera: Lymantriidae) to a baculovirus and host plant foliage constituents. *Environmental Entomology* **17**, 952–958.

Keating, S.T., McCarthy, W.J. & Yendol, W.G. (1989) Gypsy moth (*Lymantria dispar*) larval susceptibility to a baculovirus affected by selected nutrients, hydrogen ions (pH), and plant allelochemicals in artificial diets. *Journal of Invertebrate Pathology* **54**, 165–174.

King, L.A. & Possee, R.D. (1992) *The Baculovirus Expression System.* Chapman & Hall, London.

Kool, M., Ahrens, C.H., Vlak, J.M. & Rohrmann, G.F. (1995). Replication of baculovirus DNA. *Journal of General Virology* **76**, 2103–2118.

Lafferty, K.J. & Crichton, R. (1973) Immune responses of invertebrates. In: *Viruses and Invertebrates* (ed. A. J. Gibbs), pp. 301–320. North Holland Publishing Co., Amsterdam.

Laitinen, A.M., Otvos, I.S. & Levin, D.B. (1996) Genotypic variation among wild isolates of Douglas fir tussock moth (Lepidoptera: Lymantriidae) nuclear polyhedrosis virus. *Journal of Economic Entomology* **89**, 640–647.

Leather, S.R. (1990) Life-span and ovarian dynamics of the pine beauty moth, *Panolis flammea* (D&S)—the effect of low-temperatures after adult emergence on reproductive success. *Physiological Entomology* **15**, 347–353.

Leather, S.R. (1993) Influence of site factor modification on the population development of the pine beauty moth (*Panolis flammea*) in a Scottish lodgepole pine (*Pinus contorta*) plantation. *Forest Ecology and Management* **59**, 207–223.

Levene, H. (1953) Genetic equilibrium when more than one ecological niche is available. *American Naturalist* **87**, 331–333.

Levin, B.R. & Bull, J.J. (1994) Short-sighted evolution and the virulence of pathogenic microorganisms. *Trends in Microbiology* **2**, 76–81.

Levin, S. & Pimentel, D. (1981) Selection of intermediate rates of increase in parasite–host systems. *American Naturalist* **117**, 308–315.

Lindstrom L., Alatalo, R.V. & Mappes, J. (1997) Imperfect Batesian mimicry—the effects of the frequency and the distastefulness of the model. *Proceedings of the Royal Society of London Series B* **264**, 149–153.

Lipsitch, M. & Moxon, E.R. (1997) Virulence and transmissibility of pathogens: what is the relationship? *Trends in Microbiology* **5**, 31–37.

Lively, C.M. (1999) Migration, virulence, and the geographic mosaic of adaptation by parasites. *American Naturalist* **153**, S34-S47.

Lively, C.M. & Dybdahl, M.F. (2000) Parasite adaptation to locally common host genotypes. *Nature* **405**, 679–681.

Mackinnon, M.J. & Read, A.F. (1999) Genetic relationships between parasite virulence and transmission in the rodent malaria *Plasmodium chabaudi. Evolution* **53**, 689–703.

MacLeod, A., Turner, C.M.R. & Tait, A. (1999) A high level of mixed *Trypanosoma brucei* infections in tsetse flies detected by three hypervariable minisatellites. *Molecular and Biochemical Parasitology* **102**, 237–248.

May, R.M. & Nowak, M.A. (1995) Coinfection and the evolution of parasite virulence. *Proceedings of the Royal Society of London Series B* **261**, 209–215.

Maynard Smith, J. (1998) *Evolutionary Genetics*, 2nd edn. Oxford University Press, Oxford.

Meijer G. & Leuchtmann, A. (1999) Multistrain infections of the grass *Brachypodium sylvaticum* by its fungal endophyte *Epichloe sylvatica*. *New Phytologist* **141**, 355–368.

Miller, L.K. (1997) *The Baculoviruses*. Plenum Press, New York.

Muñoz, D., Castillejo, J.I. & Caballero, P. (1998) Naturally occurring deletion mutants are parasitic genotypes in a wild-type nucleopolyhedrovirus population of *Spodoptera exigua*. *Applied and Environmental Microbiology* **64**, 4372–4377.

Muñoz, D. & Caballero, P. (2000) Persistence and effects of parasitic genotypes in a mixed population of the *Spodoptera exigua* nucleopolyhedrovirus. *Biological Control* **19**, 259–264.

Muñoz, D., de Escudero, I.R. & Caballero, P. (2000) Phenotypic characteristics and relative proportions of three genotypic variants isolated from a nucleopolyhedrovirus of *Spodoptera exigua*. *Entomologia Experimentalis et Applicata* **97**, 275–282.

Nakamura, T., Konishi, T., Kawaguchi, H. & Imose, J. (1992) Estimation of relative fecundity in *Eimeria tenella* strains by a mixed infection method. *Parasitology* **104**, 11–17.

Olofsson, E. (1988) Dispersal of the nuclear polyhedrosis virus of *Neodiprion sertifer* from soil to pine foliage with dust. *Entomologia Experimentalis et Applicata* **46**, 181–186.

Paul, R.E.L, Brockman, A., Price, R.N. *et al.* (1999) Genetic analysis of *Plasmodium falciparum* infections on the north-western border of Thailand. *Transactions of the Royal Society of Tropical Medicine and Hygiene* **93**, 587–593.

Paul, R.K. (1997) *Evolution and interaction of insect pathogens*. Unpublished PhD Thesis, University of Reading, UK.

Pfennig, D.W., Harcombe, W.R. & Pfennig, K.S. (2001) Frequency-dependent Batesian mimicry—predators avoid look-alikes of venomous snakes only when the real thing is around. *Nature* **410**, 323–323.

Possee, R.D. & Kelly, D.C. (1988) Physical maps and comparative DNA hybridization of *Mamestra brassicae* and *Panolis flammea* nuclear polyhedrosis-virus genomes. *Journal of General Virology* **69**, 1285–1298.

Read, A.F. (1994) The evolution of virulence. *Trends in Microbiology* **2**, 73–76.

Ridley, M. (1993) *The Red Queen*. Viking, London.

Rodgers, P.B. (1993) Potential of biopesticides in agriculture. *Pesticide Science* **39**, 117–129.

Roelvink, P.W., Corsaro, B.G. & Granados, R.R. (1995) Characterization of the *Helicoverpa armigera* and *Pseudaletia unipuncta* granulovirus enhancing genes. *Journal of General Virology*. **76**, 2693–2705.

Santos, E.A., Niel, C., Vianna, C.O.A., De Sa, C.A.M. & Gomes, S.A. (1998) Genomic variations of hepatitis B virus strains in patients chronically coinfected with hepatitis B and human immunodeficiency viruses. *Acta Virologica* **42**, 27–33.

Sinervo, B. & Lively, C.M. (1996) The rock–paper–scissors game and the evolution of alternative male strategies. *Nature* **380**, 240–243.

Smith, H.G., Kallander, H. & Nilsson, J.-A. (1989) The trade-off between offspring number and quality in the great tit *Parus major*. *Journal of Animal Ecology* **58**, 383–402.

Stearns, S.C. 1992 *The Evolution of Life Histories*. Oxford University Press, Oxford.

Stiles, B. & Himmerich, S. (1998) *Autographica californica* NPV isolates: restriction endonuclease analysis and comparative biological activity. *Journal of Invertebrate Pathology* **72**, 174–177.

Taylor, L.H., Walliker, D. & Read, A.F. (1997) Mixed-genotype infections of the rodent malaria *Plasmodium chabaudi* are more infectious to mosquitoes than single-genotype infections. *Parasitology* **115**, 121–132.

Taylor, L.H., MacKinnon, M.J. & Read, A.F. (1998) Virulence of mixed-clone and single-clone infections of the rodent malaria *Plasmodium chabaudi*. *Evolution* **52**, 583–591.

Trudeau, D., Washburn, J.O. & Volkman, L.E. (2001) Central role of hemocytes in *Autographa californica* M nucleopolyhedrovirus pathogenesis in *Heliothis virescens* and *Helicoverpa zea*. *Journal of Virology* **75**, 996–1003.

Tucker, R.A., Johnson, P.R., Reeves, W.C. & Icenogle, J.P. (1993) Using the polymerase chain reaction to genotype human papillomavirus DNAs in samples containing multiple HPVs may produce inaccurate results. *Journal of Virological Methods* **43**, 321–334.

Turner, P.E. & Chao, L. (1999) Prisoner's dilemma in an RNA virus. *Nature* **398**, 441–443.

Tuveri, R., Rothschild, C., Pol, S. *et al.* (1997) Hepatitis C virus genotypes in haemophiliacs: kinetics and reappraisal of mixed infections. *Journal of Medical Virology* **51**, 36–41.

Vasconcelos, S.D., Williams, T., Hails, R.S. & Cory, J.S. (1996) Prey selection and baculovirus dissemination by carabid predators of Lepidoptera. *Ecological Entomology* **21**, 98–104.

Wainhouse, D. & Jukes, M.R. (1997) Geographic variation within populations of *Panolis flammea* (Lepidoptera: Noctuidae) in Britain. *Bulletin of Entomological Research* **87**, 95–99.

Watt, A.D. (1989a) The growth and survival of *Panolis flammea* larvae in the absence of predators on Scots pine and lodgepole pine. *Ecological Entomology* **14**, 225–234.

Watt, A.D. (1989b) The chemical-composition of pine foliage in relation to the population-dynamics of the pine beauty moth, *Panolis flammea*, in Scotland. *Oecologia* **78**, 251–258.

Watt, A.D. & Hicks, B.J. (2000) A reappraisal of the population dynamics of the pine beauty moth, *Panolis flammea*, on lodgepole pine, *Pinus contorta*, in Scotland. *Population Ecology* **42**, 225–230.

Weeds, P.L., Beever, R.E. & Long, P.G. (2000) Competition between aggressive and non-aggressive strains of *Botrytis cinerea* (*Botryotinia fuckeliana*) on French bean leaves. *Australasian Plant Pathology* **29**, 200–204.

Widemo, F. (1998) Alternative reproductive strategies in the ruff, *Philomachus pugnax*: a mixed ESS? *Animal Behaviour* **56**, 329–336.

Woods, S.A. & Elkinton, J.S. (1987) Bimodal patterns of mortality from nuclear polyhedrosis-virus in gypsy-moth (*Lymantria dispar*) populations. *Journal of Invertebrate Pathology* **50**, 151–157.

Chapter 16

Host dynamics and pathogen variation in insect–baculovirus interactions

C. M. Griffiths, R. McVean, R. Benmayor, S. M. Sait, D. J. Thompson, R. D. Possee, J. S. Cory and M. Begon*

Introduction

The dynamics exhibited by animal populations are both varied and complex, ranging from stable equilibria to possibly chaotic patterns in abundance, but perhaps the most intriguing are populations that exhibit cyclical dynamics. Cycles in a number of forest insect species are particularly well documented, showing regular periods between peak outbreaks in many cases (Myers 1988). 'Generation-length' cycles in stored product pest insects have also been extensively studied (Knell 1998).

Dynamics are influenced by many intrinsic, biotic factors and extrinsic, environmental factors. The impact of pathogens on the dynamics of their hosts is recognized as potentially significant, but has remained difficult to disentangle from many other important factors in the natural world (Myers 1988). Modelling studies have examined host–pathogen interactions (Anderson & May 1981; Onstad & Carruthers 1990; Anderson 1991; Briggs & Godfray 1995, 1996) and suggest that host dynamics are affected by pathogens and disease in much the same way as predator–prey interactions. Both predator and pathogen attack rates are dependent, for example, on prey density. In the case of pathogens, this may lead to a host density threshold, below which level the pathogen fails to persist (Knell *et al.* 1998a). The host density threshold is particularly relevant to pathogens whose principle route of transmission is horizontal, such as the baculoviruses studied here. Horizontal transmission requires the virus to be ingested by susceptible larvae, which, on becoming diseased, may generate subsequent rounds of transmission. As long as the number of secondary infections arising from a given infection ('R_0'; the basic reproductive rate) is greater than 1, the pathogen is able to persist. Many factors affect the basic reproductive rate, such as host susceptibility, pathogen virulence, decay rate of the pathogen in the environment and pathogen refuges; but the basic reproductive rate is likely to increase with the density of susceptible hosts, insofar as this increases the rate of uptake of, say, contaminated foliage or infected cadavers. Vertical transmission, however, may circumvent this direct relationship between the density of suscepti-

* Correspondence author: jsc@ceh.ac.uk

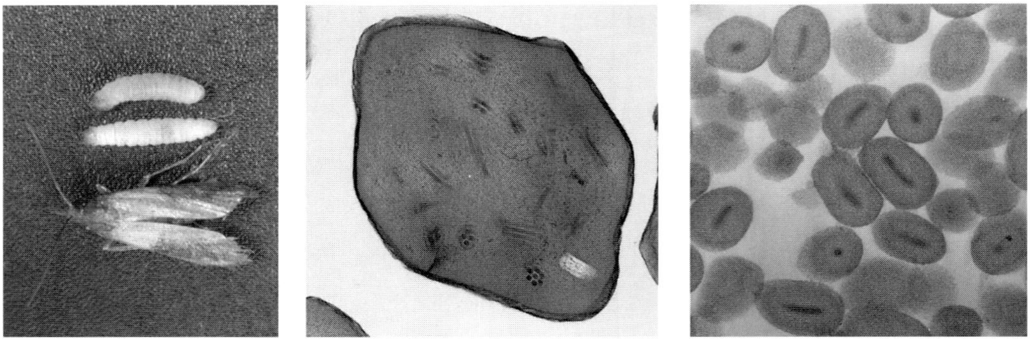

Figure 16.1 (a) 5th instar *Plodia interpunctella* larvae. The upper, healthy individual is translucent and light-brown colour, whilst the lower individual is infected with *Pi*GV and has become opaquely white or pinkish due to the accumulation of virus granules in the tissues and haemolymph. An adult is shown beneath both larvae. (b) Electron micrograph showing occlusion bodies of *Ac*MNPV (often called polyhedra), containing multiple virions. Each enveloped virion possess several nucleocapsids. (c) Electron micrograph showing occlusion bodies (granules) of *Pi*GV, each containing a single, enveloped virion.

bles and secondary infections and enable the pathogen, theoretically, to persist at a host density below which a horizontally transmitted agent would become extinct.

Insect–pathogen systems have provided some of the very few data sets on long-term host–pathogen dynamics that permit validation of theoretical work (Bird & Elgee 1957; Hofmaster 1961; Fleming *et al.* 1986). Many of these data sets come from field populations in which environmental factors, which may affect host dynamics (Myers 1988), cannot be separated from the effects caused by interactions between the host and pathogen alone. Several Lepidoptera associated with stored food products can be reared in controlled laboratory conditions on artificial diet, allowing the vagaries of light cycle, temperature, resource availability and other extraneous influences to be constrained. Insect species associated with stored food products often display short generation cycles, with multiple cycles during one year, in contrast to field populations that generally have only a single generation each year. The development of laboratory systems using stored product pests can provide long-term data sets of insect–pathogen population dynamics over relatively short experimental time frames, with limited environmental variability. The influence of many factors influencing population dynamics (diet, density, adult size and fecundity, disease, predation and parasitoid invasion) have been well studied for a range of insects that naturally feed on stored products (Park 1948; Benson 1973).

Arguably the most well studied model for examining host–pathogen systems uses caged populations of the Indian meal moth, *Plodia interpunctella,* and its baculovirus (Figure 16.1). *Plodia interpunctella* (Hübner) (Lepidoptera; Pyralidae) is a significant pest for a wide range of stored foods, especially cereals, dried fruits and nuts (Madrid & Sinha 1982; Madrid *et al.* 1983), and is one of the commonest species occurring on food imports into the UK. Although originating from warm temperate and subtropical climates, it has become established in cool temperate climates, especially where the temperature is raised artificially. *P. interpunctella* granulovirus (*Pi*GV; Figure 16.1c), the baculovirus associated with this species, was first observed in the 1960s (Arnott & Smith 1968), its high virulence and specificity making it a strong candidate for development as a biopesticide (Hunter *et al.* 1973a, 1977, 1979; Kinsinger & McGaughey 1976). Short-term experiments with *P. interpunctella* and *Pi*GV have provided much information on insect–pathogen population dynamics (Sait *et al.* 1994a, b, 1996, 1998a,b; Boots 1998; Knell *et al.* 1998a; Boots & Norman 2000). Longer-term studies have shown that the presence of the pathogen reduces the density of the host-insect population, but never causes it to near extinction (Sait *et al.* 1994c; Knell *et al.* 1998b). Essentially, while there is a positive effect of increasing host density on pathogen density, subsequent increased pathogen density does not significantly reduce host density. Further studies of this system have focused on the influence of natural enemies, such as parasitoids and baculoviruses, either singly or in combination, on *P. interpunctella*. In contrast to *Pi*GV, for example, the parasitoid *Venturia canescens* engages in a tightly coupled predator–prey interaction such that host generation cycles are driven largely by the parasitoid (Bjørnstad *et al.* 2001). Although a well-studied system, there are few precise data on the dynamics of the virus itself in coexisting populations of *P. interpunctella* and *Pi*GV. Moreover, in this system as elsewhere, very little is known regarding the influence of evolutionary change (genotypic and/or phenotypic), especially in the pathogen, on the ecological dynamics of the host–pathogen interaction; here we focus on these issues.

The Baculoviridae

Detailed descriptions of a 'wilting' or 'jaundice' disease of silkworms have been found within reports dating from the 16th and 17th centuries, but diseases caused by baculoviruses have (unknowingly) been witnessed by man for over 2000 years (Benz 1986). Several virus families infect invertebrates, including members of the Iridoviridae, Poxviridae and Reoviridae, but only the Baculoviridae are restricted solely to arthropod hosts (Huber 1990). The majority of baculovirus isolates are from lepidopteran insects (Matthews 1992), although isolates have been recorded from the Hymenoptera, Diptera, Coleoptera, Neuroptera, Trichoptera and Crustacea (Couch 1974; Granados & Federici 1986; Anderson & Prior 1992; Chang *et al.* 1993). Baculoviruses have been used successfully to control pests of agricultural crops, forests and plantations (Payne 1982, 1988; Black *et al.* 1997), and developed for high-level expression of heterologous proteins in insect cells and larvae (for reviews see Bishop 1992; Griffiths 1994; Jarvis 1997).

Classification and structure

Members of the Baculoviridae display a biphasic replication cycle and are divided into two genera, the nucleopolyhedroviruses (NPVs) and the granuloviruses (GVs), based on structural criteria (Volkman *et al.* 1995). Virus particles (virions) comprise enveloped nucleocapsids (rod-shaped protein structures) enclosing a single copy of the virus genome; a closed-circular molecule of supercoiled, double-stranded deoxyribounucleic acid (dsDNA) of between 80 and 200 kilobasepairs (kbp). At late stages of replication (from 24 hours post infection), virions are enclosed within a protective protein coat, the occlusion body (OB). The OBs of NPVs (see Figure 16.1b) are assembled in the nucleus and are large, polyhedral structures (0.15–15 μm) that occlude many virions, and hence many genome molecules. Each virion envelope may contain a single (SNPV), or multiple (MNPV, usually 4–8) nucleocapsids. Granulovirus OBs (see Figure 16.1c) are smaller (0.3 μm × 0.5 μm), generally ovocylindrical, and contain only one enveloped virion (i.e. a single genome). Baculovirus OBs are primarily composed of a single viral protein, called polyhedrin in NPVs or granulin in GVs, that is highly conserved within the family.

Infection biology

Occlusion bodies protect baculovirus particles from UV damage and desiccation, facilitating virus persistence in the environment. Horizontal transfer to new hosts is primarily via ingestion of OBs released onto foliage from infected cadavers. Virus yields can exceed 10^{10} OBs per larva, hence insects feeding in the vicinity of dead larvae may ingest massive doses of virus. Insects can also ingest OBs through cannibalism of infected larvae, or released from hosts by defecation or regurgitation (Jaques 1962; Vasconcelos 1996), but the degree of infection within natural populations due to these routes is uncertain. Vertical transmission of baculoviruses has been recorded (Melamed-Madjar & Raccah 1979; Kukan 1999) and may be important for virus persistence in insect populations at low density or that are highly mobile. Recent studies have shown vertical transfer of *Pi*GV in *P. interpunctella* from adults of either sex to their offspring (Burden *et al.* 2002). *Pi*GV was shown to persist within apparently healthy larvae, pupae and adult stages, following virus challenge, and viral genomic DNA and mRNA transcripts were also detected in first-generation offspring (up to 5th instar) of these adults. Interestingly, *Pi*GV was transmitted vertically from adults to progeny as a sublethal infection, never as an overt infection, and remained sublethal within larvae of the new generation (Burden *et al.* 2002). The concept of sublethal baculovirus infections is not new. Studies with a culture of *Mamestra brassicae* reared in the laboratory for over 20 years demonstrated a latent baculovirus infection that could be triggered by superinfection with heterologous baculoviruses (Hughes *et al.* 1993, 1997). The origin of the sublethal infection in the *M. brassicae* population is unknown; the insects had been reared in culture for many years prior to detection of virus, and may have been acquired after the insects were collected in the field. At the time of writing, the insects still carry the latent virus infection (J. Burden, personal communication), which is clearly no major hindrance to insect reproduction. Sublethal effects within *P. interpunctella* surviving *Pi*GV challenge are documented and include significant reductions in fecundity and altered host development time (Sait *et al.* 1994b). These effects may be due to the costs incurred by fighting off a lethal virus infection, such as through virus clearance from the mid-gut

(Keddie *et al.* 1989; Englehard & Volkman 1995), or to a direct consequence of the continued presence of virus, possibly as a persistent asymptomatic infection. Whatever the mechanics of sublethal infections, their effects are likely to have a major impact on the population dynamics of the insect host.

Population dynamics

P. interpunctella can be maintained on artificial diet for multiple generations, and has been used extensively to examine a range of effects (population density, resource quality, disease, predation and parasitism) on population dynamics. Long-term studies of the *P. interpunctella*–granulovirus system where resource quality is varied (compare Sait *et al.* 1994c with Knell *et al.* 1998a), have lent support to the idea that when resource is limiting, competition for food is the predominant influence on dynamics, affecting larval mortality and development rates, adult fecundity and inducing cannibalism at high densities (Benson 1973; Podoler 1974). In continuous culture on good quality food, either alone or with *Pi*GV, *P. interpunctella* adults exhibit discrete generation cycling (Sait *et al.* 1994b,c; Knell 1998). Statistical analysis of population time series (Bjornstad *et al.* 1998) and theoretical models (Briggs *et al.* 2000) suggest that the observed cycles are due to asymmetrical competition between larvae. Older larvae cannibalize and outcompete younger cohorts, before maturing to produce a peak in the number of adult moths; few younger larvae remain, the numbers of adults then crashes and the process is repeated. In contrast to cyclic host population fluctuations, the pathogen exhibits a steady increase in prevalence, with no evidence of coupled dynamics (Bjornstad *et al.* 2001), perhaps surprising given that *Pi*GV appears to be exclusive in its infection of *P. interpunctella*. In the presence of virus however, the length of host cycles is extended and mean adult densities are reduced, as the virus modulates host development time and reproduction (Sait *et al.* 1994c; Bjornstad *et al.* 1998, 2001). Given the anticipated differences in underlying host dynamics, driven by environmental contrasts, we predict that virus associated with host populations will undergo corresponding changes in both genotype and phenotype, to maximize transmission rates and persistence. Here we discuss a range of studies that have looked at this and other questions regarding host and virus dynamics under varying environmental regimes.

In our studies, direct comparisons of equivalent long-term *P. interpunctella* populations on contrasting resource regimes confirm that populations on higher quality food are significantly larger than those on poor quality diet, when measured by mean adult abundance (Figure 16.2, thick lines) or mean larval numbers (Table 16.1). Autocorrelation function (ACF) analysis of weekly counts of adults (Figure 16.3) demonstrates cyclic dynamics on good, but not on poor quality food. The absence of cyclic dynamics on poor food probably reflects a lack of significant competition for resource, due to the lower numbers of *P. interpunctella* within these populations (Sait *et al.* 1994c). Adult body size is also consistently greater when populations are maintained on higher quality food (Table 16.1). Reproductive potential is positively correlated with body size (Gage & Cook 1994; Gage 1995), and

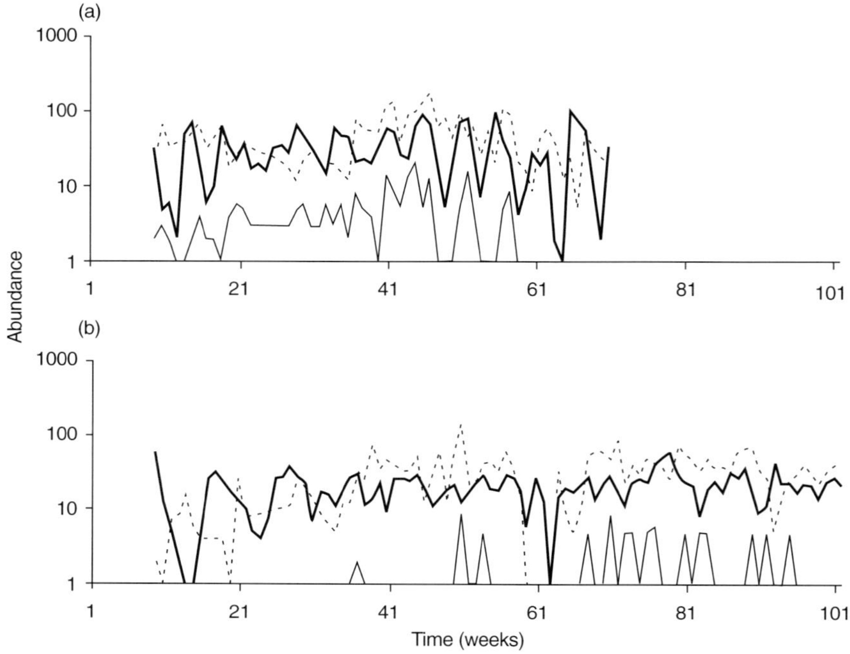

Figure 16.2 Adult and larval abundance counted for *Pi*GV-infected *P. interpunctella* populations, maintained on good (a) and poor (b) quality diet. Populations were initiated with 15 female and 15 male healthy 5th instar larvae and maintained at 28 ± 2 °C, 65 ± 5% RH with a 12 h : 12 h light : dark cycle. An equal amount of virus suspension (obtained by macerating infected 5th instar larvae into distilled water) was pipetted onto the food of each cage, to initiate infection, and the populations were allowed to reproduce undisturbed for 9 weeks. From week 10 onwards, the oldest section of food was replaced each week, a quarter being mixed with the fresh food to encourage the persistence of virus. Dead adults in each cage were removed and counted each week (bold lines), as a measure of population size for the previous week. Weekly counts were also made of the number of infected (single line) and uninfected (dashed line) larvae present in the oldest food section, as well as in subsamples of the newer food sections. All live larvae, and dead, infected larvae of instars III–IV were returned to the cages with fresh food each week.

the increased adult size on good food probably contributes to the dynamics of population cycling by increased fecundity and host density. The prevalence of *Pi*GV infection (see Figure 16.2, single line) is greater in host populations maintained on good quality diet, and the virus appears to be more readily sustained than in hosts maintained on a poor diet regime, where virus appears as disparate peaks. The proportion of infected larvae (see Figure 16.2, single line) does increase in later generations in populations maintained for long periods on poor food, whereas on good food, the virus can become cyclic or may even fall to low levels. The reasons for this

Table 16.1 Summary of the differences between populations of *P. interpunctella* and *Pi*GV, maintained on good and poor quality food.

Weekly parameters measured	Poor food			Good food			Difference	
	Replicate 1	Replicate 2	Replicate 3	Replicate 1	Replicate 2	Replicate 3	Statistic	*P*
Dead adults (no.)								
Mean ± SE	18.5 ± 1.3	19.1 ± 1.4	19.2 ± 1.1	32.6 ± 5.6	33.5 ± 3.2	45.3 ± 10.5	$F = 7.8$	<0.05
Range	0–66	0–68	0–58	0–132	0–101	0–228	—	—
Larvae (no.)								
Mean ± SE	35.4 ± 3.3	33.1 ± 3.2	28.7 ± 2.5	63.2 ± 7.7	46.3 ± 4.5	52.9 ± 7.0	$F = 31.6$	<0.001
Range	0–156	0–144	0–140	0–152	4–176	0–156	—	—
Virus-infected larvae (no.)								
Mean ± SE	0.2 ± 0.1	0.3 ± 0.1	0.7 ± 0.2	0.9 ± 0.2	3.0 ± 0.5	1.9 ± 0.8	$F = 9.8$	<0.05
Range	0–8	0–8	0–8	0–4	0–20	0–20	—	—
Larvae infected with virus (%)								
Mean ± SE	0.6 ± 0.1	0.9 ± 0.2	2.4 ± 0.3	1.3 ± 0.2	6.0 ± 0.4	3.5 ± 0.5	$\chi^2 = 122$	<0.001
Range	0–18.2	0–14.3	0–16.7	0–7.4	0–26.7	0–33.3	—	—
Length (mm) of rear right femur (mean ± SE)	1.32 ± 0.04	1.36 ± 0.04	1.32 ± 0.03	1.55 ± 0.03	1.49 ± 0.02	1.51 ± 0.03	$F = 69.1$	<0.001

All populations were initiated as described (see legend for Figure 16.2). Numbers of adult moths, infected and uninfected larvae were all recorded weekly. Femur length was used as a measure of adult size, generally accepted as an indicator of reproductive potential.

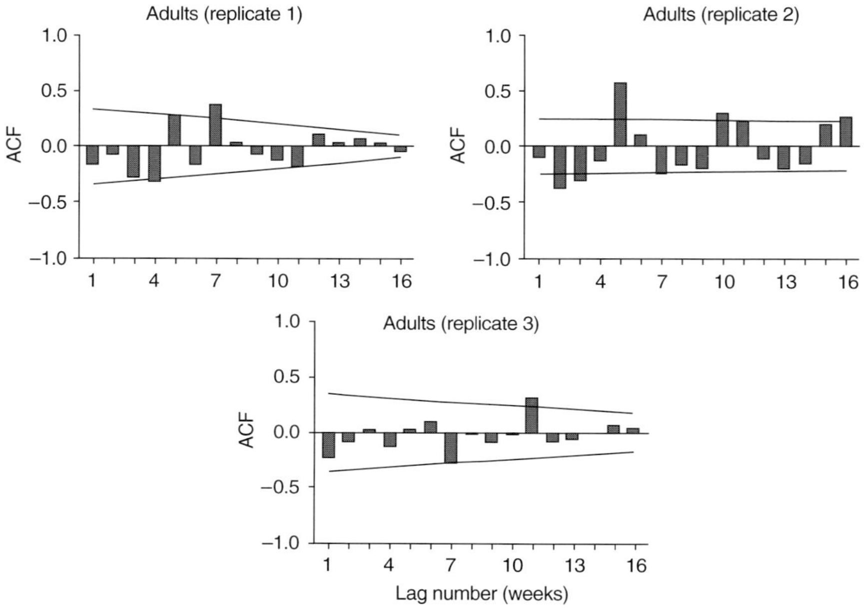

Figure 16.3 Autocorrelation function analyses (ACFs) for populations reared on good quality food. Weekly counts of dead adults from three replicate populations demonstrated significantly cyclic behaviour, with cycle lengths of either 7 weeks (replicate 1), 5 weeks (replicate 2) or 11 weeks (replicate 3). Significant ($P < 0.05$) correlation of the lag period (x-axis) is represented where the solid bars cross the horizontal indicator lines, and may be positive or negative (y-axis).

disparity are not known. Nonetheless, the persistence of *Pi*GV even in the low-density populations is consistent with the calculations of Knell *et al.* (1998a), which suggest the threshold density for *Pi*GV is very low but the basic reproductive rate increases less than linearly with host density.

Further studies have shown resistance to *Pi*GV infection varies when larval hosts are maintained on contrasting nutritional regimes. Virus-naive *P. interpunctella* maintained on good quality diet show a higher susceptibility to *Pi*GV than larvae maintained on poor food, as evidenced by a low LD_{50} (Table 16.2), compared to larvae on a poor diet. However, larvae on good food survive *Pi*GV infections for longer (greater time to death, LT_{50}) than their counterparts on poor quality food. The reasons for this are unclear, but a reduced relative fitness of insects reared on a nutritionally poor diet combined with a viral challenge may lead to more rapid deaths. An increased speed of kill ultimately reduces virus productivity, which in turn reduces the transmission rate of *Pi*GV and its potential for long-term persistence in the environment. This interaction between the host and its environment, then, may be a driving force behind any adaptive evolution on the part of the virus that results in reduced virulence of the virus; i.e. a greater tendency towards producing sublethal or

Table 16.2 Phenotype (LD_{50} and LT_{50}) of *Pi*GV in larvae maintained on contrasting food regimes.

	Good food	Poor food
Capsules/larva		
LD_{50} (log)	3.163	5.183
95% confidence limits		
Upper	3.298	6.570
Lower	3.032	4.587
Days		
LT_{50}	21.2	15.4
95% confidence limits		
Upper	24.3	17.9
Lower	18.2	13.0

Virus infectivity was assayed using 3rd instar larvae that had been continuously maintained on either good or poor quality diet (virus-naive populations) for over ten generations. Larvae were droplet fed with virus suspension, within the dose range 1.1×10^2 to 1.1×10^4 virus capsules per larva. Infectivity of the viruses is expressed as LD_{50} values (the number of virus particles required to produce 50% lethality of infected larvae) and speed of kill as LT_{50} (timeframe between the point of infection and death of 50% of infected larvae).

latent infections. These findings contrast with the work of Boots and Begon (1994), who found food quality to have no effect on the susceptibility of *P. interpunctella* larvae to a granulovirus. The discrepancy almost certainly results from the nutrient differences between good and poor quality foods used in each case. Boots and Begon used a cereal-based food for the high quality diet, the nutrient content of which was diluted by the addition of methylcellulose, an indigestible bulking agent, to create the poor quality diet. As a result, both diet regimes had the same nutrient availability, but nutrient content was more diluted within the poor food. The contrasting (good and poor quality) diets used in the study described here contained different nutrients, and therefore differ qualitatively as well as quantitatively. It is possible that this lack of nutrients, rather than the low abundance, affects susceptibility to the virus (Benz 1987).

Can resource quality influence the virus?

Several laboratory studies have shown diet as a significant factor influencing insect–baculovirus dynamics (Forschler *et al.* 1992; Duffey *et al.* 1995; Cory *et al.* 1997; Boots 2000). Resource quality in an insect–pathogen system may not only alter both host and pathogen population dynamics but also influence host and pathogen population structures and virus-induced mortality. Insects reared on contrasting regimes exhibit physiological differences that may affect their resistance to baculovirus infections, or their ability to respond immunologically to an invading virus. Our work supports the earlier evidence that the dynamics of host populations vary both quantitatively and qualitatively under the influence of resource quality (Sait

Table 16.3 Phenotype of *Pi*GV following association with *P. interpunctella* populations on good and poor quality food.

	Good food		Poor food	
	Initial virus	Virus after 2 years	Initial virus	Virus after 2 years
Capsules/larva				
LD_{50} (log)	3.631	3.221	4.802	5.044
95% confidence limits				
Upper	3.921	3.385	5.557	6.171
Lower	3.422	3.059	4.384	4.520
Days				
Time to first symptoms ± 95% confidence limits	5.4 ± 0.2	5.7 ± 0.2	5.5 ± 0.3	5.5 ± 0.3

3rd instar (virus-naive) larvae were infected with *Pi*GV isolated from the long-term *P. interpunctella* populations maintained on contrasting diet regimes. Pathogenicity (LD_{50}) of viruses was assayed as described in the notes for Table 16.2. *Pi*GV used to initiate the infected populations of *P. interpunctella* described (ancestral virus) was compared with virus extracted from these infected populations after 2 years (host-associated virus).

et al. 1994c; Knell *et al.* 1998a), and we hypothesize that resource may also impress measurable genetic and phenotypic changes upon the pathogen. There is a considerable body of evidence to support the influence of secondary chemicals on baculovirus infections (Duffey *et al.* 1995), but variation in diet quality can influence virus in more subtle ways, for example by altering the population structure of variants, influencing alterations within host dynamics or host feeding behaviour (Boots 2000), or as a direct response to the nutrients themselves.

Viruses that persist in host populations demonstrate changes in their phenotypic characteristics over time. Our experiments have compared the phenotype of *Pi*GV isolates propagated through *P. interpunctella* larvae for multiple generations (on contrasting diet regimes) to that of the parental virus from which these isolates originally derived. *Pi*GV that had persisted within host populations on high quality diet demonstrated a higher infectivity (evidenced by a lower LD_{50}) than the parental virus (Table 16.3), but curiously displayed a slower pathogenesis, taking longer to induce symptoms within infected larvae (lethargy, loss of colour). The reasons for this observation are unclear, but it may again represent an adaptive response to enhance *Pi*GV persistence in the population, given the correlation (mentioned above) between survival time of infected larvae and virus yield. By inducing a slower progression to death, high virus yield could be sustained despite an increased virulence, which is important to avoid the decline of virus levels within host populations over time. By contrast, *Pi*GV isolates in association with host populations on poor quality diet displayed no significant differences in infectivity when compared to the parental virus (Table 16.3). Clearly, the quality of host resource was able to influence the phenotypic development of the virus through multiple (evolutionary) generations.

The effects of second pathogens

In wild populations, insect hosts and pathogens rarely exist in isolation, but are influenced by interactions with other natural enemies. It has been possible to look at the effects of such a multicomponent system using *P. interpunctella*, *Pi*GV and the parasitoid wasp *Venturia canescens*. Heavily infected *P. interpunctella* make poor hosts for *V. canescens* larvae (Begon *et al.* 1997), which require that hosts develop to the fifth instar for their own full development (Harvey *et al.* 1997). Hence the presence of *Pi*GV within a population of *P. interpunctella* affects the population of *V. canescens*, and the wasp conversely influences virus transmission by providing an additional mechanism (surface contamination and spread) for dissemination throughout the habitat (Sait *et al.* 1996). The dynamics of the three-species system is strikingly different to those of two-species interactions (Begon *et al.* 1996; Sait *et al.* 2000); rather than establishing stability, both host and predator cycles lose synchrony and become multigenerational, eventually leading to extinction of the system. Insects in the wild are subject to attack by multiple pathogens, and simultaneous infection of individual larvae with more than one virus has been observed (Tanada 1959; Goto 1990; Hughes *et al.* 1993). Compared to single infections, multiple viral infections may result in lowered (virus inhibition), additive (neutral) or enhanced (synergistic) mortality levels (Hackett *et al.* 2000). Most baculoviruses are highly specific in their choice of target host, infecting only one or two species. The almond moth, *Ephestia cautella* (Lepidoptera; Phycticidae) is a considerable pest of stored products sharing many of the same habitats as *P. interpunctella*. A granulovirus (*Ec*GV) and a nucleopolyhedrovirus (*Ec*NPV) have been isolated from *E. cautella*, and each has been reported infectious to *P. interpunctella* (Hunter & Hoffman 1970; Hunter *et al.* 1973b,c). Our own long-term studies of *P. interpunctella* and *Pi*GV (on high quality diet) have shown that the introduction of *Ec*NPV as a second natural enemy can influence the dynamics of host–virus populations (Figure 16.4). Host adult cycling times (that is, generation times) in populations challenged with both viral pathogens are shorter than populations infected with *Pi*GV only. Mean adult abundance is also higher in populations containing both viruses, compared to hosts exposed to *Pi*GV alone (Benmayor 2002). Introduction of *Ec*NPV into populations that harbour *Pi*GV effectively reverses the influence of *Pi*GV infections (extended generation times and reduced population sizes, compared to uninfected hosts), suggesting that the presence of a second baculovirus interferes with *Pi*GV replication. Although the presence of replicating *Pi*GV is readily detectable in this system, further investigation is still required to confirm the continued presence of *Ec*NPV in long-term *P. interpunctella* populations. At present the full effects of a second natural enemy on *P. interpunctella* and *Pi*GV are not understood, nor is it known if the effects observed on virus–host populations is a general phenomenon, or exclusive to the particular system under study.

Baculovirus genotype variation

Several studies make it clear that baculovirus populations are heterologous, comprising a number of isolates distinguishable by their genomic differences (Smith &

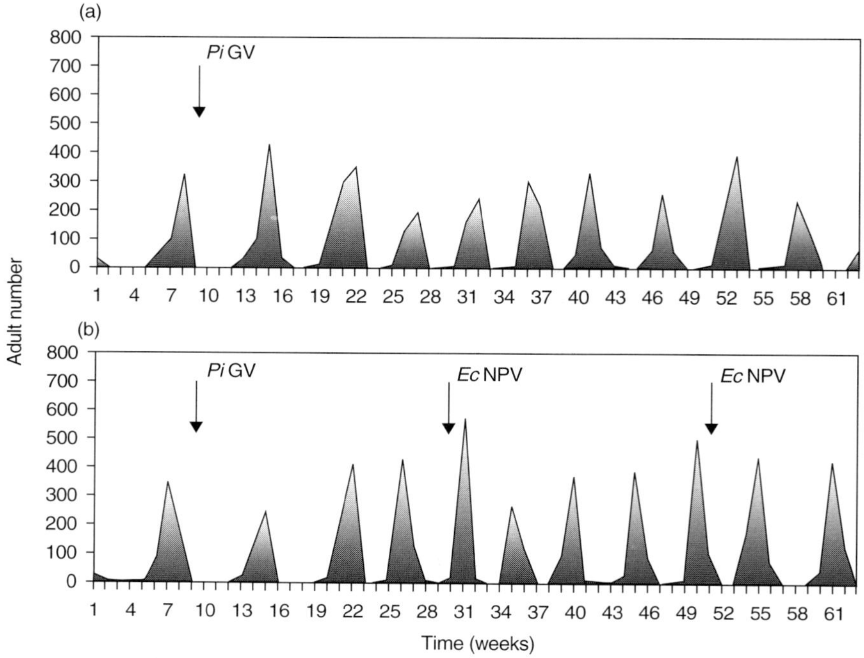

Figure 16.4 Adult abundance in *P. interpunctella* populations infected with *Pi*GV alone (a) or coinfected with *Pi*GV and *Ec*NPV (b). Populations were initiated and maintained as described for Figure 16.2, with the exception that introduction of *Pi*GV (as virus-infected cadavers introduced into the cages) was delayed until host populations had established cyclic dynamics (week 10, arrows). *Ec*NPV was introduced into mixed infection populations (arrows) at week 30 and again at week 52. As for Figure 16.2, dead adults in each cage (three replicates per treatment) were removed and counted each week, as a measure of population size for the previous week.

Summers 1979; Tweeten *et al.* 1980; Crook *et al.* 1985; McIntosh *et al.* 1987; Shapiro *et al.* 1991). Baculovirus genomes are promiscuous elements with regard to mutational change, and the ubiquitous nature of this genotypic variation strongly implies a role in biological function. Early studies *in vitro* showed rapid accumulations of mutant genomes within replicating baculovirus populations (Potter *et al.* 1976; Burand & Summers 1982; Fraser & Hink 1982; Kumar & Miller 1987). Mutations may be associated with insertion into the virus genome of insect (chromosomal) sequences (Fraser *et al.* 1983), or of mobile genetic elements known as transposons (Miller & Miller 1982; Jehle *et al.* 1995, 1997). Deletions of large regions of the virus genome have also been recorded (Carstens 1982; Kool *et al.* 1991; Lee & Krell 1992), and many deletion mutants lack the essential genomic regions encoding virus replication proteins, relying on homogenous helper viruses for effective packaging and propagation. By acting as defective interfering (DI) particles (Huang & Baltimore 1970; Holland 1990), deletion mutants can contribute to the dynamics of baculovirus persistence in cell cultures. Between closely related baculovirus genomes within an insect cell, physical exchange of genetic information (an event known as homologous recombination) occurs at high frequency (above 20%), both *in vitro* (Smith & Summers 1980; Kondo & Maeda 1991; Hajos *et al.* 2000) and *in vivo* (Croizier *et al.* 1988; Croizier & Ribeiro 1992; Munoz *et al.* 1997). The overall contribution of these factors to the genetic variability of baculoviruses has not been extensively studied, but the influence that genetic mutation can have on phenotype and pathogenesis was elegantly illustrated by the work of Kondo and Maeda (1991). Genetic exchange in infected cells, between the genomes of *Ac*MNPV and *Bm*NPV (the baculoviruses infecting the fall armyworm, *Autographa californica* and the silkworm, *Bombyx mori*, respectively), resulted in the generation of new recombinant viruses possessing different (broader) host range capacities than either parent virus. Greater information pertaining to baculovirus variability will contribute to our understanding of the evolution of baculoviruses from both a molecular and ecological viewpoint and may be important in the area of risk assessment, in connection with the use of engineered baculoviruses for use as pest control agents.

Genetic variation within *Pi*GV

Several studies have examined the biochemistry of *Pi*GV, but molecular knowledge of the virus is limited (for reviews see Consigli *et al.* 1983, 1986). With the exception of *Cp*GV, which infects the codling moth, *Cydia pomonella*, suitable laboratory cell lines with which to propagate granuloviruses are not available (Winstanley & Crook 1993). Reports of cell lines able to support *Pi*GV replication (Lynn & Oberlander 1983) have not been substantiated (D. Johnson, personal communication), and the virus is routinely propagated in larvae.

Genetic variation in original *Pi*GV isolates

The genome of *Pi*GV is a circular dsDNA molecule estimated at $73–78 \times 10^6$ molecular weight (Tweeten *et al.* 1977, 1980). Fragmentation using restriction enzymes (REs) that recognize specific DNA sequences gives rise to linear fragments that can be separated and visualized by agarose gel electrophoresis. Several REs were used to examine the *Pi*GV genome, giving rise to distinct fragment profiles (Figure 16.5) each comprising linear genomic DNA fragments at equivalent concentrations (note the relatively even intensity of adjacent bands within individual lanes). Such equimolar fragment patterns are indicative of high genetic homogeneity, since each dsDNA molecule within the population of virus genomes must contain the same number (and location) of recognition sequences for the RE to produce an identical fragment profile.

The source of *Pi*GV used in our laboratory studies has been routinely propagated over many years in controlled populations of *P. interpunctella*, and in comparison to a wild isolate that is subject to a greater variety of host and environmental parameters might be predicted to show little genetic variation. However, it may also be true that *Pi*GV populations simply do not possess great variability (Tweeten *et al.* 1980). The

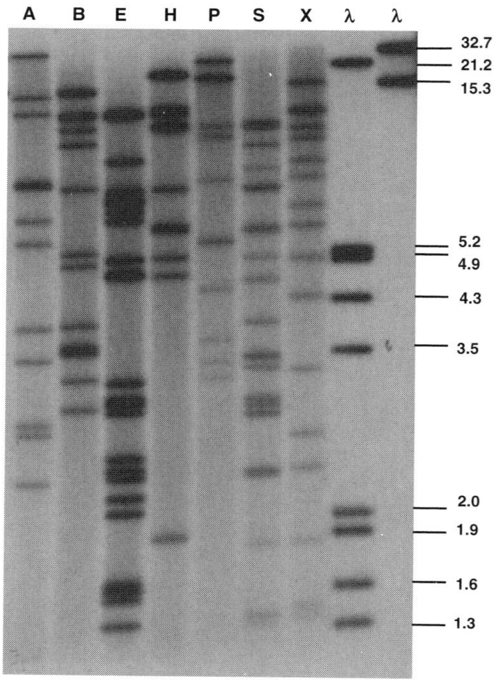

Figure 16.5 Restriction enzyme fragments for *Pi*GV genomic DNA. Virus DNA was extracted from pooled, infected *P. interpunctella* larvae (5th instars) and digested to completion with the enzymes shown. Linear genomic DNA fragments were radioactively end-labelled and separated by agarose gel electrophoresis. Once dried onto a filter paper support the DNA fragments were visualized with X-Omat film. *Pi*GV DNA was digested with *Asp* 718 (lane A), *Bam* HI (lane B), *Eco* RI (lane E), *Hin*d III (lane H), *Pst* I (lane P), *Sal* I (lane S) or *Xho* I (lane X). Molecular weight markers (lane λ) were used to provide fragment size estimates, in kb.

extent to which wild baculovirus isolates can display genetic heterogeneity is demonstrated in a related study, involving the baculovirus that infects the pine beauty moth, *Panolis flammea* (*Pf*NPV; Hodgson *et al.* this volume). Here, a single virus isolate collected from an individual cadaver contained upwards of 25 distinct (and distinguishable) genetic variants, each capable of *in vivo* infection. It is perhaps to be expected that one might detect greater variation within isolates of NPVs than GVs, given the ability of each NPV OB to encapsulate multiple genomes (see Figure 16.1).

The presence of genetic variants within *Pi*GV genome populations cannot be wholly discounted, however, and indeed we have detected genetic variation, in the form of submolar DNA fragments within RE profiles, for other isolates of the virus (data not shown). The range of REs shown in Figure 16.5 is not exhaustive, and reveals only those mutations that affect the number and relative position of these enzyme recognition sites (linear stretches of 6–8 base-pairs). Other genetic changes, which do not affect these specific sites, may remain undetected by the method used

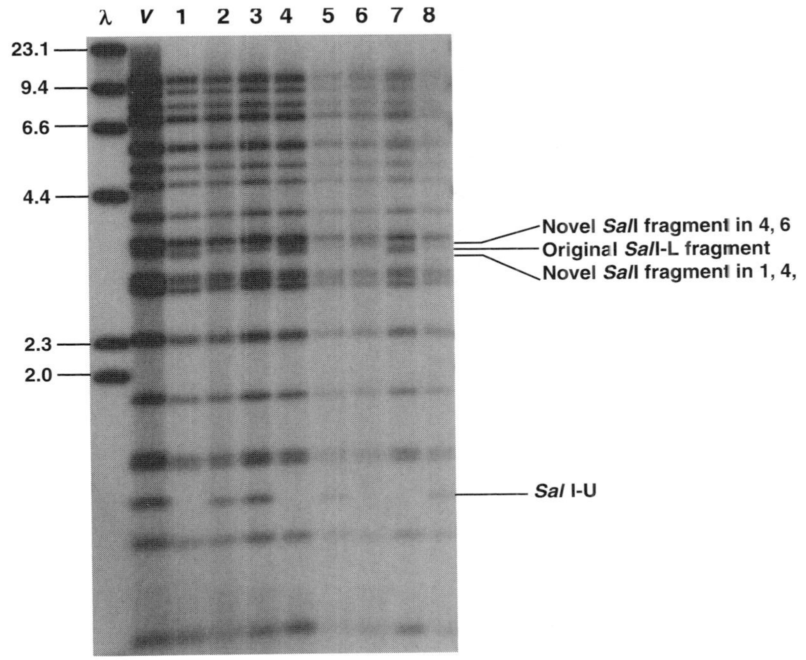

Figure 16.6 *Sal* I restriction enzyme profiles of *Pi*GV DNA isolates. *Pi*GV was extracted from individual larvae (lanes 1–8) taken from the end-point of long-term population studies. Viral DNA was examined as described previously, and restriction fragment profiles for *Sal* I were compared to the original virus (lane V) genomic population. Genomic fragments *Sal* I-L and *Sal* I-U show variation compared to the original profile (lane V) within several genome populations (lanes 1, 4, 6 and 7). Molecular weight markers (lane λ) were used to provide fragment size estimates, in kb.

here. Nevertheless, our studies have focused on the use of RE profiles to analyse *Pi*GV genomes, examining genetic changes within virus isolates persisting through long-term *P. interpunctella* populations maintained on contrasting food regimes.

Genetic change in persisting *Pi*GV populations

There are many reports of baculoviruses responding to repeated passage through an insect host by an increased virulence towards that host (Veber 1962; Tompkins *et al.* 1981, 1988; Kolodny-Hirsch & Van Beek 1997). The mechanisms for this viral response are not documented, but are likely to include recombination between viral genomes and selection of more virulent isolates. We have demonstrated (Table 16.3) that close association of *Pi*GV with its host through multiple generations can result in phenotypic (infectivity) changes. Virus populations have also been examined by RE profiles, to detect the presence of genotypic change, compared to the parental genome profile. To examine genetic heterogeneity at several levels, we examined viruses extracted from individual larva, allowing the assessment of variation across both the broad width of the entire population (the whole host cohort) and within each individual virus 'subpopulation' (in single host cohorts) that make up the whole. Figure 16.6 shows the *Sal* I genomic profiles of eight virus subpopulations, each isolated from an individual host within the long-term infected populations (described above) maintained on good quality food. Initial observation of Figure 16.6 reveals the *Sal* I profiles derived from distinct virus subpopulations (lanes 1–8) to be strikingly similar to each other and to the ancestral virus (lane V). The number and size of linear fragments generated by digestion of virus genomes has remained largely constant, even after multiple (perhaps 50) generations through successive host cycles. Strict maintenance of genome integrity is indicative of a highly stable genetic content, suggesting there is either little room for, or perhaps need of, change within the system examined.

Changes in the DNA sequence can result in the generation of new linear fragment patterns, if those changes alter the number and/or location of RE recognition sites. When all genome molecules possess the same change(s), the fragment pattern for a specific RE will alter, but both original and new fragments will remain molar. However, if only a proportion of genome molecules change, then the new linear fragments will be generated by only a subset of genome molecules. These so-called 'submolar' fragments, detected by agarose gel electrophoresis, appear at lower intensities compared to molar fragments that make up the majority of the genome profile. Four of the eight virus subpopulations examined in Figure 16.6 demonstrate differences within their *Sal* I profiles, when compared to the parental virus (V). Within these novel viruses (see lanes 1, 4, 6 and 7), genetic changes were found to reside in two specific regions, the *Sal* I-L (3.3 kbp) and *Sal* I-U (1.2 kbp) fragments. The smaller *Sal I*-U fragment has been completely lost from all four of the new viruses, indicating that every genome in these isolates possesses the specific mutation causing this change. This indicates genetic heterogeneity between the subpopulations that make up part of the larger *Pi*GV population, with distinct virus 'lineages' emerging in separate hosts. A further level of variation is clear if we examine the mutations that have arisen to affect the *Sal* I-L fragments of these viruses. In the parental virus (V) and subpopulations 2, 3, 5 and 8, all genomes possess a molar *Sal*I-L fragment, whilst subpopulations 1, 4, 6 and 7 possess new fragments at submolar intensities. Within lanes 1, 6 and 7 a second *Sal* I-L fragment is now visible, approximately 100 bp larger (isolate 6) or smaller (isolates 1 and 7) than the parental *Sal*I-L. The virus subpopulation represented by lane 4 contains both of these new fragments, so that three *Sal*I-L fragments now exist. The genetic changes that have given rise to the new *Sal*I-L fragments are not present in every genome copy of these subpopulations, since the new fragments are submolar compared to the main fragment pattern. Viewing these fragments simply by eye, they appear to be equimolar to each other, implying that the two or three variant genome types within these subpopulations are of approximately equal prevalence. A more definitive estimate of variant ratios could be gained by examining the relative intensity of fragments using sensitive imaging equipment.

Genotypic variation then, exists within virus subpopulations that together make up the total *Pi*GV population, and the development of genetic change at this level

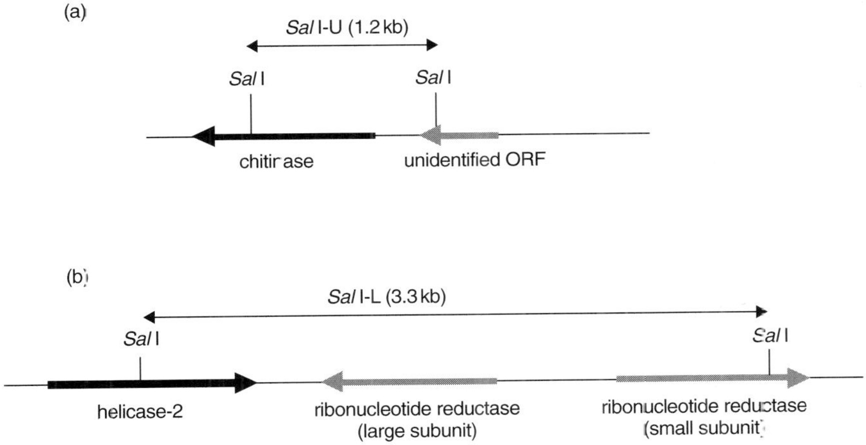

Figure 16.7 Predicted gene content of *Pi*GV *Sal* I-U (a) and *Sal* I-L (b) genomic fragments. The *Sal* I-U and *Sal* I-L fragments were identified as showing variation in several *Pi*GV populations following long-term interactions with host insects on good quality diet. DNA sequence data for these regions reveals putative open reading frames (ORFs) for chitinase, helicase-2, both subunits of a ribonucleotide reductase, and an unidentified ORF. Arrows indicate the direction of transcription with respect to the granulin gene (positive (genome) sense, left to right).

gives rise to greater variation at the global level, as novel genotypes (arising within an individual larva) mix within, and are disseminated through, the larger population during subsequent rounds of infection. It should perhaps be highlighted that the variants detected within our virus subpopulations were not detected in the ancestral virus isolate, and the evolution of multiple virus generations through long-term association in hosts may have given rise to these variants by genetic mutations. However, it is also possible that these variants were in fact present in the parental virus isolate, perhaps as a very low proportion such that they were below the detection limits of the methods used, and that multiple rounds of replication through a large number of hosts allowed them to be separated and amplified as distinct lineages. It is likely that both these possibilities have contributed to the genetic heterogeneity observed within our studies. The degree of influence that host environment has to drive virus evolution and change is unclear, but it is interesting that, just as for phenotypic change, no genotypic changes were detected within virus subpopulations taken from hosts on poor quality diet. Clearly there is a link between resource quality and virus variation, but the details are as yet elusive.

Linking genotypic and phenotypic variation

We have determined the DNA sequence for much of the *Pi*GV genome, and the putative coding capacities of the *Sal* I-L and *Sal* I-U fragments are shown in Figure 16.7. The smaller *Sal* I-U fragment (1.2 kbp) contains a portion of the *Pi*GV chitinase gene homologue and part of a second, as yet unidentified, open reading frame (ORF). The chitinase gene of *Pf*NPV resides in a genetic locus known to be subject to hypervariation between virus isolates (Paul 1997; Hitchman 2002; see Chapter 15), and a similar situation may exist for *Pi*GV. Chitinase does not form part of the 'core' set of gene products essential for baculovirus replication, and although widespread through the Baculoviridae, is not ubiquitous. Where present, chitinase plays an important role in the egress and dissemination of baculoviruses, acting in conjunction with a second virus enzyme, cathepsin, to facilitate host cuticle degradation and tissue liquefaction (Hawtin *et al.* 1997), thereby permitting virus release and exposure to new hosts. *Pi*GV does not liquefy infected *P. interpunctella* following death of the host, despite an ability to encode for a chitinase protein. To date, we have been unable to identify a gene encoding cathepsin within the *Pi*GV genome, although a metalloproteinase gene, adjacent to the chitinase gene, may potentially serve the same function. Cannibalism of infected insects is a major factor contributing to the persistence of several baculoviruses, including *Pi*GV and *Spodoptera frugiperda* NPV. In both cases, successful horizontal transmission via cannibalism provides a means of circumventing the need to liquefy hosts for virus dissemination, which in turn may permit mutation of the chitinase gene without serious detriment to the transmission of virus. Studies to investigate the functionality of *Pi*GV are under way, but at present nothing is known regarding the *Sf*NPV chitinase.

The *Sal* I-L fragment (3.3 kbp) of *Pi*GV contains three ORFs, the *Pi*GV homologues for helicase-2 (Lu & Carstens 1991; Laufs *et al.* 1997; Kuzio *et al.* 1999) and for the large and small subunits of ribonucleotide reductase (van Strien *et al.* 1997). Like chitinase, ribonucleotide reductase is also an auxiliary protein, not essential for baculovirus replication and the genes encoding its subunits have only been found in a few baculovirus genomes (Ahrens *et al.* 1997; Ijkel *et al.* 1999; Kuzio *et al.* 1999). The coding sequences of proteins not crucial for the replication or transmission of *Pi*GV may accumulate mutations at a higher rate than essential virus genes, which are constrained by selective pressure to maintain sequence integrity. DNA sequences at these loci may contribute to a greater level of variation than at other sites within the genome, providing areas of hypervariability that drive genetic change and evolution for the virus population.

Discussion and closing remarks

P. interpunctella, maintained on contrasting diet regimes, demonstrate different population dynamics. We confirm that in the presence or absence of *Pi*GV, a baculovirus pathogen, larvae on good quality food maintain cyclic dynamics, supporting earlier modelling studies. However, similar dynamics are not seen when diet is of low quality, suggesting a strong influence of environment on insect population dynamics. Host susceptibility to *Pi*GV differs between populations perpetuated with high or low food quality, and virus lineages persisting under contrasting food regimes also behave differently with regard to their pathogenicity, resulting in an increased infectivity on good food, but not on poor food. Recent mathematical studies have shown that *Pi*GV does not engage in coupled dynamics with *P. interpunctella*, and that changes in host abundance do not appear to affect changes in pathogen density (Bjornstad *et al.* 1998, 2001). It is perhaps not surprising, then, that changes in the underlying processes that bring about alternative dynamic patterns in larval and adult host abundance, as a result of contrasting environments, do not act as a major selective force on the virus.

Overall, we have found the degree of genetic change occurring over multiple generations of *Pi*GV replication to be low, indicating a high degree of stability for *Pi*GV genomes. However, both inter- and intra-population variation exists, and our study has shown that change is a feature of virus dynamics during long-term association with the host. Several regions of the *Pi*GV genome are subject to significant variation, accumulating mutational changes as they continue in association with the host on good food, although no such changes were detectable when hosts were maintained on poor quality diet. The variation that was found within *Pi*GV when examined under different environmental regimes suggests that both the host and pathogen engage in an interaction that possesses subtleties not previously recognized or detected in the experimental and theoretical examination of the dynamics of *P. interpunctella* and its granulovirus.

Key parameters crucial to host–pathogen interactions include pathogenicity, speed of kill and pathogen productivity (see Cory *et al.* 1997). These factors are critical components of the rate of horizontal transmission and persistence of the pathogen, and can interact in a variety of complex ways that largely depend on host age-structure and size of initial pathogen inoculum. If the baculoviruses undergo genotypic changes as a result of potential coevolutionary processes, which are driven by intrinsic and extrinsic factors, it is possible that one or more of these parameters may be affected. In addition to natural factors, the use of modern pesticides has introduced new elements into the plethora of influences that act on insect populations. Full knowledge and understanding of all these parameters is necessary before satisfactory and sustained control can be possible, particularly if control is to be effected by natural or engineered baculoviruses.

Acknowledgements

We thank T. Heyes (Liverpool) and W. Tyne (Oxford) for technical support, S. Turner for assistance with bioinformatic manipulations, C. Hatton for photography and A. Patmanidi for preparation of electron micrographs. This work was funded by the Natural Environment Research Council's thematic programme EDGE (Ecological Dynamics and Genes).

References

Ahrens, C.H., Russell, R.L.Q., Funk, C.J. *et al.* (1997) The sequence of the *Orgyia pseudotsugata* multinucleocapsid nuclear polyhedrosis virus genome. *Virology* **229**, 381–399.

Anderson, I.G. & Prior, H.C. (1992) Baculovirus infections in the mud crab *Scylla serrata*, and a freshwater crayfish, *Cherax quadricarinolus*, from Australia. *Journal of Invertebrate Pathology* **60**, 265–273.

Anderson, R.M. (1991) Populations and infectious diseases: ecology or epidemiology. *Journal of Animal Ecology* **60**, 1–50.

Anderson, R.M. & May, R.M. (1981) The population dynamics of microparasites and their invertebrate hosts. *Philosophical Transactions of the Royal Society of London Series B* **291**, 451–524.

Arnott, H.J. & Smith, K.M. (1968) An ultrastructural study of the development of a granulosis virus in the cells of the moth *Plodia interpunctella* (Hbn). *Journal of Ultrastructure Research* **21**, 251–268.

Begon, M., Sait, S.M. & Thompson, D.J. (1996) Predator–prey cycles with period shifts between two- and three-species systems. *Nature* **381**, 311–315.

Begon, M., Sait, S.M. & Thompson, D.J. (1997) Two's company, three's a crowd: host–pathogen–parasitoid dynamics. In: *Multitrophic Interactions in Terrestrial Systems* (eds A. C. Gage & V. K. Brown), pp. 307–332. Blackwell Science, Oxford.

Benmayor, R. (2002) *The population dynamics of the Indian meal moth,* Plodia interpunctella, *with two baculoviruses.* PhD thesis, Liverpool University.

Benson, J. F. (1973) The biology of Lepidoptera infesting stored products, with special reference to population dynamics. *Biological Reviews* **48**, 1–26.

Benz, G. A. (1986) Introduction: historical perspectives. In: *The Biology of Baculoviruses: Biological Properties and Molecular Biology* (eds R. R. Granados & B. A. Federici), pp. 1–35. CRC Press, Boca Raton, FL.

Benz, G. (1987) Environment. In: *Epizootiology of Insect Diseases* (eds J. R. Fuxa & Y. Tanada), pp. 177–214. John Wiley, New York.

Bird, F.T. & Elgee, D.E. (1957) A virus disease and introduced parasites as factors controlling the European spruce sawfly, *Diprion hercyniag* (Htg.), in central New Brunswick. *Canadian Entomology* **89**, 371–322.

Bishop, D.H.L. (1992) Baculovirus expression vectors. *Seminars in Virology* **3**, 253–264.

Bjornstad, O.N., Begon, M., Stenseth, N.C. *et al.* (1998) Population dynamics of the Indian meal moth: demographic stochasticity and delayed regulatory mechanisms. *Journal of Animal Ecology* **67**, 110–126.

Bjornstad, O.N., Sait, S.M., Stenseth, N.C., Thompson, D.J. & Begon, M. (2001) The impact of specialized enemies on the dimensionality of host dynamics. *Nature* **409**, 1001–1006.

Black, B.C., Brennan, L.A., Dierks, P. M. & Gard, I.E. (1997) Commercialization of baculoviral insecticides. In: *The Baculoviruses* (ed. L. K. Miller), pp. 341–387. Plenum Press, New York.

Boots, M. (1998) Cannibalism and the stage-dependent transmission of a viral pathogen of the Indian meal moth, *Plodia interpunctella. Ecological Entomology* **23**, 118–122.

Boots, M. (2000) Density-independent resource limitation and the transmission of an insect pathogen. *Oecologia* **124**, 172–175.

Boots, M. & Begon, M. (1994) Trade-offs with resistance to a granulovirus virus in the Indian meal moth, examined by a laboratory evolution experiment. *Functional Ecology* **7**, 528–534.

Boots, M. & Norman, R. (2000) Sublethal infection and the population dynamics of host–microparasite interactions. *Journal of Animal Ecology* **69**, 517–524.

Briggs, C.J. & Godfray, H.C.J. (1995) The dynamics of insect–pathogen interactions in stage-structured populations. *American Nature* **145**, 855–887.

Briggs, C.J. & Godfray, H.C.J. (1996) The dynamics of insect–pathogen interactions in seasonal environments. *Theoretical Population Biology* **50**, 149–177.

Briggs, C.J., Sait, S.M., Begon, M., Thompson, D.J. & Godfray, H.C.J. (2000) What causes generation cycles in populations of stored product moths? *Journal of Animal Ecology* **69**, 352–366.

Burand, J.P. & Summers, M.D. (1982) Alteration of *Autographa californica* nuclear polyhedrosis virus DNA upon serial passage in cell culture. *Virology* **119**, 223–229.

Burden, J.P., Griffiths, C.M., Cory, J.S., Smith, P. &

Sait, S.M. (2002). Vertical transmission of sublethal granulovirus infection in the Indian meal moth, *Plodia interpunctella*. *Molecular Ecology*, **11**, 547–555.

Carstens, E.B. (1982) Mapping the mutation site of an *Autographa californica* Nuclear Polyhedrosis Virus polyhedron morphology mutant. *Journal of Virology* **43**, 809–818.

Chang, P.S., Lo, C.L., Kou, G.H., Lu, C.C. & Chen, S.N. (1993) Purification and amplification of DNA from *Penaeus monodon* type baculovirus (MBV). *Journal of Invertebrate Pathology* **62**, 116–120.

Consigli, R.A., Tweeten, K.A., Anderson, D.K. & Bulla, L.A., Jr (1983) Granulosis viruses, with emphasis on the GV of the Indian Meal Moth, *Plodia interpunctella*. *Advances in Virus Research* **28**, 141–173.

Consigli, R.A., Russell, D.L. & Wilson, M.E. (1986) The biochemistry and molecular biology of the granulosis virus that infects *Plodia interpunctella*. *Current Topics in Microbiology and Immunology* **131**, 69–100.

Cory, J.S., Hails, R.S. & Sait, S.M. (1997) Baculovirus ecology. In: *The Baculoviruses* (ed. L. K. Miller), pp. 301–339. Plenum Press, New York.

Couch, J.A. (1974) Free and occluded virus, similar to baculovirus, in hepatopancreas of pink shrimp. *Nature* **247**, 229–231.

Croizier, G. & Ribeiro, H.C.T. (1992) Recombination as a possible major cause of genetic heterogeneity in *Anticarsia gemmatalis* nuclear polyhedrosis virus wild populations. *Virus Research* **26**, 183–196.

Croizier, G., Croizier, L., Quiot, J.M. & Lereclus, D. (1988) Recombination of *Autographa californica* and *Rachiplusia ou* nuclear polyhedrosis viruses in *Galleria mellonella* L. *Journal of General Virology* **69**, 177–185.

Crook, N.E., Spencer, R.A., Payne, C.C. & Leisy, D. J.(1985) Variation in *Cydia pomonella* granulosis virus isolates and physical maps of the DNA from three variants. *Journal of General Virology* **66**, 2423–2430.

Duffey, S.S., Hoover, K., Bonning, B. & Hammock, B.D. (1995) The impact of host plant on the efficacy of baculoviruses. *Reviews in Pest Toxicology* **3**, 137–275.

Englehard, E.K. & Volkman, L.E. (1995) Developmental resistance in fourth instar *Trichoplusia ni* orally inoculated with *Autographa californica* M nuclear polyhedrosis virus. *Virology* **209**, 384–389.

Fleming, S.B., Kalmakoff, J., Archibald, R.D. & Stewart, K.M. (1986) Density-dependent virus mortality in populations of *Wisnea* (Lepidoptera: Hepialidae). *Journal of Invertebrate Pathology* **48**, 193–198.

Forschler, B.T., Young, S.Y. & Felton, G.W. (1992) Diet and the susceptibility of *Helicoverpa zea* (Noctuidae: Lepidoptera) to a nuclear polyhedrosis virus. *Environmental Entomology* **21**, 1220–1223.

Fraser, M.J. & Hink, W.F. (1982) The isolation and characterization of the MP and FP plaque variants of *Galleria mellonella* polyhedrosis virus. *Virology* **117**, 366–378.

Fraser, M.J., Smith, G.E. & Summers, M.D. (1983) Acquisition of host cell DNA sequences by baculoviruses: relationship between host insertions and FP mutants of *Autographa californica* and *Galleria mellonella* nuclear polyhedrosis viruses. *Journal of Virology* **47**, 287–300.

Gage, M.J.G. (1995) Continuous variation in reproductive strategy as an adaptive response to population-density in the moth *Plodia interpunctella*. *Proceedings of the Royal Society of London Series B* **261**, 25–30.

Gage, M.J.G. & Cook, P.A. (1994) Sperm size or numbers? Effects of nutritional stress upon eupyrene and apyrene sperm production strategies in the moth *Plodia interpunctella* (Lepidoptera: Pyralidae). *Functional Ecology* **8**, 594–599.

Goto, C. (1990) Enhancement of a nuclear polyhedrosis virus (NPV) infection by a granulosis virus (GV) isolated from the spotted cutworm, *Xestia c-nigrum* L. (Lepidoptera: Noctuidae). *Applied Entomology and Zoology* **25**, 135–137.

Granados, R.R. & Federici, B.A. (1986) *In vivo* infection and replication of baculoviruses.

In: *Biology of Baculoviruses* (eds R. Granados & B. Federici), pp. 89–108. CRC Press, Boca Raton, FL.

Griffiths, C.M. (1994) Baculovirus expression vectors: advances and applications. *Expert Opinion in Therapeutic Patents* **4**, 1065–1082.

Hackett, K.J., Boore, A., Deming, C. *et al.* (2000) *Helicoverpa armigera* granulovirus interference with progression of *H. zea* nucleopolyhedrovirus disease in *H. zea* larvae. *Journal of Invertebrate Pathology* **75**, 99–106.

Hajos, J.P., Pijenburg, J., Usnamy, M. *et al.* (2000) High frequency recombination between homologous baculoviruses in cell culture. *Archives of Virology* **145**, 159–164.

Harvey, J.A., Harvey, I.F. & Thompson, D.J. (1997) Flexible larval growth allows the use of a range of host sizes by a parasitoid wasp. *Ecology* **75**, 1420–1428.

Hawtin, R.E., Zarkowska, T., Arnold, K. *et al.* (1997) Liquefaction of *Autographa californica* nucleopolyhedrovirus-infected insects is dependent on the integrity of virus-encoded chitinase and cathepsin genes. *Virology* **238**, 243–253.

Hitchman, R.B. (2002) *The evolution and molecular biology of baculoviruses.* PhD thesis, Oxford Brookes University, Oxford.

Hofmaster, R.N. (1961) Seasonal abundance of the cabbage looper as related to light trap collections, precipitation, temperature and the incidence of a nuclear polyhedrosis virus. *Journal of Economic Entomology* **54**, 796–798.

Holland, J.J. (1990) Defective viral genomes. In: *Virology*, Vol. 1. (eds B.N. Fields & D.M. Knipe), 2nd edn, pp. 151–165. Raven Press, New York.

Huang, A.S. & Baltimore, D. (1970) Defective viral particles and viral disease processes. *Nature* **226**, 325–327.

Huber, J. (1990) Viral insecticides: Potential, problems and prospects. In: *Conference on Crop Protection*, pp. 601–606. Brighton.

Hughes, D.S., Possee, R.D. & King, L.A. (1993) Activation and detection of a latent baculovirus resembling *Mamestra brassicae* nuclear polyhedrosis virus in *M. brassicae* insects. *Virology* **194**, 608–615.

Hughes, D.S., Possee, R.D. & King, L.A. (1997) Evidence for the presence of a low-level, persistent baculovirus infection of *Mamestra brassicae* insects. *Journal of General Virology* **78**, 1801–1805.

Hunter, D.K. & Hoffman, D.F. (1970) A granulosis virus of the almond moth, *Cadra cautella. Journal of Invertebrate Pathology* **16**, 400–407.

Hunter, D.K., Collier, S.J. & Hoffman, D.F. (1973a) Effectiveness of a granulosis virus of the Indian meal moth as a protectant for stored inshell nuts: preliminary observations. *Journal of Invertebrate Pathology* **22**, 4–10.

Hunter, D.K., Hoffman, D.F. & Collier, S.J. (1973b) Pathogenicity of a nuclear polyhedrosis virus of the almond moth, *Cadra cautella. Journal of Invertebrate Pathology* **21**, 282–286.

Hunter, D.K., Hoffman, D.F. & Collier, S.J. (1973c) Cross-infection of a nuclear polyhedrosis virus of the almond moth to the Indian meal moth. *Journal of Invertebrate Pathology* **22**, 186–192.

Hunter, D.K., Collier, S.J. & Hoffman, D.F. (1977) Granulosis virus of Indian meal moth as a protectant for stored inshell almonds. *Journal of Economic Entomology* **70**, 493–494.

Hunter, D.K., Collier, S.J. & Hoffman, D.F. (1979) The effect of a granulosis virus on *Plodia interpunctella* (Lepidoptera: Pyralidae) infestations occurring in raisins. *Journal of Stored Products Research* **15**, 65–79.

Ijkel, W.F.J., van Strien, E.A., Heldens, J.G.M. *et al.* (1999) Sequence and organization of the *Spodoptera exigua* multicapsid nucleopolyhedrovirus genome. *Journal of General Virology* **80**, 3289–3304.

Jaques, R.P. (1962) The transmission of nuclear polyhedrosis virus in laboratory populations of *Trichoplusia ni* (Hubner). *Journal of Insect Pathology* **4**, 433–445.

Jarvis, D.L. (1997) Baculovirus expression vectors. In: *The Baculoviruses* (ed. L.K. Miller), pp. 389–431. Plenum Press, New York.

Jehle, J.A., Fritsch, E., Nickel, A., Huber, J. & Backhaus, H. (1995) TCI4.7: a novel Lepidopteran transposon found in *Cydia pomonella* granulosis virus. *Virology* **207**, 369–379.

Jehle, J.A., van der Linden, I.F.A., Backhaus, H. & Vlak, J.M. (1997) Identification and sequence analysis of the integration site of transposon TCp3.2 in the genome of *Cydia pomonella* granulovirus. *Virus Research* **50**, 151–157.

Keddie, B.A., Aponte, G.W. & Volkman, L.E. (1989) The pathway of infection of *Autographa californica* nuclear polyhedrosis virus in an insect host. *Science* **243**, 1728–1730.

Kinsinger, R.A. & McGaughey, W.H. (1976) Stability of *Bacillus thuringiensis* and a granulovirus of *Plodia interpunctella* on stored wheat. *Journal of Economic Entomology* **69**, 149–154.

Knell, R.J. (1998) Generation cycles. *Trends in Ecology and Evolution* **13**, 186–190.

Knell, R.J., Begon, M. & Thompson, D.J. (1998a) Host–pathogen dynamics, basic reproductive rates and threshold densities. *Oikos* **81**, 299–308.

Knell, R.J., Begon, M. & Thompson, D.J. (1998b) Transmission of *Plodia interpunctella* granulosis virus does not conform to the mass action model. *Journal of Animal Ecology* **67**, 592–599.

Kolodny-Hirsch, D.M. & Van Beek, N.A.M. (1997) Selection of a morphological variant of *Autographa californica* nuclear polyhedrosis virus with increased virulence following serial passage in *Plutella xylostella*. *Journal of Invertebrate Pathology* **69**, 205–211.

Kondo, A. & Maeda, S. (1991) Host range expansion by recombination of the baculoviruses *Bombyx mori* nuclear polyhedrosis virus and *Autographa californica* nuclear polyhedrosis virus. *Journal of Virology* **65**, 3625–3632.

Kool, M., Voncken, J.W., van Lier, F.L.J., Tramper, J. & Vlak, J.M. (1991) Detection and analysis of *Autographa californica* nuclear polyhedrosis virus mutants with defective interfering properties. *Virology* **183**, 739–746.

Kukan, B. (1999) Vertical transmission of nucleopolyhedrovirus in insects. *Journal of Invertebrate Pathology* **74**, 103–111.

Kumar, S. & Miller, L.K. (1987) Effects of serial passage of *Autographa californica* nuclear polyhedrosis virus in cell culture. *Virus Research* **7**, 335–349.

Kuzio, J., Pearson, M.N., Harwood, S.H. *et al.* (1999) Sequence and analysis of the genome of a baculovirus pathogenic for *Lymantria dispar*. *Virology* **253**, 17–34.

Laufs, S., Lu, A., Arrell, K. & Carstens, E.B. (1997) *Autographa californica* nuclear polyhedrosis virus *p143* gene product is a DNA-binding protein. *Virology* **228**, 98–106.

Lee, H.Y. & Krell, P.J. (1992) Generation and analysis of defective genomes of *Autographa californica* nuclear polyhedrosis virus. *Journal of Virology* **66**, 4339–4347.

Lu, A. & Carstens, E.B. (1991) Nucleotide sequence of a gene essential for viral DNA replication in the baculovirus *Autographa californica* Nuclear Polyhedrosis Virus. *Virology* **181**, 336–347.

Lynn, D.E. & Oberlander, H. (1983) The establishment of cell lines from imaginal wing discs of *Spodoptera frugiperda* and *Plodia interpunctella*. *Journal of Insect Pathology* **29**, 591–596.

McIntosh, A.H., Rice, W.C. & Ignoffo, C.M. (1987) Genotypic variants in wild-type populations of baculovirus. In: *Biotechnology in Invertebrate Pathology and Cell Culture* (ed. M. Maramorosch), pp. 305–325. Academic Press, New York.

Madrid, F.J. & Sinha, R. (1982) Feeding damage of three stored-product moths (Lepidoptera: Pyralidae) on wheat. *Journal of Economic Entomology* **75**, 1017–1020.

Madrid, F.J., White, N.D.G. & Sinha, R.N. (1983) Effects of malathion dust on Indian meal moth and almond moth (Lepidoptera: Phycitidae) infestation on stored wheat. *Journal of Economic Entomology* **76**, 1401–1404.

Matthews, R.E.F. (1992) Classification and nomenclature of viruses. *Intervirology* **17**, 1–199.

Melamed-Madjar, V. & Raccah, B. (1979) The trans stadial and vertical transmission of a granulosis virus from the corn borer *Sesamia nonagroides*. *Journal of Invertebrate Pathology* **33**, 259–264.

Miller, D.W. & Miller, L.K. (1982) A virus mutant with an insertion of a *copia*-like transposable element. *Nature* **299**, 562–564.

Munoz, D., Vlak, J.M. & Caballero, P. (1997) Recombination between two strains of the nucleopolyhedrovirus of *Spodoptera exigua* (SeMNPV) in its natural host. *Applied and Environmental Microbiology* **63**, 3025–3031.

Myers, J.H. (1988) Can a general hypothesis explain population cycles of forest Lepidoptera? *Advances in Ecology Research* **18**, 179–242.

Onstad, D.W. & Carruthers, R.I. (1990) Epizootiological models of insect diseases. *Annual Review of Entomology* **35**, 399–419.

Park, T. (1948) Inter-species competition in *Trifolium confusum* Duval and *Tribolium castaneum* Herbst. *Ecological Monographs* **18**, 265–307.

Paul, R.K. (1997) *Evolution and interaction of insect pathogens.* PhD thesis, University of Reading.

Payne, C.C. (1982) Insect viruses as control agents. *Parasitology* **84**, 35–67.

Payne, C.C. (1988) Pathogens for the control of insects: Where next? *Philosophical Transactions of the Royal Society of London Series B* **318**, 225–248.

Podoler, H. (1974) Effects of intraspecific competition in the Indian meal moth, *Plodia interpunctella* (Hubner) (Lepidoptera: Phycitidae) on populations of the moth and its parasite *Nemeritis canescens* (Gravenhorst) (Hymenoptera: Ichnemonidea). *Journal of Animal Ecology* **43**, 641–651.

Potter, K.N., Faulkner, P. & Mackinnon, E.A. (1976) Strain selection during serial passage of *Trichoplusia ni* nuclear polyhedrosis virus. *Journal of Virology* **18**, 1040–1050.

Sait, S.M., Begon, M. & Thompson, D.J. (1994a) The influence of larval age of the response of *Plodia interpunctella* to a granulosis virus. *Journal of Invertebrate Pathology* **63**, 107–110.

Sait, S.M., Begon, M. & Thompson, D.J. (1994b) The effects of a sublethal baculovirus infection in the Indian meal moth, *Plodia interpunctella. Journal of Animal Ecology* **63**, 541–550.

Sait, S.M., Begon, M. & Thompson, D.J. (1994c) Long-term population dynamics of the Indian meal moth, *Plodia interpunctella*, and its granulosis virus. *Journal of Animal Ecology* **63**, 861–870.

Sait, S.M., Begon, M., Thompson, D.J. & Harvey, J.A. (1996) Parasitism of baculovirus-infected *Plodia interpunctella* by *Venturia canescens* and subsequent virus transmission. *Functional Ecology* **10**, 586–591.

Sait, S.M., Begon, M. & Thompson, D.J. (1998a). Complex population dynamics in a host–pathogen–parasitoid interaction. In: *Population and Community Ecology for Insect Management and Conservation* (eds J. Baumgartner, P. Brandmayr & B.F.J. Manly), pp. 183–190. Balkema Press, Rotterdam.

Sait, S.M., Gage, M.J.G. & Cook, P.A. (1998b) Effects of a fertility-reducing baculovirus on sperm numbers and sizes in the Indian meal moth, *Plodia interpunctella. Functional Ecology* **12**, 56–62.

Sait, S.M., Liu, W-C., Thompson, D.J., Godfray, H.C.J. & Begon, M. (2000) Invasion sequence affects predator–prey dynamics in a multi-species interaction. *Nature* **405**, 448–450.

Shapiro, D.I., Fuxa, J.R., Braymer, H.D. & Pashley, D.P. (1991) DNA restriction polymorphism in wild isolates of *Spodoptera frugiperda* nuclear polyhedrosis virus. *Journal of Invertebrate Pathology* **58**, 96–105.

Smith, G.E. & Summers, M.D. (1979) Restriction maps of five *Autographa californica* MNPV variants, *Trichoplusia ni* MNPV, and *Galleria mellonella* MNPV DNAs with endonucleases *Sma*I, *Kpn*I, *Bam*HI, *Sac*I, *Xho*I, and *Eco*RI. *Journal of Virology* **30**, 828–838.

Smith, G.E. & Summers, M.D. (1980) Restriction map of *Rachiplusia ou* and *Rachiplusia ou–Autographa californica* nuclear polyhedrosis virus recombinants. *Journal of Virology* **33**, 311–319.

van Strien, E.A., Faktor, O., Hu, Z.H. *et al.* (1997) Baculoviruses contain a gene for the large subunit of ribonucleotide reductase. *Journal of General Virology* **78**, 2365–2377.

Tanada, Y. (1959) Synergism between viruses of the armyworm, *Pseudaletia unipuncta* (Haworth) (Lepidoptera, Noctuidae). *Journal of Invertebrate Pathology* **1**, 215–231.

Tompkins, G.J., Vaughn, J.L., Adams, J.R. & Reichelderfer, C.F. (1981) Effect of propagating *Autographa californica* nuclear polyhedrosis virus and its *Trichoplusia ni* variant in different hosts. *Environmental Entomology* **10**, 801–806.

Tompkins, G.J., Dougherty, E.M., Adams, J.R. & Diggs, D. (1988) Changes in the virulence of nuclear polyhedrosis viruses when propagated in alternative noctuid (Lepidoptera: Noctuidae) cell lines and hosts. *Journal of Economic Entomology* **81**, 1027–1032.

Tweeten, K.A., Bulla, L.A., Jr & Consigli, R.A. (1977) Supercoiled circular DNA of an insect granulosis virus. *Proceedings of the National Academy of Sciences of the United States of America* **74**, 3574–3578.

Tweeten, K.A., Bulla, L.A., Jr & Consigli, R.A. (1980) Restriction enzyme analysis of the genomes of *Plodia interpunctella* and *Pieris rapae* granulosis viruses. *Virology* **104**, 514–519.

Vasconcelos, S.D. (1996) Alternative routes for horizontal transmission of a nucleopolyhedrovirus. *Journal of Invertebrate Pathology* **68**, 269–274.

Veber, J. (1962) Virulence of an insect virus increased by repeated passages. Coloquialisms Internationales des Pathologiques Insectes, Paris **3**, 403–405.

Volkman, L. E., Blissard, G. W., Friesen, P. *et al.* (1995) Baculoviridae. In: *Virus Taxonomy: Classification and Nomenclature of Viruses* (eds F. A. Murphy, C. M. Fauquet, D. H. L. Bishop *et al.*), pp. 104–113. Springer-Verlag, New York.

Winstanley, D. & Crook, N.E. (1993) Replication of *Cydia pomonella* granulosis virus in cell cultures. *Journal of General Virology* **74**, 1599–1609.

Chapter 17

The ecological significance of symbiotic microorganisms in animals: perspectives from the microbiota of aphids

A. E. Douglas, A. C. Darby, L. M. Birkle and K. F. A. Walters*

Introduction

Certain habitats and diets are exploited exclusively by animals with symbiotic microorganisms. For example, virtually all reef-building corals in illuminated, nutrient-poor waters have photosynthetic algae, all insects feeding on vertebrate blood or plant sap through their life cycle have symbiotic bacteria or yeasts, and almost all herbivorous mammals have cellulolytic bacteria. These symbiotic microorganisms possess biochemically and genetically complex metabolic capabilities absent from their animal partners. By forming a symbiosis, the corals have acquired access to photosynthetic carbon fixation, the blood- and plant-sap-feeding insects have gained access to vitamins and essential amino acids, respectively, and the mammalian herbivores have acquired the capacity to degrade plant cell wall polysaccharides such as cellulose. Animals with symbiotic microorganisms are of great ecological importance. Taking the examples cited above, corals are the architectural foundation to all modern shallow-water reefs, ecosystems of exceptional biodiversity and productivity (Wood 1998) and herbivores play a crucial role in shaping terrestrial ecosystems (e.g. Olff *et al.* 1999).

Genes and molecular biology are central to our understanding of the ecological significance of symbiotic microorganisms in animals for two very different reasons. The first is fundamental to the animal 'condition'. As well as including taxa of remarkable morphological and behavioural complexity, the animals, as a group, are metabolically impoverished. The nutritional ecology of most animals is shaped by their inability to fix either inorganic carbon or nitrogen or to synthesize various specific compounds, including the essential amino acids (nine of the 20 amino acids that make up protein) and vitamins (coenzymes required for activity of enzymes central to metabolism). Certain ecologically successful phyla have additional metabolic limitations, e.g. arthropods cannot synthesize sterols and chordates cannot

* Correspondence author: aed2@york.ac.uk

degrade cellulose. As the examples in the opening paragraph of this chapter illustrate, various animals have circumvented these metabolic limitations by forming symbioses with microorganisms. Through symbiosis, animals have acquired not only the genes for complex metabolic capabilities, but also the molecular and cellular machinery for the regulated expression of those genes.

The second reason for the importance of genes and molecular biology to the study of the ecology of symbioses in animals is that many of the symbiotic microorganisms are unculturable and therefore not amenable to traditional microbiological methods to study either their taxonomic diversity or metabolic capabilities. Even among apparently culturable symbiotic microorganisms, approaches based on cultivation can be misleading. This can be illustrated by the research of Goulet and Coffroth (1997) on the symbiotic dinoflagellate algae in the gorgonian coral *Plexaura kuna*. A DNA fingerprinting analysis revealed that the algae brought into culture from this animal were allied to, but distinct from, the dominant forms in the symbiosis. Detailed analysis of the characteristics of the algae cultured from *P. kuna* would have provided insight only into the properties of genotypes that are quantitatively unsuccessful in the symbiosis.

The focus of this chapter is the bacterial symbiosis in a phloem-feeding insect, the aphids. The insect–microbial symbioses were described in the first half of the 20th century and have been summarized by Buchner (1966). However, for decades these associations have been considered as experimentally intractable, in comparison to the photosynthetic symbioses in corals, cellulolytic microbiota of mammals and especially microbial symbioses in plants. This is because the insect symbioses are obligate (neither insect nor microbial partner can be maintained in long-term culture in isolation) and they cannot be synthesized experimentally from the isolated partners. These limitations have been circumvented, at least in part, by molecular techniques, and further molecular opportunities are offered by the availability of the complete genome sequence of the dominant symbiotic bacterium in aphids (Shigenobu *et al.* 2000).

The purpose of this article is to provide a critical review of the diversity, significance and function of the symbiotic microorganisms in aphids, as deduced from an integrated molecular and ecological approach.

The diversity of microorganisms in aphids

Microorganisms in the gut and cells of aphids

Most animals bear a great diversity of microorganisms, and the greatest abundance is in the gut, especially the hindgut (Cazemier *et al.* 1997; Falk *et al.* 1998). Aphids are unusual in that they appear to be associated with relatively few microorganisms, most of which are predominantly or exclusively intracellular. One important contributory reason for the dearth of gut-associated microorganisms is the anatomical simplicity of the aphid gut, which is a tube of relatively uniform diameter and short length (ca. four times the length of the aphid body) without caeca or other diverticula. Most microorganisms ingested by aphids pass directly through the gut with the rapid flow of the liquid food (Harada & Ishikawa 1997).

The gut microbiota of aphids in laboratory culture has been investigated. The guts of some aphid cultures are apparently microbiologically sterile, as indicated by parallel microscopical and molecular analysis (Wilkinson *et al.* 1997), and others contain populations of one or several different taxa, including species of the genera *Erwinia* and *Staphylococcus* (Grenier *et al.* 1994; Harada *et al.* 1996). However, the composition of the gut microbiota of insects can be influenced by laboratory culture conditions, as is illustrated by the dramatic changes in the microbial community in termites brought into the laboratory (Lovelock *et al.* 1985) and the realization that the actinomycete *Nocardia rhodnii*, originally described as the sole and obligate symbiont of the blood-sucking bug *Rhodnius prolixus*, is just one of many different gut-associated microorganisms and absent from some populations of this insect species (Dasch *et al.* 1984; Durvasula *et al.* 1999). To our knowledge, the microbiology of the guts of aphids in natural populations has not been investigated.

Buchnera: the dominant symbiont in most aphids

As Buchner (1966, p. 298) comments, the microorganisms in various aphids 'have an extraordinary resemblance to one another'. They are coccoid bacteria, 2–4 μm in diameter and located exclusively in the cytoplasm of large insect cells, called mycetocytes, in the aphid body cavity. Furthermore, as the microscopical analyses of Uichanco (1924) and Hinde (1971) reveal, these bacteria are transmitted from the maternal mycetocytes to the eggs or developing embryos in the female aphid ovary. Analysis of the 16S rRNA gene sequence in various aphids has revealed that the primary symbionts are a phylogenetically coherent group within the γ-*Proteobacteria*, distinct from all other described bacteria and with a phylogeny strictly congruent with that of aphids (Munson *et al.* 1991a; Moran *et al.* 1993). The bacteria have been assigned formally to a novel genus, *Buchnera* (Munson *et al.* 1991b), and it can reasonably be concluded from the sequence data that all *Buchnera* are derived from a single evolutionary transition to the symbiotic life style in aphids, probably between 160 and 280 million years ago (Moran *et al.* 1993). These data are consistent with the possibility that *Buchnera* was acquired by the first aphids in the late Palaeozoic.

The diversity of *Buchnera* can be summarized as three linked elements, all shaped ultimately by the obligate vertical transmission of these bacteria from mother aphid to her offspring.

1 The *Buchnera* cells in an individual aphid are of low diversity and closely related by descent. This is principally because relatively small numbers of bacteria (probably 50–500) are transferred from the mycetocytes of a female aphid to each egg or embryo in the ovary (Douglas 1998) and the bacteria in each offspring divide relatively few times (up to 7–15 rounds of doubling) before transmission to the grand-daughters of the original aphid (Whitehead & Douglas 1992; Baumann *et al.* 1994; Humphreys & Douglas 1997). The diversity of *Buchnera* cells transmitted to each offspring aphid is reduced further by their derivation from just one of 60–120 maternal mycetocytes (Hongoh & Ishikawa 1994; Wilkinson & Douglas 1998), containing the descendants of a subpopulation (probably <10%) of the *Buchnera* acquired by the mother. It is widely accepted that the selection pressure on the insect to maintain *Buchnera* populations of high relatedness is the risk of selfish mutations in the *Buchnera* cells (reviewed in Rispe & Moran 2000). Selfish mutations (e.g. elevated division rates) are of immediate selective advantage to the mutant and detrimental to other *Buchnera* cells and the aphid. The selective advantage of selfish mutations is low in populations closely related by descent, and the very small numbers of bacteria transmitted to each offspring would additionally ensure that any selfish mutation is exposed to selection relatively quickly.

2 The global diversity of *Buchnera* across all aphids is very high, in terms of sequence diversity. The substitution rate of 16S rRNA genes has been studied intensively and is two to three times faster than in related bacteria (Moran *et al.* 1995); and elevated rates of sequence evolution are also evident in all other genes investigated to date (e.g. Moran 1996; Brynnel *et al.* 1998). This characteristic is a direct consequence of the population structure of *Buchnera* at the level of the individual aphid lineage. As a result of the small numbers of *Buchnera* transmitted, mutations are predicted to become fixed by drift (Moran 1996; Rispe & Moran 2000). Mutations with slightly deleterious effects, e.g. in genes that are not absolutely required for the symbiosis or in regulatory sequences, are expected to accumulate. These mutations are unlikely to be purged by recombination because the bacterial cells are very closely related (see above). Once the genes for DNA repair and recombination start to 'decay', then the genome-wide rates of sequence evolution would be accelerated further.

3 The total genome size of *Buchnera* is very uniform across the aphid species studied to date: 0.64 Mb in *Acyrthosiphon pisum*, *Schizaphis graminum* and *Rhopalosiphum padi* and 0.63 Mb in *Melaphis rhois* (Shigenobu *et al.* 2000; Wernegreen *et al.* 2000). These values are very small, compared to 4–5 Mb in related free-living bacteria, and close to the predicted minimal size of a cell genome (Maniloff 1996). Shigenobu *et al.* (2000) aptly describe the gene complement of *Buchnera* in the pea aphid *A. pisum* (Table 17.1) as a 'subset of the genome of *E. coli*'. This condition is consistent with the predicted decay and deletion of genes not essential for the symbiotic life style. A key outstanding issue is the variation in gene content among *Buchnera* from different aphid taxa. Perhaps the genome reduction to the minimal complement occurred rapidly after the origin of the symbiosis, before the diversification of all modern aphids, and the genomes of modern *Buchnera* are relatively uniform because of very strong stabilizing selection (Clark *et al.* 1999) and isolation from horizontal transfer of genes through decay of genes involved in recombination (Moran 1996; but see Van Ham *et al.* (2000) for a candidate example of lateral acquisition of genes by *Buchnera*). Alternatively, the genome may vary between aphid taxa depending on, and possibly coevolving with, the gene complement in different aphids. In particular, the population structure of aphids may influence the gene content and function of *Buchnera*; aphid mutations

Table 17.1 Distribution of cellular functions of gene products of *Buchnera* and *E. coli*.

	Number of genes	
Functional category	*E. coli*	*Buchnera* sp.
Biosynthesis of small molecules		
Amino acid synthesis*	119	53
Other	237	68
Intermediate metabolism†	383	74
Macromolecule metabolism	419	209
Other	607	213
Total	1765	617

Data from Shigenobu *et al.* (2000) and Riley (1993).
* Includes *trpEG* and *leuA–D*, which are plasmid-borne in *Buchnera*, and *trpG*, excluded from the genes listed for *E. coli* in Riley (1993).
† Genes functional in sulfur metabolism (*cysC, cysD, cysE, cysH, cysI, cysN, cysQ* in both *E. coli* and *Buchnera* sp., and *cysP* in *E. coli*) are included in the functional category 'intermediate metabolism', and not 'amino acid synthesis'.

that compensate for deleterious mutations in *Buchnera* (Douglas 1997a) are more likely to arise in species with large population sizes.

The opportunity in the coming years for comparative genomics of *Buchnera* in aphids of different families, ecology and life-history traits will offer new insight into the molecular ecology of *Buchnera* and its symbiosis with aphids. For the present, however, two general points emerge from these considerations of the diversity of *Buchnera*. First, the determinants of *Buchnera* diversity are well-understood, at least in outline: *Buchnera* diversity is shaped by aphid control over the *Buchnera* population structure, as mediated by the pattern of vertical transmission. Second, the perceived diversity of *Buchnera* depends on the criterion of diversity adopted and the phylogenetic scale over which it is determined. From the molecular perspective, sequence diversity is high, but the gene complement may be uniform across the full range of *Buchnera*. This sequence diversity is highly structured, with little or no variation at the level of the individual aphid but very high divergence between phylogenetically-distant aphids.

Accessory bacteria

In many aphids, *Buchnera* is not the sole bacterium in the aphid tissues (Buchner 1966). Aphids from both laboratory cultures and natural populations commonly bear bacteria morphologically distinct from *Buchnera*, either associated with the mycetocytes or free in the haemolymph (insect 'blood') and generally transmitted maternally via the aphid ovary. These bacteria are much less abundant than *Buchnera*. For example, *Buchnera* accounts for 95–99% of the total bacterial population in both the black bean aphid *Aphis fabae* (Wilkinson *et al.* 2001) and the pea aphid *A. pisum* (L.M. Birkle *et al.*, unpublished observations). Sequence analysis of

Table 17.2 The diversity of accessory bacteria in aphids.

Accessory bacteria	Aphid species	Reference
γ-Proteobacteria		
PASS/R-type	*Acyrthosiphon pisum*	Chen & Purcell (1997)
	Macrosiphum rosae	Fukatsu *et al.* (2000)
	Uroleucon caligatum	Sandström *et al.* (2001)
PABS/T-type	*A. pisum*	Darby *et al.* (2001)
	Uroleucon ambrosiae	Sandström *et al.* (2001)
	U. rudbeckiae	
	U. astronomus	
U-type	*A. pisum*	Sandström *et al.* (2001)
	Macrosiphoniella ludovicianae	
	Uroleucon astronomus	
	U. solidagensis	
	U. aeneum	
	U. helianthicola	
α-Proteobacteria		
Rickettsia	*A. pisum*	Chen *et al.* (1996)
Spiroplasma	*A. pisum*	Fukatsu *et al.* (2001)

PABS, pea aphid *Bemesia*-like symbiont; PASS, pea aphis secondary symbiont.

16S rRNA genes, obtained by PCR amplification with universal bacterial primers and total aphid DNA as template, has revealed various different taxa in aphids (Table 17.2 and Figure 17.1). These bacteria are often described as 'secondary symbionts' but we prefer to call them 'accessory bacteria', a term that cannot be interpreted to presuppose benefit to the aphid.

Although the data sets are small and fragmentary, they do reveal that the incidence of accessory bacteria varies within and between aphid species. For example, 50/57 (80%) of the pea aphids (*A. pisum*) collected by Chen and Purcell (1997) in California bore PASS, and of the 35 pea aphids from natural populations in UK studied by Darby *et al.* (2001) 13 (37%) bore PABS and none bore PASS.

One might anticipate that there is an upper limit to the total diversity of bacteria borne by an individual aphid or that certain taxa may be incompatible, either because of extensive niche overlap between different taxa and competitive exclusion or because certain bacteria display virulent characteristics in mixed infections resulting in selection against aphids bearing a high bacterial diversity. Competitive exclusion and incompatibility among different microbial taxa have been described in other symbioses, notably those involving *Chlorella* algae, but the outcome of these interactions vary widely. In hydras, the 'resident' algal genotype is not replaced when challenged by any other genotypes, independent of the relative infectivity or capacity to support hydra growth of the resident and challenging genotypes in single infections (McAuley & Smith 1982). The protist *Paramecium bursaria* can be colonized by yeast, *Chlorella* and bacteria but stable associations invariably involve

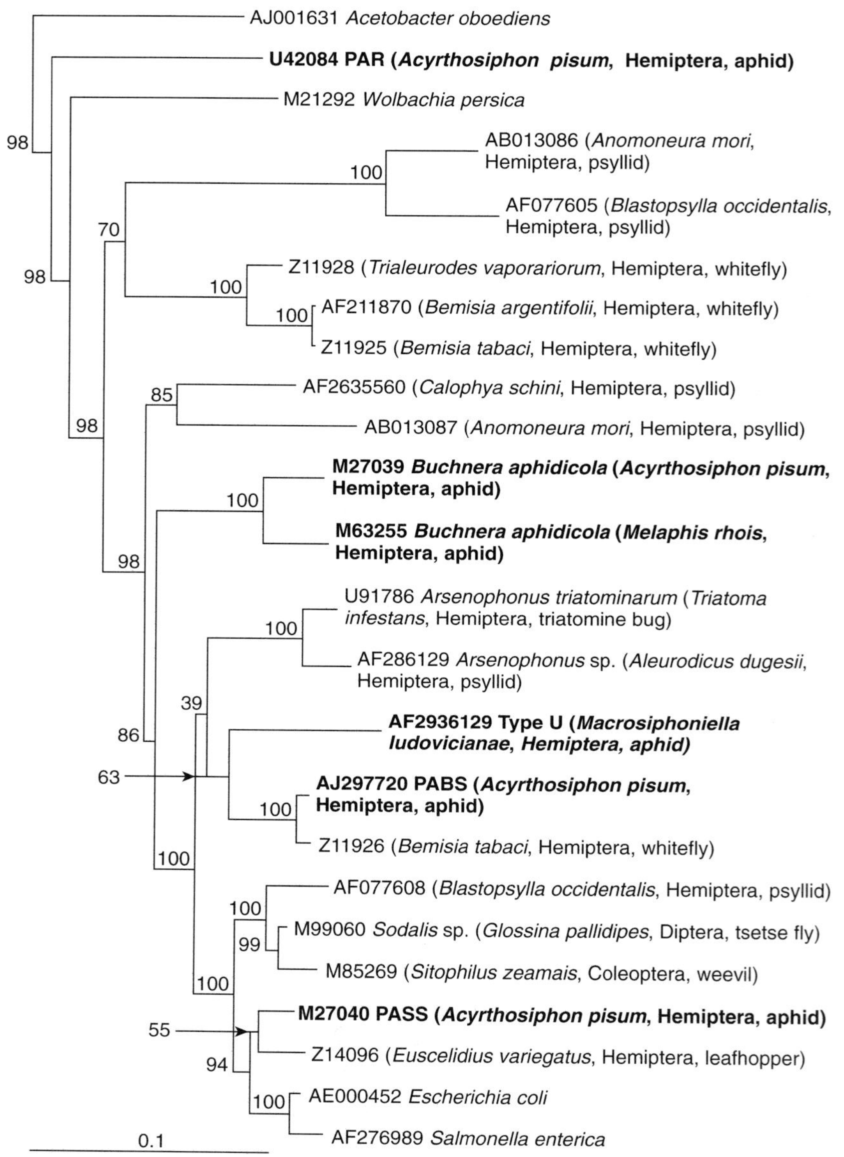

Figure 17.1 Phylogenetic tree of 16S rDNA sequences of symbiotic bacteria in aphids (bold) and other insects. Numbers above branches are %bootstrap values for nodes calculated by the neighbour-joining method. Branch labels show GenBank accession number, bacterial species where known and, in parentheses, the species, order and informal name of the insect host of the bacteria. PAR, pea aphid rickettsia. (Redrawn from figure 2 in Darby *et al.* 2001.)

just one taxon, with a well-defined hierarchy, from the superior yeast, through *Chlorella* to inferior bacteria, independent of the order in which the *Paramecium* may be exposed to the different microorganisms (Görtz 1982). In other algal symbioses, different algal genotypes may coexist in apparent harmony (e.g. Rowan *et al.* 1997). For the accessory bacteria in aphids, more extensive sampling is required to establish whether certain forms may be incompatible or whether there is an upper limit to the total diversity of bacteria accommodated in an individual aphid.

Various laboratory studies have confirmed that accessory bacteria are transmitted vertically and can be transferred horizontally between aphids by injection into the haemolymph (e.g. Chen *et al.* 2000; Fukatsu *et al.* 2001). There is also persuasive phylogenetic evidence for horizontal transmission of accessory bacteria in aphids (Sandström *et al.* 2001). Both the occasional horizontal transmission and failure of vertical transmission may contribute to the observed prevalence of accessory bacteria in natural populations, although laboratory studies suggest that horizontal transmission is rare and vertical transmission close to assured (e.g. Darby *et al.* 2001; Sandström *et al.* 2001). It has been suggested that the oral route and aborted parasitoid attack may mediate horizontal transmission of accessory bacteria (Chen *et al.* 2000; Darby *et al.* 2001; Sandström *et al.* 2001), but neither has been demonstrated definitively for any aphid species.

Significance of symbiotic bacteria to aphids

The impact of eliminating bacteria on aphid fitness

Buchnera is absolutely required for sustained growth and reproduction of all aphid species possessing these bacteria. When the bacteria are eliminated, for example by treatment with antibiotics, the growth of larvae is inhibited and reproductive output is suppressed (reviewed in Douglas 1998). There are a few reports of larviposition by antibiotic-treated aphids, but the offspring are stillborn or die within a few days without growing or developing. In natural aphid populations, *Buchnera* is present in all morphs apart from well-defined dwarf males and soldiers of certain taxa (Buchner 1966; Fukatsu & Ishikawa 1992).

The variable incidence of accessory bacteria in both natural populations and laboratory cultures of aphids (e.g. Chen & Purcell 1997; Fukatsu *et al.* 2000, 2001; Darby *et al.* 2001; Sandström *et al.* 2001) indicates that these bacteria are not required by aphids. Two complementary approaches have been adopted to explore the significance of accessory bacteria for aphids: analysis of the fitness of single parthenogenetic lines of aphids lacking and experimentally infected with accessory bacteria in the laboratory; and comparisons of fitness of aphids in natural populations bearing and lacking the bacteria. To date, all the studies have been conducted on *A. pisum*. As the data in Figure 17.2a,b illustrate, *Spiroplasma* sp. depresses the fecundity of the one *A. pisum* clone studied, and the impact of PASS and PAR on aphid fecundity varies with temperature. Further experiments have revealed that the

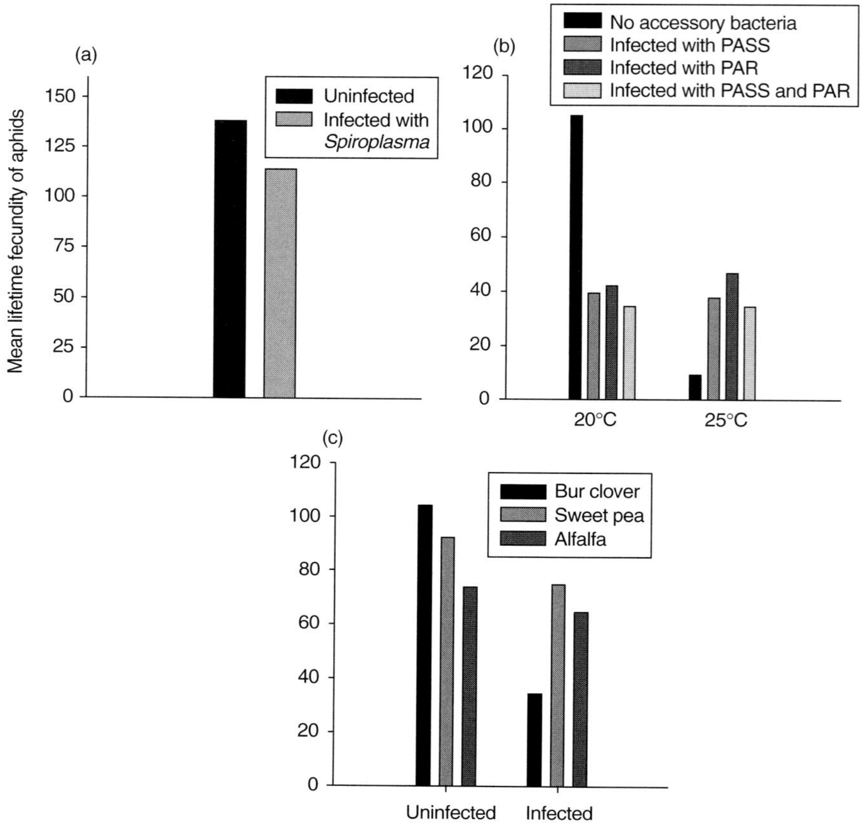

Figure 17.2 The fitness of pea aphids, *Acyrthosiphon pisum*, experimentally infected with accessory bacteria: (a) *Spiroplasma*; (b) PASS and PAR (either singly or as double infection) on bur clover at 20°C and 25°C; (c) double infection with PASS and PAR on different host plants at 20°C. (Redrawn from data in Fukatsu *et al.* (2001) and Chen *et al.* (2000).)

impact of PASS and PAR on aphid fitness varies among parthenogenetic lines of the aphid (Chen *et al.* 2000). The effect of PABS on aphid fitness in the laboratory has not been investigated, but the fitness of aphids in natural populations is not significantly affected by PABS whether quantified as fecundity (Figure 17.3), survivorship or larval development time (Darby 2002).

To summarize, the datasets on the significance of accessory bacteria to aphids are far from comprehensive, but the information available suggests that these bacteria have a small but variable effect on aphid fitness, at least as defined narrowly by survival, reproductive output, etc.

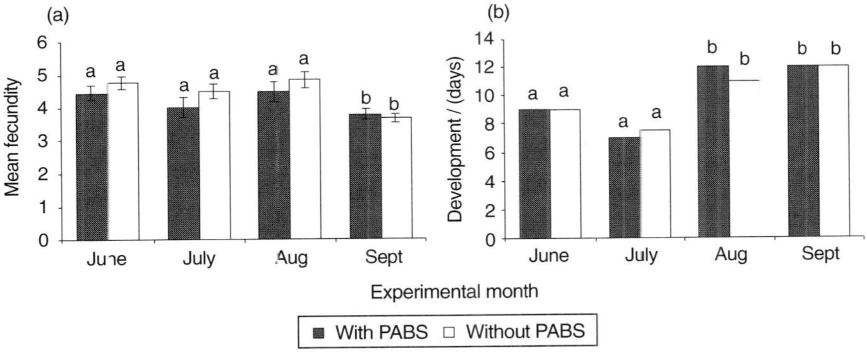

Figure 17.3 The fitness of natural *Acyrthosiphon pisum* populations in North Yorkshire, UK under field conditions, in 1999. (a) Mean fecundity of aphids and (b) development time of nymphs to adulthood for aphids with and without PABS. (Error bars indicate standard error of means; treatments with the same superscript letters are not significantly different.) (Redrawn from Darby 2002.)

Symbiotic bacteria and plant range of aphids

From the ecological perspective, a particularly interesting aspect to the significance of symbiotic microorganisms to aphids is their possible role in shaping insect plant range (e.g. Jones 1984; Southwood 1985; Dowd 1991; Adams & Douglas 1997). It has been argued that the microorganisms may enhance the capacity of phytophagous insects to utilize plants with certain allelochemical or nutrient profiles. For the vertically transmitted microorganisms in aphids, two interacting evolutionary processes could be involved: selection pressure on the aphid to specialize on plants most compatible with their complement of microorganisms; and selection on the microorganisms to maximize fitness of the insect on its principal host plants.

Chen *et al.* (2000) provide the first experimental evidence that accessory bacteria may influence the relative suitability of different plant species for aphids. Under laboratory conditions, infection of a parthenogenetic line of *A. pisum* with PASS and PAR depressed aphid fecundity on bur clover by 67%, but reduced mean aphid fecundity on the alternative host plants, sweet pea and alfalfa, by just 19% and 12%, respectively (see Figure 17.2c).

It is more difficult to assess the importance of *Buchnera* in shaping plant range of aphids because there is no method to construct experimental symbioses between the aphid and *Buchnera* partners of symbioses with different plant ranges; *Buchnera* cells are lysed when experimentally transferred between aphids by the injection method routinely used for accessory bacteria (unpublished results). However, two lines of evidence are consistent with the interpretation that *Buchnera* is not generally an important determinant of the plant range of aphids. First, the *Buchnera* population in *Aphis fabae* persists when the aphids are cultured on plant species that fail to support survival to adulthood (Wilkinson *et al.* 2001), indicating that collapse

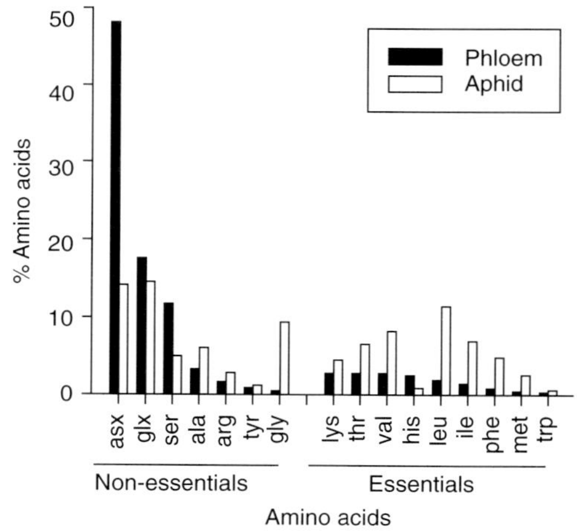

Figure 17.4 Amino acid composition of the phloem sap of *Vicia faba* and the black bean aphid *Aphis fabae*. ala, Alanine; arg, arginine; asx, asparagine + aspartic acid; glx, glutamine + glutamic acid; gly, glycine; his, histidine; ile, isoleucine; leu, leucine; lys, lysine; met, methionine; phe, phenylalanine; ser, serine; thr, threonine; trp, tryptophan; tyr, tyrosine; val, valine. Proline and cysteine (both non-essential amino acids) were not quantified in this analysis. (Data from Douglas *et al.* (2001) and C. R. Tosh, K. F. A. Walters, A. E. Douglas (unpublished results).)

of the symbiosis does not contribute to poor aphid performance. Second, *A. pisum* uses cues at the plant surface to discriminate between host and non-host plant species, and these responses are unaffected by elimination of the bacteria from the aphids (Wilkinson & Douglas 1995, 1998b).

The functions of bacteria in aphids

On the functions of *Buchnera*

The functions of symbiotic bacteria can be defined as traits of selective advantage to their partners (e.g. aphids). There is overwhelming evidence that the principal, and possibly sole, function of *Buchnera* is to provide aphids with essential amino acids, nutrients that the aphid cannot synthesize and that are in short supply in the aphid diet of plant phloem sap (Douglas 1998). The nutritional value of essential amino acid provisioning by *Buchnera* to aphids can be illustrated by a comparison of amino acid composition of the aphid *Aphis fabae* and one of its host plants, the broad bean *Vicia faba* (Figure 17.4). As is general among animals, the essential amino acids account for ca. 50% of the total amino acid content of the aphid (more than 98% of the amino acids in aphids is in protein). Free amino acids are the dominant nitrogenous compounds in plant phloem sap, but less than 20% are essential amino acids in most plants (Douglas 1993; Sandström *et al.* 2000).

The direct evidence that symbiotic bacteria provide essential amino acids is twofold. Aphids containing *Buchnera*, but not those experimentally deprived of their complement of *Buchnera*, first, can be maintained through multiple generations on diets lacking individual essential amino acids (reviewed in Douglas 1998) and, second, can synthesize essential amino acids from a variety of precursors, including sucrose and glutamic acid (Febvay *et al.* 1995, 1999; Douglas *et al.* 2001). Molecular data support these nutritional and metabolic studies. Despite massive reduction in the genome, including loss of the capacity to synthesize all non-essential amino acids, *Buchnera* in *A. pisum* have the full complement of genes required to synthesize most essential amino acids (Shigenobu *et al.* 2000). Furthermore, for the *Buchnera* in many aphids, the genes *leuA-D*, coding for enzymes in the leucine biosynthetic pathway, and the genes *trpEG*, coding for the enzyme anthranilate synthase of the tryptophan biosynthetic pathways, are amplified on plasmids, known as *p*leu and *p*trp, respectively (Lai *et al.* 1994; Bracho *et al.* 1995) (Figure 17.5).

In recent years, insight into *Buchnera* function has been gained from genomic analysis. Technological developments, especially relating to the availability of the complete genome sequence for *Buchnera* in *A. pisum* (Shigenobu *et al.* 2000), offer opportunities to explore the wider implications of *Buchnera* function in the ecology of aphids. There are indications, however, that function cannot reliably be deduced from sequence data alone and requires corroboration by phenotypic and fitness analyses. Three issues illustrating this point are considered below.

Copy number of *trpEG* genes in *Buchnera*

The number of functional copies of *trpEG* genes on the *p*trp plasmid of *Buchnera* relative to the number of *Buchnera* chromosomal genes, varies among *Buchnera* from different aphid lineages. Up to three molecular mechanisms may be involved (varying between aphid species): variation in the number of *trpEG* repeats per plasmid; variation in the number of plasmids per *Buchnera* chromosome; and, in certain species, silencing of one or more copies of *trpEG* through nonsense mutations or deletions (Lai *et al.* 1996; Baumann *et al.* 1997). From the sequence of *trpE*, anthranilate synthase activity of *Buchnera* is predicted to be subject to feedback inhibition by tryptophan. From this, it can be argued that the rate of tryptophan synthesis would increase with anthranilate synthase protein and, by implication, with the number of functional *trpEG* genes, relative to the chromosomal *trp* genes (Lai *et al.* 1994). The concentration of tryptophan in the free amino acid fraction of the aphids has been adopted as an index of tryptophan biosynthetic capability of *Buchnera*. Only one of the two interspecific comparisons summarized in Table 17.3 is consistent with the predicted positive relationship between *trpEG* copy number and tryptophan biosynthetic capability. Furthermore, a detailed analysis of 12 *A. pisum* clones revealed no consistent relationship between *trpEG* copy number and rates of tryptophan biosynthesis. Perhaps, the controls over the rates of tryptophan synthesis are distributed across multiple enzymes in the tryptophan biosynthetic pathway in some aphid–*Buchnera* symbioses (see Fell 1996) and factors other

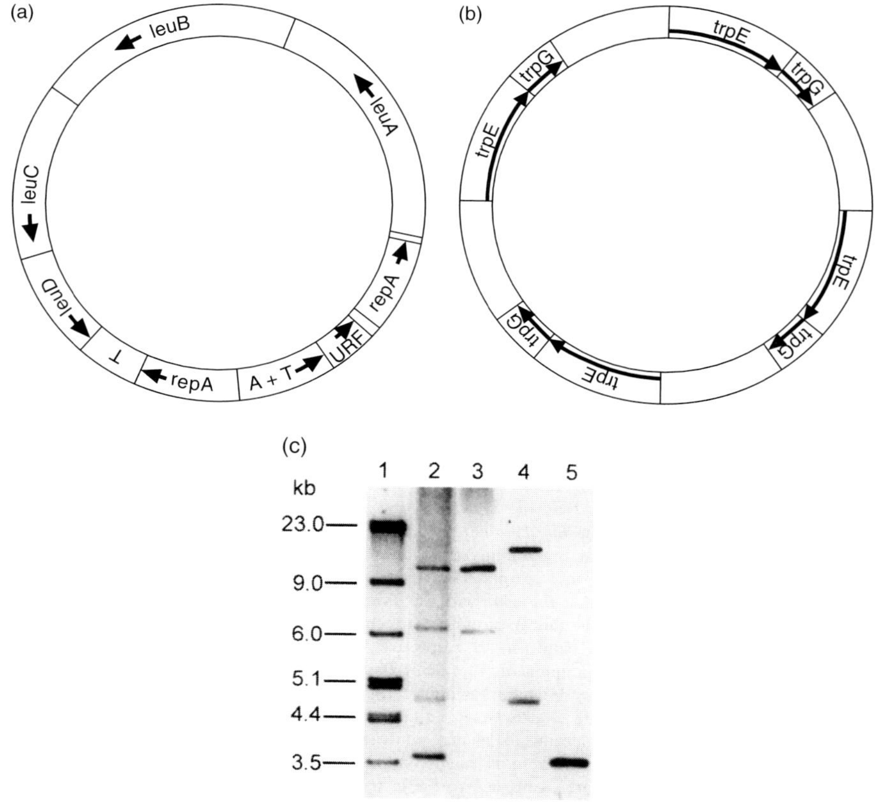

Figure 17.5 The plasmids of *Buchnera* from members of the family Aphididae. (a) Gene map of *p*leu from *Rhopalosiphum padi* (redrawn from Bracho *et al.* 1995). (b) Gene map of *p*trp from *Schizaphis graminum* (redrawn from Lai *et al.* 1994). (c) Products of restriction digestion with *Bgl*II of covalently closed circular DNA molecules isolated from *Acyrthosiphon pisum* by alkali lysis. Lane 1, molecular weight markers; lane 2, stained with ethidium bromide; lanes 3–5, Southern blots with mtDNA probe, *p*leu probe and *trpE* probe, respectively. (Note that *Bgl*II does not cut *p*leu, which is in the rapidly migrating supercoiled form and slowly migrating open relaxed form; and *p*trp is cut once per tandem repeat of 3.6 kb by *Bgl*II.) (Reproduced from Douglas 1997a.)

than aphid demand for tryptophan may dictate the copy number of *trpEG* (Moran 1996).

How is essential amino acid synthesis by *Buchnera* regulated?

Molecular analyses have revealed the complete absence of transcriptional controls over all *Buchnera* genes involved in amino acid biosynthesis and of post-translational controls over the activity of most biosynthetic pathways (Jimenez *et al.* 2000; Shigenobu *et al.* 2000). (The retention of feedback inhibition of anthranilate

Table 17.3 Tryptophan content of the free amino acid fraction of aphids bearing *Buchnera* with different levels of *trpEG* amplification, as determined by ratio of number of functional plasmid-borne copies of the gene *trpE* to number of copies of the chromosomal gene *trpB*.

Aphid species	*trpE*: *trpB*	% tryptophan in aphid-free amino acid fraction
Diuraphis* species		
D. mexicana	14:1	3
D. noxia	1.8:1	1
Aphid species affiliated to *Vicia faba*†		
Acyrthosiphon pisum	16:1	0.9 ± 0.03
Aphis fabae	4:1	1.1 ± 0.03

* Data of Thao *et al.* (1998) and Wernegreen and Moran (2000).
† Data of Birkle *et al.* (unpublished data), mean ± SE of 15 repetitions, for aphids reared to 4th larval stadium at 18°C.

synthase activity by tryptophan, considered under 'Copy number of *trpEG* genes in *Buchnera*' above, is exceptional.) Metabolic studies, by contrast, have demonstrated elevated rates of *Buchnera* synthesis of amino acids that are in short supply in the aphid diet (Febvay *et al.* 1999; Douglas *et al.* 2001). Perhaps *Buchnera* cells are genetically constitutive overproducers of essential amino acids and their realized rates of essential amino acid synthesis are regulated by the supply of precursors from the aphid, as has been suggested for other types of symbioses (reviewed in Douglas 1994).

It is important to our understanding of the ecology of this symbiosis to resolve the mismatch between the apparently incompatible scenarios of *Buchnera* as a constitutive overproducer of essential amino acids (molecular studies) and as a co-operative symbiont that scales amino acid to aphid demand (metabolic studies). The essential amino acid composition of plant phloem sap varies widely between plant species and with plant developmental age and environmental conditions (e.g. Blackmer *et al.* 1999; Brugière *et al.* 1999). The responsiveness of essential amino acid synthesis rates by *Buchnera* to aphid demand may be an important factor influencing the capacity of aphids to utilize plants of different amino acid profiles—and defining the limits to that capability.

Gene content of *Buchnera*

The metabolic capabilities of the aphid symbiosis appear to exceed the genes identified from complete genome sequence of *Buchnera*. In particular, key genes in the biosynthesis of methionine and isoleucine, *metC* and *ilvA* respectively, are apparently missing from the genome of *Buchnera* from on parthenogenetic line of pea aphids (Shigenobu *et al.* 2000), but *Buchnera* in other lines of pea aphids and other aphid species can synthesize both of these amino acids and some aphids can be maintained indefinitely on diets lacking either methionine or isoleucine (e.g. Srivastava *et al.* 1985; Douglas 1988; Febvay *et al.* 1999; Douglas *et al.* 2001). At present, these discrepancies are open to multiple interpretations. Perhaps these functions are

Table 17.4 Intraspecific variation in the essential amino acid requirement of aphids reared on chemically defined diets.

Aphid	his	ile	leu	lys	met	phe	thr	trp	val
Acyrthosiphon pisum*									
Biotype C	0		0	0	0		0	0	
Biotype J	0				0	0	0		0
***Aphis fabae*†**									
HR91/3	0				0				
AED94/81	0				0		0		
AED94/82	0				0				
AED94/84					0				
DY94/109	0				0				
AED95/19									
AED95/31	0				A		A	A	
AED95/33	0								0

0, Aphids cannot be maintained through multiple generations on diet; A, the cultures produced predominantly alate (winged) adults which did not settle on the diets or produce offspring.
Abbreviations as in legend to Figure 17.4.
* Data of Srivastava *et al.* (1985).
† Data of T.L. Wilkinson and A.E. Douglas (unpublished data).

mediated in *Buchnera* by novel genes (or sequences with low similarity to *metC* and *ilvA* of other α-*Proteobacteria*) or by substrate ambiguity of other enzymes (Cordwell 1999). For example, LeuA, which can mediate the reaction usually catalysed by IlvA when expressed at high levels in *Escherichia coli* (Umbarger 1996), may be expressed strongly in *Buchnera* because *leuA* is amplified on a plasmid (see 'On the functions of *Buchnera*' above). Alternatively, the gene complement may vary between different *Buchnera* lineages or the accessory bacteria may provide the aphid with certain essential amino acids, complementing or supplementing the supply from the *Buchnera*. An indication that the bacteria in aphids are not functionally equivalent across different aphid lineages comes from the several demonstrations of intraspecific variation in aphid requirements for dietary essential amino acids (Table 17.4).

If future research confirms variation in the array of essential amino acids synthesized in different aphids, whether as a consequence of variation in the gene complement of *Buchnera* or identity of accessory bacteria, the implications for the fitness of aphids on different plants could be profound. These considerations also raise the issue of the function of accessory bacteria, which is addressed next.

On the functions of accessory bacteria in aphids

As addressed above (see 'Accessory bacteria' and 'The impact of eliminating bacteria on aphid fitness'), the accessory bacteria of aphids are of variable incidence and have either no discernible or small impacts on aphid fitness. It is entirely possible that the relationship between some accessory bacteria and aphids can be best described in terms of the bacteria exploiting the aphid habitat and transmission routes that have evolved in relation to the *Buchnera* symbiosis. These accessory bacteria are 'hitching a ride'.

Under certain environmental circumstances, however, the aphid may derive benefit from these 'hitch-hiking' accessory bacteria. For example, they may protect the insect from pathogens, as has been demonstrated many times for microorganisms in the digestive tract of mammals (e.g. van der Waaij & van der Waaij 1990; Berg 1996) or, as discussed in the section above, provide nutrients to the insect. More generally, we cannot exclude the possibility that the profile of accessory bacteria may contribute to intraspecific differences between aphids, including plant affiliation, thermal tolerance, pathogen susceptibility and morph production (e.g. Chen *et al.* 2000). Furthermore, certain multigenerational shifts in the plant range or other characteristics of parthenogenetic aphids (e.g. Douglas 1997b, 2000) could reflect changes in the identity or abundance of accessory bacteria.

Concluding remarks

Our understanding of the biology of symbioses between animals and microorganisms has been transformed over the last decade by the application of molecular techniques; and the aphid–bacterial symbiosis has played a major part in these advances. Most of the fundamental issues that early investigators considered intractable have been resolved. These include the taxonomic identity of the microorganisms, their abundance in aphids and the mechanistic basis of aphid dependence on the symbiosis. We now have a real insight into the complement of genes and functions that aphids have acquired through symbiosis, and good prospects to identify the gene content and expression patterns of symbiotic microorganisms that underpin the phenotype and fitness of aphids.

There are, however, indications that the relationship between the bacterial genome and ecologically important traits of the aphid may be complex. This can be illustrated by the apparent mismatch of *Buchnera* gene content and regulatory sequences with symbiosis function (see pages 319–320). Other issues remaining to be resolved include the role of symbiotic bacteria in defining plant utilization patterns of aphids and the impact of accessory bacteria on ecologically important traits of the aphid, such as fecundity and thermal tolerance. We anticipate that molecular approaches will continue to make a vital contribution to our understanding of the ecological significance of symbiosis, a route by which phylogenetically distant genomes are brought together to form a predominantly co-operative unit.

Acknowledgements

This work was funded by the Natural Environment Research Council's thematic programme EDGE (Ecological Dynamics and Genes).

References

Adams, D. & Douglas, A.E. (1997) How symbiotic bacteria influence plant utilization by the polyphagous aphid, *Aphis fabae. Oecologica* **110**, 528–532.

Baumann, L., Clark, A., Rouhhakhsh, D. *et al.* (1997) Endosymbionts (*Buchnera*) of the aphid *Uroleucon sonchi* contain plasmids with *trpEG* and remnants of *trpE* pseudogenes. *Current Microbiology* **35**, 18–21.

Baumann, P., Baumann, L. & Clark, M.A. (1994) Levels of *Buchnera aphidicola* chaperonin GroEL during growth of the aphid *Schizaphis graminum. Current Microbiology* **32**, 279–285.

Berg, R.D. (1996) The indigenous gastrointestinal microflora. *Trends in Microbiology* **4**, 430–435.

Blackmer, J.L. & Byrne, D.N. (1999) Changes in the amino acids in *Cucumis melo* in relation to life-history traits and flight propensity of *Bemisia tabaci. Entomologia Experimentalis et Applicata* **93**, 29–40.

Bracho, A.M., Martinez-Torres, D., Moya, A. & Latorre, A. (1995) Discovery and molecular characterization of a plasmid localized in *Buchnera* sp. bacterial endosymbiont of the aphid *Rhopalosiphum padi. Journal of Molecular Evolution* **41**, 67–73.

Brugière, N., Dubois, F., Limami, A.M. *et al.* (1999) Glutamine synthetase in the phloem plays a major role in controlling proline production. *Plant Cell* **11**, 1995–2001.

Brynnel, E.U., Kurland, C.G., Moran, N.A. & Andersson, S.G.E. (1998) Evolutionary rates for *tuf* genes in endosymbionts of aphids. *Molecular Biology and Evolution* **15**, 574–582.

Buchner, P. (1966) *Endosymbiosis of Animals with Plant Microorganisms.* Wiley, London.

Cazemier, A.E., Hackstein, J.H.P, Op den Camp, H.J.M, Rosenberg, J. & van der Drift, C. (1997) Bacteria in the intestinal tract of different species of arthropods. *Microbial Ecology* **33**, 189–197.

Chen, D-Q. & Purcell, A.H. (1997) Occurrence and transmission of facultative endosymbionts in aphids. *Current Microbiology* **34**, 220–225.

Chen, D-Q., Campbell, B.C. & Purcell, A.H. (1996) A new rickettsia from a herbivorous insect, the pea aphid *Acyrthosiphon pisum* (Harris). *Current Microbiology* **33**, 123–128.

Chen, D-Q., Montllor, C.B. & Purcell, A.H. (2000) Fitness effects of two facultative endosymbiotic bacteria on the pea aphid, *Acyrthosiphon pisum*, and the blue alfalfa aphid, *A. kondoi. Entomologia Experimentalis et Applicata* **95**, 315–323.

Clark, M.A., Moran, N.A. & Baumann, P. (1999) Sequence evolution in bacterial endosymbionts having extreme base compositions. *Molecular Biology and Evolution* **16**, 1586–1598.

Cordwell, S.J. (1999) Microbial genomes and 'missing' enzymes: redefining biochemical pathways. *Archives of Microbiology* **172**, 269–279.

Darby, A.C. (2002) *The microbiota of the pea aphid.* Acyrthosipon pisum. PhD thesis, University of York.

Darby, A.C., Birkle, L.M., Turner, S.L. & Douglas, A.E. (2001) An aphid-borne bacterium allied to the secondary symbionts of whitefly. *FEMS Microbiology Ecology*, in press.

Dasch, G.A., Weiss, E. & Chang, K.P. (1984) Endosymbionts of insects. In: *Bergey's Manual of Systematic Bacteriology*, Vol. 1 (ed. N.R. Krieg), pp. 881–883. Williams & Wilkins, Baltimore.

Douglas, A.E. (1988) Sulphate utilization in an aphid symbiosis. *Insect Biochemistry* **18**, 599–605.

Douglas, A.E. (1993) The nutritional quality of phloem sap utilized by natural aphid populations. *Ecological Entomology* **18**, 31–38.

Douglas, A.E. (1994) *Symbiotic Interactions.* Oxford University Press, Oxford.

Douglas, A.E. (1997a) Parallels and contrasts between symbiotic bacteria and bacterial-derived organelles: evidence from *Buchnera*, the bacterial symbiont of aphids. *FEMS Microbiology Ecology* **24**, 1–9.

Douglas, A.E. (1997b) Provenance, experience and host plant affiliation in the polyphagous aphid, *Aphis fabae. Entomologia Experimentalis et Applicata* **83**, 161–170.

Douglas, A.E. (1998) Nutritional interactions in insect–microbial symbioses. *Annual Reviews of Entomology* **43**, 17–37.

Douglas, A.E. (2000) Reproductive diapause and the bacterial symbiosis in the sycamore aphid *Drepanosiphum platanoidis* (Schr.). *Ecological Entomology* **25**, 256–261.

Douglas, A.E., Minto, L.B. & Wilkinson, T.L. (2001)

Quantifying nutrient production by the microbial symbiosis in an aphid. *Journal of Experimental Biology* **204**, 349–358.

Dowd, P.F. (1991) Symbiont-mediated detoxification in insect herbivores. In: *Microbial Mediation of Plant–Herbivore Interactions* (eds P. Barbosa, V. A. Krischik & C. G. Jones), pp. 411–440. Wiley, New York.

Durvasula, R.V., Kroger, A. Goodwin, M. *et al.* (1999) Strategy for introduction of foreign genes into field populations of Chagas disease vectors. *Annals of the Entomological Society of America* **92**, 937–943.

Falk, P.G., Hoper, L.V., Midtvedt, T. & Gordon, J.I. (1998) Creating and maintaining the gastrointestinal ecosystem: what we know and what we need to know from gnotobiology. *Microbiology and Molecular Biology Reviews* **62**, 1157–1170.

Febvay, G., Liadouze, I., Guillaud, J. & Bonnot, G. (1995) Analysis of energetic amino acid metabolism in *Acyrthosiphon pisum*: a multidimensional approach to amino acid metabolism in aphids. *Archives of Insect Biochemistry and Physiology* **29**, 45–69.

Febvay, G., Rahbe, Y., Rynkiewiecz, M., Guillaud, J. & Bonnot G. (1999) Fate of dietary sucrose and neosynthesis of amino acids in the pea aphid, *Acyrthosiphon pisum*, reared on different diets. *Journal of Experimental Biology* **202**, 2639–2652.

Fell, D.A. (1996) *Understanding the Control of Metabolism*. Portland Press, London.

Fukatsu, T. & Ishikawa, H. (1992) Soldier and male of an eusocial aphid *Colphina arma* lack endosymbiont—implications for physiological and evolutionary interaction between host and symbiont. *Journal of Insect Physiology* **38**, 1033–1042.

Fukatsu, T., Nikoh, N., Kawai, R. & Koga, R. (2000) The secondary endosymbiotic bacterium of the pea aphid *Acyrthosiphon pisum* (Insecta: Homoptera) *Applied and Environmental Microbiology* **66**, 2748–2758.

Fukatsu, T., Tsuchida, T., Nikoh, N. & Koga, R. (2001) *Spiroplasma* symbiont of the pea aphid, *Acyrthosiphon pisum* (Insecta: Homoptera). *Applied and Environmental Microbiology* **67**, 1284–1291.

Görtz, H.D. (1982) Infections of *Paramecium bursaria* with bacteria and yeasts. *Journal of Cell Science* **58**, 445–453.

Goulet, T.L. & Coffroth, M.A. (1997) A within colony comparison of zooxanthella genotypes in the Caribbean gorgonian *Plexaura kuna*. In: *Proceedings of the 8th International Coral Reef Symposium*, Vol. 2, pp. 1331–1334.

Grenier, A-M., Nardon, C., Rahbe, Y. (1994) Observations on the microorganisms occurring in the gut of the pea aphid *Acyrthosiphon pisum*. *Entomologia Experimentalis et Applicata* **70**, 91–96.

Harada, H. & Ishikawa, H. (1997) Experimental pathogenicity of *Erwinia aphidicola* to pea aphid, *Acyrthosiphon pisum*. *Journal of General and Applied Microbiology* **43**, 363–367.

Harada, H., Oyaizu, H. & Ishikawa, H. (1996) A consideration about the origin of aphid intracellular symbiont in connection with gut bacterial flora. *Journal of General Applied Microbiology* **42**, 17–26.

Hinde, R. (1971) The control of the mycetome symbiotes of the aphids *Brevicoryne brassicae*, *Myzus persicae*, and *Macrosiphum rosae*. *Journal of Insect Physiology* **17**, 1791–1800.

Hongoh, Y. & Ishikawa, H. (1994) Changes of mycetocyte symbiosis in response to flying behaviour of alatiform aphid (*Acyrthosiphon pisum*). *Zoological Science* **11**, 731–735.

Humphreys, N.J. & Douglas, A.E. (1997) The partitioning of symbiotic bacteria between generations of an insect: a quantitative study of *Buchnera* in the pea aphid (*Acyrthosiphon pisum*) reared at different temperatures. *Applied and Environmental Microbiology* **63**, 3294–3296.

Jimenez, N., Gonzalez-Candelas, F. & Silva, F.J. (2000) Prephenate dehydratase from the aphid endosymbiont (*Buchnera*) displays changes in the regulatory domain that suggest its desensitization to inhibition by phenylalanine. *Journal of Bacteriology* **182**, 2967–2969.

Jones, C.G. (1984) Microorganisms as mediators of plant resource exploitation by insect herbivores. In: *A New Ecology: Novel Approaches to Interactive Systems* (eds P. W. Price, C. N. Slobodchikoff & W. S. Gand), pp. 53–99. Wiley, New York.

Lai, C-Y., Baumann, L. & Baumann, P. (1994)

Amplification of *trpEG*: adaptation of *Buchnera aphidicola* to an endosymbiotic association with aphids. *Proceedings of the National Academy of Sciences of the United States of America* **91**, 3819–3823.

Lai, C-Y., Baumann, P. & Moran, N.A. (1996) The endosymbiont (*Buchnera* sp.) of the aphid *Diuraphis noxia* contains plasmids consisting of *trpEG* and tandem repeats of *trpEG* pseudogenes. *Applied and Environmental Microbiology* **62**, 332–339.

Lovelock, M., O'Brien, R.W. & Slaytor, M. (1985) Effect of laboratory containment on the nitrogen metabolism of termites. *Insect Biochemistry* **15**, 503–509.

McAuley, P.J. & Smith, D.C. (1982) The green hydra symbiosis. VII. Conservation of the host cell habitat by the symbiotic algae. *Proceedings of the Royal Society of London Series B* **216**, 415–426.

Maniloff, J. (1996) The minimal cell genome: 'on being the right size'. *Proceedings of the National Academy of Sciences of the United States of America* **93**, 10004–10006.

Moran, N.A. (1996) Accelerated evolution and Muller's rachet in endosymbiotic bacteria. *Proceedings of the National Academy of Sciences of the United States of America* **93**, 2873–2878.

Moran, N.A., Munson, M.A., Baumann, P. & Ishikawa, H. (1993) A molecular clock in endosymbiotic bacteria is calibrated using the insect hosts. *Proceedings of the Royal Society of London Series B* **253**, 167–171.

Moran, N.A, von Dohlen, C.D. & Baumann, P. (1995) Faster evolutionary rates in endosymbiotic bacteria than in cospeciating insect hosts. *Journal of Molecular Evolution* **41**, 727–731.

Munson, M.A., Baumann, P., Clark, M.A. *et al.* (1991a) Evidence for the establishment of aphid–eubacterium endosymbiosis in an ancestor of four aphid families. *Journal of Bacteriology* **173**, 6321–6324.

Munson, M.A., Baumann, P. & Kinsey, M.G. (1991b) *Buchnera* gen. nov. and *Buchnera aphidicola* sp. nov., a taxon consisting of the mycetocyte-associated, primary endosymbionts of aphids. *International Journal of Systematic Bacteriology* **41**, 566–568.

Olff, H., Brown, V.K. & Drent, R.H. (1999) *Herbivores: Between Plants and Predators. 38th Symposium of the British Ecological Society.* Blackwell Science, Oxford.

Riley, M. (1993) Functions of the gene products of *Escherichia coli*. *Microbiology Reviews* **57**, 862–952.

Rispe, C. & Moran, N.A. (2000) Accumulation of deleterious mutations in endosymbionts: Muller's ratchet with two levels of selection. *American Naturalist* **156**, 425–441.

Rowan, R., Knowlton, N., Baker, A. & Jara, A. (1997) Landscape ecology of algal symbionts creates variation in episodes of coral bleaching. *Nature* **388**, 265–269.

Sandström, J.P., Telang, A. & Moran, N.A. (2000) Nutritional enhancement of host plants by aphids—a comparison of three aphid species on grasses. *Journal of Insect Physiology* **46**, 33–40.

Sandström, J.P., Russell, J.A., White, J.P. & Moran, N.A. (2001) Independent origins and horizontal transfer of bacterial symbionts of aphids. *Molecular Ecology* **10**, 217–228.

Shigenobu, S., Watanabe, H., Hattori, M., Sakaki, Y. & Ishikawa, H. (2000) Genome sequence of the endocellular bacterial symbiont of aphids *Buchnera* sp. APS. *Nature* **407**, 81–86.

Southwood, T.R.E. (1985) Interactions of plants and animals: patterns and processes. *Oikos* **44**, 5–11.

Srivastava, P.N., Gao, Y., Levesque, J. & Auclair, J.L. (1985) Differences in amino acid requirements between two biotypes of the pea aphid, *Acyrthosiphon pisum*. *Canadian Journal of Zoology* **63**, 603–606.

Thao, M.L., Baumann, L., Baumann, P. & Moran, N.A. (1998) Endosymbionts (*Buchnera*) from the aphids *Schizaphis graminum* and *Diuraphis noxia* have different copy numbers of the plasmid containing the leucine biosynthetic genes. *Current Microbiology* **36**, 238–240.

Uichanco, L.B. (1924) Studies on the embrogeny and postnatal development of the Aphididae, with special reference to the history of the 'symbiotic organ' or 'mycetom'. *Phillippine Journal of Science* **24**, 143–247.

Umbarger, H.E. (1996) Biosynthesis of branched-chain amino acids. In: *Escherichia coli* and *Salmonella: Cellular and Molecular Biology* (ed. F. C. Neidhardt), pp. 442–457. American Society for Microbiology.

Van Ham, R.C.H.J., Gonzalez-Candelas, F. *et al.* (2000) Postsymbiotic plasmid acquisition and evolution of the *repA1*-replicon in *Buchnera aphidicola. Proceedings of the National Academy of Sciences of the United States of America* **97**, 10855–10860.

van der Waaij, D. & van der Waaij, B.D. (1990) The colonization resistance of the digestive tract in different animal species and in man: a comparative study. *Epidemiology and Infection* **105**, 237–243.

Wernegreen, J.J. & Moran, N.A. (2000) Decay of mutualistic potential in aphid endosymbionts through silencing of biosynthetic loci: *Buchnera* of *Diuraphis. Proceedings of the Royal Society of London Series B* **267**, 1423–1431.

Wernegreen, J.J., Ochman, H., Jones, I.B. & Moran, N.A. (2000) Decoupling of genome size and sequence divergence in a symbiotic bacterium. *Journal of Bacteriology* **182**, 3867–3869.

Whitehead, L.F. & Douglas, A.E. (1992) Populations of symbiotic bacteria in the parthenogenetic pea aphid (*Acyrthosiphon pisum*) symbiosis. *Proceedings of the Royal Society of London Series B* **254**, 29–32.

Wilkinson T.L. & Douglas A.E. (1995) Aphid feeding, as influenced by the disruption of the symbiotic bacteria. *Journal of Insect Physiology* **41**, 635–640.

Wilkinson, T.L. & Douglas, A.E. (1998a) Host cell allometry and regulation of the symbiosis between pea aphids, *Acyrthosiphon pisum*, and bacteria, *Buchnera. Journal of Insect Physiology* **44**, 629–635.

Wilkinson, T.L. & Douglas, A.E. (1998b) Plant penetration by pea aphids (*Acyrthosiphon pisum*) of different plant range. *Entomologia Experimentalis et Applicata* **87**, 43–50.

Wilkinson, T.L., Ashford, D.A., Pritchard, J. & Douglas, A.E. (1997) Honeydew sugars and osmoregulation in the pea aphid *Acyrthosiphon pisum. Journal of Experimental Biology* **200**, 2137–2143.

Wilkinson, T.L., Adams, D., Mintro, L.B. & Douglas, A.E. (2001) The impact of hostplant on the abundance and function of symbiotic bacteria in an aphid. *Journal of Experimental Biology* **204**, 3027–3028.

Wood, R. (1998) *Reef Evolution.* Oxford University Press, Oxford.

Index

Note: page numbers in *italics* refer to figures; those in **bold** refer to tables.